T0345136

Springer Texts in Statistics

Advisors:
Stephen Fienberg   Ingram Olkin

Springer Texts in Statistics

H.T. Nguyen   G.S. Rogers

Fundamentals of Mathematical Statistics

Volume I

Probability for Statistics

With 23 Illustrations

Springer-Verlag
New York Berlin Heidelberg
London Paris Tokyo Hong Kong

Hung T. Nguyen
Gerald S. Rogers
Department of Mathematical Sciences
New Mexico State University
Las Cruces, NM 88003-0001, USA

Mathematics Subject Classification: 60-01

Library of Congress Cataloging-in-Publication Data
Nguyen, Hung T., 1944–
     Fundamentals of mathematical statistics / H.T. Nguyen and G.S.
  Rogers.
        p.   cm. — (Springer texts in statistics)
     Bibliography: v. 1,
     Includes indexes.
     Contents: v. 1. Probability for statistics — v. 2. Statistical
  inference.
     ISBN 0-387-97014-2 (v. 1 : alk. paper). — ISBN 0-387-97020-7 (v.
  2 : alk. paper)
     1. Mathematical statistics.   I. Rogers, Gerald Stanley, 1928–
  II. Title.   III. Series.
  QA276.12.N49 1989
  519.5—dc20   89-6426

Printed on acid-free paper.

Camera-ready copy prepared by the authors.
Printed and bound by R.R. Donnelley & Sons, Harrisonburg, Virginia.
Printed in the United States of America.

9 8 7 6 5 4 3 2 1

ISBN 0-387-97014-2 Springer-Verlag New York Berlin Heidelberg
ISBN 3-540-97014-2 Springer-Verlag Berlin Heidelberg New York

Preface

This is the first half of a text for a two semester course in mathematical statistics at the senior/graduate level for those who need a strong background in statistics as an essential tool in their career. To study this text, the reader needs a thorough familiarity with calculus including such things as Jacobians and series but somewhat less intense familiarity with matrices including quadratic forms and eigenvalues.

For convenience, these lecture notes were divided into two parts: *Volume I*, Probability for Statistics, for the first semester, and *Volume II*, Statistical Inference, for the second.

We suggest that the following distinguish this text from other introductions to mathematical statistics.

1. The most obvious thing is the layout. We have designed each lesson for the (U.S.) 50 minute class; those who study independently probably need the traditional three hours for each lesson. Since we have more than (the U.S. again) 90 lessons, some choices have to be made. In the table of contents, we have used a * to designate those lessons which are "interesting but not essential" (INE) and may be omitted from a general course; some exercises and proofs in other lessons are also "INE". We have made lessons of some material which other writers might stuff into appendices. Incorporating this freedom of choice has led to some redundancy, mostly in definitions, which may be beneficial.

For the first semester, Parts I, II, III of this Volume I contain 49 lessons of which 10 are INE; for the second semseter (Volume II) Parts IV, V, VI, contain 36 lessons (1 INE) and Part VII contains 11 (partially independent) choices.

2. Not quite so obvious is the pedagogy. First, most of the exercises are integrated into the discussions and cannot be omitted. Second, we started with elementary probability *and* statistics (Part I) because most students in a first course have little, if any, previous experience with statistics as a mathematical discipline. Just as importantly, in this part, the discussions begin with (modified) real examples and include some simple proofs. We think that our line of discussion leads naturally to the consideration of more general formulations.

3. In this vein, we believe that today's students of mathematical statistics need exposure to measure theoretic aspects, the "buzz words", though not the truly fine points.

Part II is such an introduction and is more difficult than Part I but, as throughout these notes, we blithely skip over really involved proofs with a reference to more specialized texts. In teaching, we have at times omitted some of the proofs that are included herein. Otherwise, we have tried to emphasize concepts and not details and we recognize that some instructors prefer more "problems". Part III contains special pieces of analysis needed for probability and statistics.

4. The topics in Volume II, Parts IV, V, VI are rather traditional but this listing does not show the care with which we have stated the theorems even when the proof is to be limited to a reference. The immense breadth of statistics forces all writers to be selective; we think we have a good choice of major points with some intersting sidelights. Moreover, our INE materials (in particular Part VII) give the reader (instructor) additional variety.

The following general comments on the contents summarize the detailed overviews at the beginning of each part.

Volume I: Probability For Statistics.

Part I: *Elementary Probability and Statistics.* We begin with the (Western) origins of probability in the seventeenth century and end with recognition of the necessity of more sophisticated techniques. In between, there are some practical problems of probability and examples of two principles of statistics: testing hypotheses and finding confidence intervals. As the title says, the notions are elementary; they are not trivial.

Part II: *Probability and Expectation.* As the reader is asked to take a broader view of functions and integration, this material is a good bit harder than that in Part I. Although we present most of the theory, we do so with a minimum of detailed proofs and concentrate on appreciation of the generalization of concepts from Part I. In particular, "almost surely" comes to be understood and not overlooked as some authors suggest.

Part III: *Limiting Distributions.* Herein, we complete the basic results of probability needed for "statistics". We have not addressed the problem specifically but Parts II and III should be adequate preparation for a course in "stochastic processes". Again, the emphasis is on appreciation of concepts and not all details of all proofs of all theorems are given. However, we do include an almost complete proof of one central limit theorem.

Volume II: Statistical Inference.

Part IV: *Sampling and Distributions.* Now we pursue the concepts of random sample and statistics as the bases of the theory of statistical inference; we include some sampling distributions in normal populations. We explain some useful properties of statistics such as sufficiency and completeness, in particular, with respect to exponential families.

Part V: *Statistical Estimation.* This is a discussion of some of the problems of point estimation and interval estimation of (mostly) real parameters. In some cases, the results are exact for a given sample size; in others, they are only approximations based on asymptotic (limiting) distributions. Criteria for comparing estimators such as bias, variance, efficiency are discussed. Particular care is given to the presentation of basic results regarding the maximum likelihood method.

Part VI: *Testing Hypotheses* In this part, we discuss three variations of likelihood (colloquially, probability) ratios used to derive tests of hypotheses about parameters. Most of the classical examples about normal populations are included.

Part VII: *Special Topics.* We call the topics in this part special because each is the merest introduction to its area of statistics such as decision theory, linear models, non–parametrics, goodness–of–fit, classified data. The reader is invited to choose freely and often.

We extend our thanks to: various groups of students, particularly those in our course in 1986–87, for their patience and comments; our department head, Carol Walker, for her patience and encouragement; to Valerie Reed for the original typing in T^3.

Hung T. Nguyen and Gerald S. Rogers
Las Cruces, March 10, 1989.

PS. We especially want to recognize the people at TCI Software Research, Inc. Without their generous assistance and T^3 word processing system, we would not have been able to present this work in our own style. HTN/GSR

CONTENTS

Volume I: Probability for Statistics

Contents

* indicates a lesson which is "interesting but not essential" and may be omitted.

PART I: ELEMENTARY PROBABILITY
AND STATISTICS

Overview

Lesson 1 contains the old notion of probability as a limiting relative frequency in repeated trials. There are other interpretations which assign the values of "probability" by taking account of other kinds of previous experience but the rules for subsequent calculations are the same. Graphing these relative frequencies brings out their existence as functions.

In order to appreciate the contemporary form of probability, it is necessary to understand the concepts of sets and operations with sets like union $\cup$, intersection $\cap$; some of this algebra of sets is explained in Lessons 2 and 3.

In Lesson 4, the notion of counting is formalized as a set function on finite sets. This leads to (Lesson 5) the basic rules for a probability function:

$$0 \leq P(A) \leq 1 \text{ for all events (subsets)}$$

$$P(A_1 \cup A_2) = P(A_1) + P(A_2) \text{ for disjoint events.}$$

The next two lessons contain basic definitions and formulas for (6) permutations and (7) combinations. These are used (Lesson 8) in classical examples of probability for cards, coins, dice, etc. Some discussion is included to suggest that other "real" problems may have the same kinds of sample space structure but not, of course, the same P values.

A much used result in probability is that of independence: the occurrence of one event does not affect the probability of occurrence of another event; this is explained in Lesson 9. Then Lesson 10 contains discussion (and applications) of the intimately related concept of conditional probability:

$$P(A \mid B) = P(A \cap B)/P(B) \text{ when } P(B) > 0 .$$

To a large extent, statistics is based on "numbers and their averages"; the corresponding technical terms – random variables and expected value – are introduced in Lessons 11 and 12.

Lesson 13 introduces the hypergeometric distribution for sampling from a finite population of "defectives and non–defectives". Some techniques of sampling are discussed in Lesson 14.

The principle idea – testing – introduced in Lesson 15 pervades a good deal of statistics; all of Part VI and some of Part VII (Volume II) are devoted to the theory and practice of testing.

Lesson 16 completes a statistical discussion of the hypergeometric begun in Lessons 13 and 15.

At the beginning of Lesson 17, the binomial distribution turns up as an approximation for the hypergeometric; the second part of this lesson considers the binomial in its own right as one mathematical model dealing with repeated trials.

The discussion of Lesson 18 concerns only one case of the inclusion–exclusion formula: the probability of

the occurrence of at least one of the events $A_1, A_2, \cdots, A_n$.

Other cases are outlined in the exercises.

A second general topic of statistics – estimation, specifically, confidence sets for parameters – is introduced for Bernoulli θ in Lesson 19. All of Part V and some of Part VII (Volume II) are devoted to the theory and practice of estimation.

In Lesson 20, it is seen that one approximation for the binomial density leads to the Poisson formula which in turn leads to the necessity of an extension of the basic axioms of probability to countable sets and countable unions. In Lesson 21, another view of Bernoulli trials – waiting times – also leads to this necessity and, hence, Part II.

LESSON 1. RELATIVE FREQUENCY

We have all done some experimenting in our time; even grade school children have their science projects. If an *experiment* is merely the making of an observation, then our lives are filled with experiments in all kinds of subject matter: biology, physics, medicine, accounting, chemistry, agriculture, education, even eating, swimming, and on and on. We are familiar too with the repetition of experiments, usually in science laboratories; yet even though the experiment is performed "in the same way", the *outcome* is not always the same. Such repeated trials of an experiment are easily perceived in the traditional examples (tossing dice or coins) but may be more elusive in other cases.

Examples: a) Toss a coin. The possible outcomes are usually denoted by H(ead) and T(ail). We can toss the coin as often as we wish, repeating the experiment, with some outcomes H and some outcomes T.

b) Toss an ordinary die—a cube made of homogeneous material with the faces labeled 1, 2, 3, 4, 5, 6 . The outcome of the experiment (the toss) is the value on the upface of the die.

c) Fry an egg. A second egg may taste like the first but there are always some differences in shape, color, consistency, · · ·

d) A doctor's treatment of the same symptoms in different patients is often considered as a sequence of repeated trials. In broad terms, the outcome on each trial (patient) might be one of symptoms relieved, symptoms unchanged, symptoms aggravated.

e) Plants of the same kind, grown in the same environment (not the same pot of course) may be likened to repeated trials—of growing a plant. Each outcome could be one or more of plant height, leaf color, seed yield, bugs, · · ·

f) An assembly line is an attempt to repeat certain essential elements in each automobile, case of cola, chair, lightbulb, personal computer, · · ·

g) An auditor might consider a collection of invoices as repeated trials in accuracy.

As we will see again, which experimental results are selected to be "outcomes" depends on the subject matter and purpose of the experiment. For example, you know of courses in *micro*–economics and *macro*–economics. A selective point of view is maintained even in the classical cases; for example,

both represent the upface 6 and the orientation or size is unimportant.

Exercise 1: Describe some "repeatable" experiments in areas of your own interests.

We will now examine a particular case in some detail. When we ignore the different types of illnesses, ages, personalities and such of the patients, we may consider their lengths of stay in a hospital as repeated trials. Suppose our first few observations are 15, 14, 13, 13, 16, 12; i.e., patient one stayed 15 days; patient two stayed 14 days; patient three stayed 13 days, and so on. In column 2 of the table below is a record of 50 such observations. We concentrate on one property of these, namely, that indicated by the statement, "The length of stay was less than 14 days"; for brevity, we denote this by A.
For the first two patients, A is false; for patients 3 and 4, A is true. Let us denote the number of times A is true for the first "n" patients by $f_n(A)$. Thus,

$$f_1(A) = 0 = f_2(A); \; f_3(A) = 1; \; f_4(A) = 2;$$

$$f_5(A) = 2; \; f_6(A) = 3; \text{ etc.}$$

The table includes these *frequencies* $f_n(A)$ and the *relative frequencies* $f_n(A)/n$.

As you look down the column of relative frequencies in this table (and the one you are to complete as the next exercise), note that there are "large fluctuations" in these values among the first ten or fifteen trials but much less variation among these values in the last ten or fifteen trials. With much larger numbers of repetitions, this settling down of the relative frequencies could be even more pronounced; a later exercise in this section contains such a record.

Record of Hospital Confinements

Trial number (Patient) n	Outcome (Length in day)	Is A true	Frequency of A $f_n(A)$	Relative frequency $f_n(A)/n$
1	15	no	0	0
2	14	no	0	0
3	13	yes	1	1/3 = .333
4	13	yes	2	2/4 = .500
5	16	no	2	2/5 = .400
6	12	yes	3	3/6 = .500
7	11	yes	4	4/7 = .571
8	11	yes	5	5/8 = .625
9	15	no	5	5/9 = .556
10	14	no	5	5/10 = .500
11	13	yes	6	6/11 = .545
12	11	yes	7	7/12 = .583
13	12	yes	8	8/13 = .616
14	11	yes	9	9/14 = .643
15	16	no	9	9/15 = .600
16	18	no	9	9/16 = .563
17	13	yes	10	10/17 = .588
18	14	no	10	10/18 = .556
19	12	yes	11	11/19 = .580
20	12	yes	12	12/20 = .600
21	16	no	12	12/21 = .571
22	13	yes	13	13/22 = .591
23	11	yes	14	14/23 = .606
24	11	yes	15	15/24 = .625
25	10	yes	16	16/25 = .640
26	17	no	16	16/26 = .616
27	8	yes	17	17/27 = .629
28	15	no	17	17/28 = .607
29	15	no	17	17/29 = .586
30	6	yes	18	18/30 = .600
31	11	yes	19	19/31 = .613
32	13	yes	20	20/32 = .625
33	14	no	20	20/33 = .606
34	13	yes	21	21/34 = .618
35	12	yes	22	22/35 = .629
36	11	yes	23	23/36 = .639
37	11	yes	24	24/37 = .649
38	14	no	24	24/38 = .632
39	13	yes	25	25/39 = .641

40	12	yes	26	26/40 = .650
41	11	yes	27	27/41 = .656
42	17	no	27	27/42 = .642
43	13	yes	28	28/43 = .651
44	14	no	28	28/44 = .636
45	17	no	28	28/45 = .622
46	9	yes	29	29/46 = .630
47	12	yes	30	30/47 = .639
48	16	no	30	30/48 = .625
49	12	yes	31	31/49 = .633
50	13	yes	32	32/50 = .640

Exercise 2: Complete columns 3 and 4 of the table following. The code is for eggs ordered at a lunch counter: S–scrambled OH–overhard   OE–overeasy   SS–sunnysideup   B–softboiled P–poached. Focus on the frequency $f_n(S)$ and the relative frequency $f_n(S)/n$ of scrambled egg orders in the first n customers.

Record of Egg Choices at a Lunch Counter

Trial (customer) frequency n)	Egg choice $f_n(S)$	Frequency of scrambled	Relative $f_n(S)/n$
1	OH	0	0
2	OH	0	0
3	OH	0	0
4	S	1	1/4 = .250
5	OH	1	1/5 = .200
6	OE	1	1/6 = .167
7	SS	1	1/7 = .143
8	S	2	2/8 = .250
9	S	3	3/9 = .333
10	OH		
11	OE		
12	OE		
13	SS		
14	SS		
15	OE		
16	B		
17	S		
18	P		

19	OH
20	B
21	S
22	S
23	OH
24	B
25	S
26	OE
27	P
28	OH
29	OH
30	OE
31	SS
32	P
33	S
34	OE
35	OE
36	SS
37	S
38	OH
39	S
40	OE
41	SS
42	B
43	S
44	S
45	OH
46	OH
47	OE
48	P
49	OE
50	OH

Exercise 3: Repeat some experiment about 50 times, recording the outcomes of each trial; calculate the frequencies and relative frequencies for some special event. E.g., observe the number of conjunctions on each line (trial) of a magazine article and compute the relative frequencies of lines without conjunctions. Record the hair color of classmates and count the number of blondes. If you really can't think of anything more interesting, toss 10 thumbtacks and count the number which fall "point up".

It may be easier to see the settling down of the relative frequencies in the more traditional graph below.

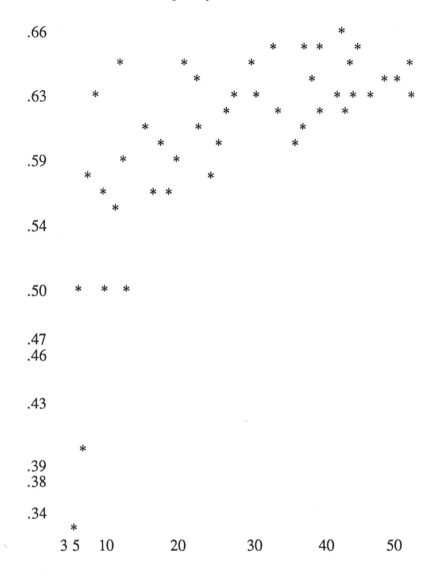

The horizontal axis is the trial number n and the vertical axis is the relative frequency f_n/n . For the hospital confinements, we plot the pairs $(n, f_n/n)$ {omitting $(1, 0)$, $(2, 0)$ for convenience}: $(3, .333)$ $(4, .500)$ $(5, .400)$ $\cdots$.

You should recognize each collection of pairs as a *function*: for each point [n] in one set,

there is one point $[f_n(A)/n]$ in another set.

The domain is the collection of first components [1, 2, 3, ..., 50] and the range is the collection of second components [0, 1/3, 1/2, ..., 32/50] .

These sentences [omitting the bracketed stuff] are the definitions of function, domain and range. Other phraseology is used: "Let y be a function of x; let f(x) be a function of x ; let y = f(x)."

Actually, y = f(x) or the like specifies only the values of the function *at* x. In an expression such as "$f(x) = x^{1/2}$", specifying the values, the domain is assumed to be understood from the context; if only real values for f are of interest, $x \geq 0$. As we shall see, much of the theory of mathematics and statistics involves properties of and relations among functions.

Exercise 4 :   a) Plot the pairs $(n, f_n(S)/n)$ for the egg orders.

b)   Plot the corresponding function for your own experiment. What are the domain and range?

In the hospital example, we have $f_{46}(A)/50 = .630$, $f_{47}(A)/50 = .639$, $f_{48}(A)/50 = .625$, $f_{49}(A)/50 = .633$, and $f_{50}(A)/50 = .640$. These values are all "close" to each other and (barring some catastrophe,) it is not unreasonable to assume that the values of $f_n(A)/n$ will deviate less and less from each other as we make more and more observations (trials). Then all the latter values will be close to some common value which we call the probability of A: P(A). In this example, P(A) is the probability that a stay in the hospital will be less than two weeks duration. Based on the 50 trials that we observed, we may guess that P(A) is close to the empirical value $f_{50}(A)/50 = .640$ but we have no way of knowing how accurate or inaccurate this may be unless we make observations on *all* patients — past, present, and future. One of the purposes of statistics is to devise techniques for assessing the accuracy without having to wait until the end.

These comments apply to other repeatable experiments: if A is some particular outcome, an experimenter may have enough knowledge of the field to suggest that $f_n(A)/n$ approaches a constant — called P(A) — as the number of repetitions increases. Note that this is a *physical* not a mathematical notion; the article by Keller, "The probability of heads", 1986, has an excellent discussion of this aspect. The fact that such a physical limit appears to be reasonable in so many different kinds of experiments is what makes certain applications of statistics and probability possible.

Exercise 5:   Consider the following record of repeated tosses of a four–sided die—a regular tetrahedron; the outcome is the value on the "down–face": 1, 2, 3, 4. On four separate axes, plot the pairs $(n, f_n(i)/n)$ for each outcome $i = 1,2,3,4$.

n	$f_n(1)$	$f_n(2)$	$f_n(3)$	$f_n(4)$
10	2	3	3	2
50	12	10	15	13
100	21	29	28	22
500	110	135	133	132
1000	250	246	253	251
5000	1232	1253	1251	1264
10000	2498	2472	2536	3750
15000	3717	3777	3756	3750
20000	5100	4906	4927	5067

LESSON 2. SAMPLE SPACES

To be realistic, we should consider large numbers of trials in our examples; certainly, a doctor would not judge the efficacy of salve tried on just three patients. But for the moment, such details are burdensome while many patterns and structures may appear within a few observations

A doctor is going to test a new salve on three people with itchy skin. At first, he is going to focus on whether or not the salve relieves the itch; for each patient, he will record R(elief) or N(o relief). He can use a "tree diagram" to describe the outcomes; each fork $\langle$ represents the possible outcomes at that stage of the experiment. The doctor can list all the possible outcomes by following the branches of the {fallen} tree:

Patient	One	Two	Three	Outcome
		R	R	RRR
			N	RRN
	R		R	RNR
		N	N	RNN
		R	R	NRR
			N	NRN
	N	N	R	NRR
			N	NNN

Of course, the doctor is interested in the treatment of individual patients and hence in which one of these outcomes will occur when he does the experiment. But it is likely that he is also interested in the general efficacy of the salve, say, in the number of patients whose itching was relieved. So he may consider the outcome as 0 if no patients felt relief; 1 if exactly one of the three felt relief; 2 if exactly two of the three felt relief; 3 if all patients felt relief. In other words, the personal physician will look at one kind of outcome and the medical researcher at another.

But there is another lesson to be learned here. When the doctor keeps his records in terms of the outcomes RRR, RRN,···, he can always find the number of patients who felt relief: RRR   corresponds to 3

              RRN RNR NRR   correspond to 2

              NRN NNR RNN   correspond to 1

              NN   corresponds to 0 .

      Obviously, if he records only the number of patients relieved, he cannot tell the effect on individual patients (cases 0 and 3 excepted.) The moral is that one should always consider outcomes which contain all the relevant information available. In fact, a doctor may have additional interests such as the presence or absence of side effects. Then he might record also A(adverse side effects present) or K(no adverse side effects) for each patient. For one patient a tree diagram is:

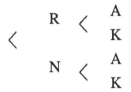

and the outcome is one of RA, RK, NA, NK. Note that RA and RK together make up an R in the left hand fork of the diagram; NA and NK together make up the N in that fork. For three patients, there will be $4 \times 4 \times 4 = 64$ distinguishable outcomes.

Exercise 1: You can see all these (and how 64 came up) by completing the tree diagram on the next page.

Exercies 2: Consider the first sample space at the beginning of this lesson with the eight triples. Rewrite this sample space using 1 to represent relief and 0 to represent no relief.

Exercise 3: List two kinds of outcomes for each experiment.
       a)   Toss a pair of coins one time.
       b)   Toss a pair of coins twice.
       c)   Test four light bulbs—each bulb either "works" or "does not work".
       d)   Two workers are each asked to classify their foreman as either Bad or Indifferent or Good.

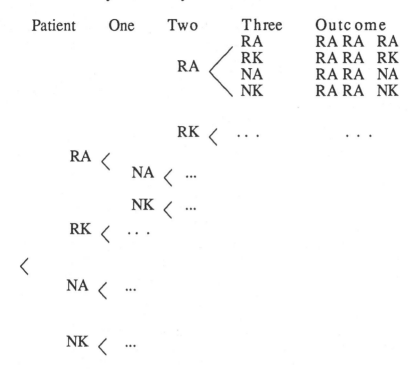

Patient	One	Two	Three	Outcome
			RA	RA RA RA
		RA	RK	RA RA RK
			NA	RA RA NA
			NK	RA RA NK

RK ⟨ ...

RA ⟨

NA ⟨ ...

NK ⟨ ...

RK ⟨ ...

⟨

NA ⟨ ...

NK ⟨ ...

Definition: *(Once the kind of outcome is decided upon) the collection of all possible outcomes for a given experiment is called a sample space.*

Note that in line with the previous discussion, we do not say *the* sample space. Because of the mixed history of the developement of probability (mathematical and applied), other terms are also in use.

A collection of distinct objects is called a *set*; the "objects" in the set are called *points* or *elements*. Thus a sample space is the set of *all sample points*. We shall use S as an abbreviation for a sample space and braces { } as an abbreviation for "The set ...".

Example: For the doctor and his itchy patients, we have at least three sample spaces only one of which would be chosen in a particular analysis.

$$S = \{RRR \ RRN \ RNR \ RNN \ NRR \ NRN \ NNR \ NNN\}$$
or $\{111 \ 110 \ 101 \ 100 \ 011 \ 010 \ 011 \ 000\}$

$$S_1 = \{0 \ 1 \ 2 \ 3\}$$

We should note that the elements of a set are not ordered among themselves. For example,

$$\{a\ b\ c\} = \{a\ c\ b\} = \{c\ a\ b\} = \{c\ b\ a\} = \{b\ a\ c\} = \{b\ c\ a\}$$

as they are all different "names" for the same set of three letters. Similarly, as sets,

$$\{0\ 2\ 3\ 1\},\ \{1\ 3\ 2\ 0\},\ \{3\ 2\ 1\ 0\}\ \cdots$$

all represent the same S_1 although for other purposes, we might want to recognize $0 < 1 < 2 < 3$ and the like.

In connection with repeated trials, we usualy need idealizations. For example, the actual toss of a coin may result in one of the ordinary points H T but also in one of the extraordinary points "falling off the table, disintegrating, standing on edge, ...". If we lump all these other points into "O" , a sample space is $\{H\ O\ T\}$. It is the collective experience of coin tossers that O occurs rather infrequently so that $f_n(O)/n$ is close to zero for large n. When we assign $P(O) = 0$, probability statements about a coin reduce to statements about $P(H)$ or $P(T)$. In practice, we ignore the extraordinary points and consider $S = \{H\ T\}$. You have done the equivalent thing in playing childhood games with "Slips goes over!" In medical trials, one would discount patients killed in car accidents when studying effects of a new cancer treatment.

Exercise 4:   Discuss some experiments with extraordinary outcomes and how they might be set aside.

The following discussion illustrates the introduction of "repeated trials" at a different stage. A "free test box" of a new candy contains chocolates — 3 with creme centers and 2 with hard centers. Mary is going to eat two pieces.
a)   If Mary is concerned only with eating a creme C or a hard center H, a tree diagram is

$$\text{First piece}\quad\text{Second piece}$$

with a corresponding sampl e space $S = \{CC\ CH\ HC\ HH\}$

b)   If Mary is concerned also with the flavor of the centers, she may consider the five pieces as distinct: C_1 C_2 C_3 H_1 H_2 . Then, a tree diagram is

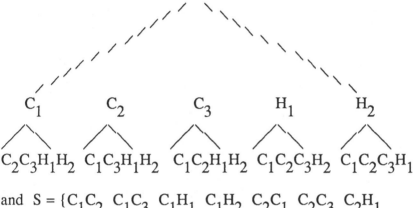

$$C_1 \qquad C_2 \qquad C_3 \qquad H_1 \qquad H_2$$

$$C_2C_3H_1H_2 \quad C_1C_3H_1H_2 \quad C_1C_2H_1H_2 \quad C_1C_2C_3H_2 \quad C_1C_2C_3H_1$$

and $S = \{C_1C_2$ C_1C_3 C_1H_1 C_1H_2 C_2C_1 C_2C_3 C_2H_1
C_2H_2 C_3C_1 C_3C_2 C_3H_1 C_3H_2 H_1C_1
H_1C_2 H_1C_3 H_1H_2 H_2C_1 H_2C_2 H_2C_3 $H_2H_1\}$

In the case of the doctor and the salve, the outcome R on one patient does not preclude the same or different outcome on the previous patient or the next one; similarly, N is non-preclusive. But in eating candy, once Mary has eaten the peach creme first, she cannot then consider eating that same piece second. To apply the principle of repeated trials (estimating probabilities), we need to think of eating two pieces from the test box as one trial; given a large number of test boxes (and Mary's), we may be able to say something about the selection of pairs:

$$P(C_1C_2), P(C_2H_2), \cdots$$

and, with more information, even something about which pieces were preferred, etc.

In a different vein, there are five candidates for two jobs; three are female [C_1 is Black, C_2 is Anglo, C_3 is Hispanic] and two are male [H_1 is Black, H_2 is Anglo]. Long run relative frequencies of such events might be used to examine possibilities of unfair labor practices.

"Distinguishability" can also be a source of confusion. Consider the case of a four–sided die wherein we look at the down–face on each toss: 1, 2, 3, 4. In the sequence first toss, second toss, it is obvious that there are 16 ordered pairs:

$$S = \{(1,1)\ (1,2)\ (1,3)\ (1,4)\ (2,1)\ (2,2)\ (2,3)\ (2,4)$$
$$(3,1)\ (3,2)\ (3,3)\ (3,4)\ (4,1)\ (4,2)\ (4,3)\ (4,4)\}.$$

When we toss a pair of "identical dice", we cannot distinguish (1,2) from (2,1), (2,3) from (3,2) etc. We see the tosses as if there were only ten pairs:

(1,1)   (1,2) = (2,1)   (1,3) = (3,1)   (1,4) = (4,1)   (2,2)

(2,3) = (3,2)   (2,4) = (4,2)   (3,3)   (3,4) = (4,3)   (4,4) .

But the dice {which do not know that} are in different physical positions and the ordered space of 16 pairs which is "finer" should be used. Our imagination must not be limited by our vision.

Exercise 5:   List the elements of a sample space for each experiment.
  a)   Toss a pair of six–sided dice.
  b)   Choose a family with three children (single births) so that the sample points may be represented by bbg (first a boy , then another boy, then a girl), bgb, $\cdots$.
  c)   Select three out of five invoices.
  d)   Select the first and second place winners from the four semifinalists Alice, Bill, Chris, Leslie.

Exercise 6:   Describe a sample space for the following experiments. Make note of the simplifications you are using.
  a)   Tossing a coin until a head appears or ten tosses have been made.
  b)   Finding the number of smokers in a group of 100 people.
  c)   Finding the number of delinquent charge accounts as of May 1 in six branches of one store.

Since our first discussions will be concerned with finite sample spaces, we should be rather specific about the term *finite*.

Definition: *The empty set (with no points) is finite with size 0. The non-empty set A is finite if there is a positive integer n such that the elements of A can be paired one to one with the set {1,2,3,...,n} (the set of positive integers less than or equal to n); then A has size n.*

Thus we have been talking about finite sample spaces all along— our experiments have only a finite number of distinct outcomes. Strictly speaking, as we are limited in our physical capacity to measure things precisely, all experiments have only a finite number of distinct outcomes; in some instances, it is convenient to idealize these in other ways and we will see some of this later.

Incidentally, pairing in one to one fashion is precisely the process of counting though we don't usually think of it in this way.

Example: a) {Jack Jill Hill Pail Water Crown Tumbling} has size 7 because we can form the pairs (1, Jack) (2, Jill) (3, Hill) (4, Pail) (5, Water) (6, Crown) (7, Tumbling).
b)   {Brigitte Clare Catherine} has size 3.
c)   At this particular moment in 1988, there are a finite number of people living in the United States, about 247 million; we can imagine the pairing even though we cannot do it physically.

If we start pairing even integers with positive integers (1,2) (2,4) (3,6) $\cdots$, there will be no end to it; both sets are not finite and are said to be *countably infinite*. It turns out that the set of rational numbers is also countably infinite but the set of real numbers is *uncountably infinite* — even greater . We shall return to these later.

INE–Exercises:
1.   Describe a sample space for the following experiments.
        a)   tossing a coin until the first head appears, then stopping;
        b)   finding the number of smokers in groups of 100 people;
        c)   counting the number of long–distance calls on a WATS line;
        d)   classifying one assembly of an article (car, dryer, $\cdots$) as satisfactory or unsatisfactory;
        e)   classifying three assemblies as in d).
2.   Hot dogs are inspected for shape (round and straight, round and curved, out of round), for color (light, medium, dark), and for taste (superior, ordinary, bad). How many "different kinds" of hot dogs are there?
3.   Cans of soda from a case are checked for proper filling by a weighing process. For each case, the process is stopped as soon as two underweight cans are found (in which event the case is returned for further checking) or five cans have been found with the proper weight (in which event the case

is sent to the shipping department). What is a sample space for this experiment if a case contains 12 cans? 24 cans?

4.   Certain stocks are rated   A+ or A– or B+ or B– or C. In how many ways could three different stocks be rated? In how many of these ways would exactly one of the three stocks be rated C ?

LESSON *3. SOME RULES ABOUT SETS

We begin with a traditional example, part of which you did as an exercise. A sample space for the tossing of a pair of six–sided dice is:

$$S = \{\begin{matrix} 11 & 12 & 13 & 14 & 15 & 16 \\ 21 & 22 & 23 & 24 & 25 & 26 \\ 31 & 32 & 33 & 34 & 35 & 36 \\ 41 & 42 & 43 & 44 & 45 & 46 \\ 51 & 52 & 53 & 54 & 55 & 56 \\ 61 & 62 & 63 & 64 & 65 & 66 \end{matrix}\} .$$

In many dice games, the rules are based not on the individual elements (pairs in this example) but on the sum of their upfaces. The following table shows these sums with the associated sample points.

Sum	Sample points with this sum
2	11
3	12 21
4	13 22 31
5	14 23 32 41
6	15 24 33 42 5
7	16 25 34 43 52 61
8	26 35 44 53 62
9	36 45 54 63
10	46 55 64
11	56 65
12	66

Each of the lists on the right hand side of this table is part of the sample space S; we can read these as:

the part of S for which the sum is 2 is the set of points 11;
the part of S for which the sum is 3 is the set of points 12 21;
the part of S for which the sum is 4 is the set of points 13 22 32 ; etc.

Of course, it is simpler to abbreviate these statements as:

Sum 2 = {11}

Sum 3 = {12 21}

Sum 4 = {13 22 31}, etc.

Now for some formalities.

Definition: *Parts of a sample space are called events or subsets of S. The points or elements of a set are said to be in the set or to belong to the set; the set contains the points. When set A is part of set B, "A is a subset of B", symbolically, $A \subset B$. In particular, atoms or singletons are sets of individual points. If point s is in set A , $s \in$ (belongs to) A but $\{s\} \subset A$.*

Example: a) Continuing the above discussion with dice, we have: $\{11\} \subset S$; $\{12 \ 21\} \subset S$; $12 \in \{\text{Sum 3}\}$; 22 is in $\{\text{Sum 4}\}$;

$$\{22\} \subset \{\text{Sum 4}\}; \ \{14 \ 23\} \subset \{14 \ 23 \ 32 \ 41\}; \quad \text{etc.}$$

b)   The integers are a subset of the rational numbers which are in turn a subset of the real numbers, itself a subset of the complex numbers.

c)   The U.S. Senators are a subset of the U.S. Congressmen.

Exercise 1:   Recall the sample space from the previous section where Mary is going to eat two of the five pieces of candy — C_1 C_2 C_3 H_1 H_2 ; S has 20 points (pairs). The event that Mary eats two cremes is $\{C_1C_2 \ C_2C_1 \ C_1C_3 \ C_3C_1 \ C_2C_3 \ C_3C_2\}$.

Similarly, you can list the points in the subsets of S described by the following:

a)   Mary eats exactly one creme.
b)   Mary eats no hard centers.
c)   Mary eats a creme first and a hard center second.
d)   Mary eats a creme first.

Example: For families with three children, ggb means that the oldest child is a girl, the next oldest child is a girl and the youngest child is a boy; etc. A sample space is

$$S = \{\text{bbb bbg bgb gbb bgg gbg ggb ggg}\}.$$

If A is the event that the oldest child is a boy, then

$$A = \{\text{bbb bbg bgb bgg}\}.$$

The event that the oldest child is a girl is, say,

$$B = \{\text{gbb gbg ggb ggg}\}.$$

Similarly, let $C = \{\text{bbb bgb gbb ggb}\}$ be the event that the youngest child is a boy and let $D = \{\text{bbg bgg gbg ggg}\}$ be the event that the youngest child is a girl.

    a)   The event that the youngest child is a girl and the oldest child is a boy must be part (a subset) of both A and D , that is, the part common to A and D . We write this as $A \cap D$ where $\cap$ is read intersect or meet. Thus, $A \cap D = \{bbg \ \ bgg\}$.

               Similarly, $B \cap C = \{gbb \ \ ggb\}$ is the event that the oldest child is a girl and the youngest is a boy .

    b)   The event that the youngest is a boy or the oldest is a girl is symbolized by $C \cup B$ where $\cup$ is read *union* or join. $C \cup B$ is the set of points in C or in B or in both; thus, $C \cup B = \{bbb \ bgb \ gbb \ ggb \ gbg \ ggg\}$.

       There are two things to note in a union. First the use of "or" in C or B or both is different from its usual use in English. The announcement "Brigitte or Catherine will sing tonight" usually precludes hearing a duet; "mathematically", this might occur. Secondly, "points" are not repeated; we do not allow $\{1 \ 2 \ 3\} \cup \{2 \ 3 \ 4\} = \{1 \ 2 \ 3 \ 2 \ 3 \ 4\}$ but only $\{1 \ 2 \ 3\} \cup \{2 \ 3 \ 4\} = \{1 \ 2 \ 3 \ 4\}$. Thus in the above example, $C \cup B$ has 6 symbols rather than eight since gbb and ggb are in both these sets: $C \cap B = \{gbb \ \ ggb\}$.

       General statements about union and intersection and some other vocabulary are contained in the following:

Definition: *Let A and B be events in a sample space S. Then $A \cap B$ is the event containing points in both A and B; $A \cup B$ is the event containing points in A or B or both. A^C is the set of points in S which are not in A; A^C is read A complement. If A and B have no points in common, $A \cap B = \phi$ is the usual symbol for the empty (null, vacuous) set; then A and B are said to be disjoint or mutually exclusive.*

       It may be an abuse of language to refer to ϕ as an empty *set* when it has no points, but it is convenient mathematics. Similarly, we allow a set to be a subset of itself so that $A \subset A$ is permissible; some writers use $A \subseteq A$. Also, ϕ may be a subset of any set.

Example: (continuing with families of three children)

    a) $A^C = \{ggg \ \ ggb \ \ gbg \ \ gbb\}$ so $A \cup A^C = S$ and

$A \cap A^c = \phi$. Note here that $A^c = B$ so $A \cup B = S$ as it should since the oldest child is either a boy (A) or a girl (B). Similarly, $C \cap D = \phi$.

b) $\{bbb \ ggg\}^c = \{bbg \ bgb \ gbb \ bgg \ gbg \ ggb\}$.

The ideas in the definition above can be displayed neatly in *Venn diagrams*. This geometric view should also clarify the lack of order and duplication: there is no "first" point on the page and each point appears only once. In interpreting these models, we ignore the "boundaries".

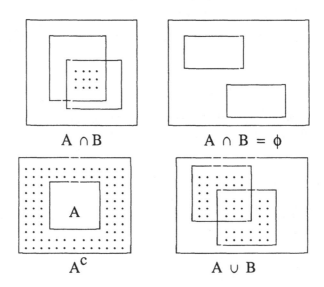

$A \cap B$ $A \cap B = \phi$

A^c $A \cup B$

Exercise 2:   Complete each statement.
a)   If $A = \{1\ 2\ 3\ 4\ 5\}$ and $B = \{6\ 7\ 5\}$ then $A \cup B =$
b)   If $A = \{a\ b\ c\ d\}$ and $B = \{c\ d\ e\}$,
        then $A \cap B = \{cd\}$.
c)   If $S = \{\ 2\ 3\ 4\ 5\ 6\ 7\ 8\ 9\ 10\ 11\ 12\}$ and B is the subset whose integers are divisible by 2, C is the subset whose integers are divisible by 3 and D is the subset whose integers are divisible by 6, then:

$C = \cdots , B = \cdots , D = \cdots , C \cap B = \cdots ,$

$C \cap D = \cdots , B \cap D = \cdots , D \cap B = \cdots ,$

$C \cup D = \cdots , B \cup C = \cdots , C^c = \cdots , D^c \cap C = \cdots ,$

$(C \cap D)^c = \cdots .$

In part c) of the above exercise, note that $D \cup C = C \cup D$ and $C \cap D = D \cap C$ as well. Indeed, these are general rules, valid for all (sub)sets. This is very easy to see in the Venn diagrams — what is common to two sets does not depend on the order of labeling the two sets; etc. Since you undoubtedly recognized the equations at the beginning of this paragraph as statements of commutative laws, you may anticipate that there are also distributive laws (see the theorem below). We also see the common part of three sets in the associative law:

$$A \cap (B \cap C) = (A \cap B) \cap C \text{ written as } A \cap B \cap C \;;$$

similarly, $A \cup (B \cup C) = (A \cup B) \cup C = A \cup B \cup C$. For completeness, we include the following recursive forms.

Theorem: *For* $A, A_1, A_2, \cdots$ *subsets of a sample space* S,

$$\bigcup_{i=1}^{n} A_i = A_1 \text{ for } n = 1 \text{ and}$$

$$\bigcup_{i=1}^{n} A_i = (\bigcup_{i=1}^{n-1} A_i) \cup A_n \text{ for } n \geq 2 \;;$$

$$\bigcap_{i=1}^{n} A_i = A_1 \text{ for } n = 1 \text{ and}$$

$$\bigcap_{i=1}^{n} A_i = (\bigcap_{i=1}^{n-1} A_i) \cap A_n \text{ for } n \geq 2 \;.$$

Then $A \cap (\bigcup_{i=1}^{n} A_i) = \bigcup_{=1}^{n} (A \cap A_i)$ *for* $n \geq 1 \;;$

$$A \cup (\bigcap_{i=1}^{n} A_i) = \bigcap_{i=1}^{n} (A \cup A_i) \text{ for } n \geq 1 \;.$$

Exercise 3: The reader may use mathematical induction to prove the theorem just as for the distributive law in the algebra of real numbers.

Exercise 4: Complete the expansions and note the use of the basic $A \cap B$ or $A \cup B$ at each step.

a) $\quad \bigcup_{i=1}^{5} A_i = (\bigcup_{i=1}^{4} A_i) \cup A_5 = \cdots$

$$= A_1 \cup A_2 \cup A_3 \cup A_4 \cup A_5$$

b) $\displaystyle\bigcap_{i=1}^{5} A_i = (\bigcap_{i=1}^{4} A_i) \cap A_5 = \cdots$

$$= A_1 \cap A_2 \cap A_3 \cap A_4 \cap A_5$$

Exercise 5:   Verify the following laws using Venn diagrams {draw pictures and color appropriate regions}; of course, this is not a proof.

De Morgan: $(A \cap B)^c = A^c \cup B^c$

$(A \cup B)^c = A^c \cap B^c$

Distributivity: $A \cap (B \cup C) = (A \cap B) \cup (A \cap C)$

$A \cup (B \cap C) = (A \cup B) \cap (A \cup C)$.

Definition: *Let $B_1, B_2, \cdots, B_n$ be subsets of a sample space S such that $S = \displaystyle\bigcup_{i=1}^{n} B_i$ and $B_i \cap B_j = \phi$ for $i \neq j$. We say that $\{B_i\}$ partitions S .*

For example, àla Venn,

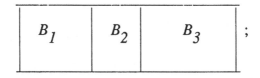

$$B_1 \quad\quad B_2 \quad\quad B_3 \quad\quad ;$$

or, perhaps

$$S = \{\, a\ e\ i\ o\ u \,\} = \{a\ e\} \cup \{i\ u\} \cup \{o\} \, .$$

Exercise 6:   a) Look at the example which began this lesson and write down the corresponding $\{B_i\}$ for the sums.

b)   Let A be a subset of S and let $A_i = A \cap B_i$ with $\{B_i\}$ as in the definition above. Show that $\{A_i\}$ partitions A. (The formalities of induction may be "imagined" if you want to work with a specific n.)

Exercise 7: Three subsets partition a sample space S into at most eight regions or subsets as in the Venn diagram. List the numbers of the regions in each event. Note: A = {1 3 4 7}, B = {1 2 4 5}, C = {1 2 3 6} .

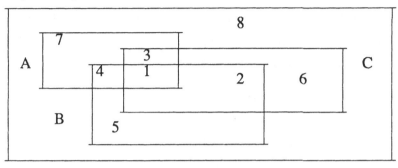

A, B, C, A ∪ B, A ∪ C, B ∪ C, A ∩ B, A ∩ C,

B ∩ C, A ∩ B^c, A^c ∪ B, (A ∩ B)^c, (A ∩ B)^c,

A ∪ C^c, (A ∩ B ∩ C)^c, A ∩ (B ∪ C), A ∩ (B ∪ C)^c.

Exercise 8 : A sample space S can be partitioned into 16 subsets by four given subsets A, B, C, D.

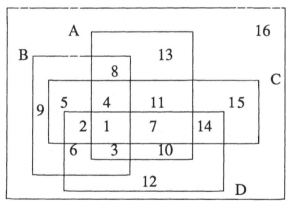

a) List the numbers of the regions corresponding to: A, B, C, D, A ∩ B, A ∩ C, A ∩ D, B ∩ C, B ∩ D, C ∩ D, A ∩ B ∩ C, A ∩ B ∩ D, A ∩ C ∩ D, B ∩ C ∩ D, A ∩ B ∩ C ∩ D .

b) Each numbered region is the intersection of four sets. For example, 1 = A ∩ B ∩ C ∩ D ,

$$2 = A^c ∩ B ∩ C ∩ D ,$$

$$16 = A^c \cap B^c \cap C^c \cap D^c .$$
Complete the list for 3 thru 15.

INE–Exercises:
1.   Insured postal packages have been inspected at the destination with the following results. "C" denotes damage to the contents; "A" denotes major damage to the packaging, and "D" denotes minor damage to the packaging.

Condition	Percentage
only C	3.5
only A	3
only D	2.5
only (C and A)	2
only (C and D)	1
only (A and D)	1.5
only (C and A and D)	.5

   a)   What percent of such packages have none of these damages?
   b)   Owners of "C" packages are reimbursed; what percentage is this?
   c)   What percent have damage to the package only?

2.   Let A_1, A_2, A_3 be events. Find expressions in terms of unions, intersections and complements for the event that: only A_2 occurs; A_1 and A_2 occur but A_3 does not occur; at least one of the three events occurs; at least two of the three events occur; not more than one of the three events occurs; exactly two of the three events occur.

3.   Let A, B, C be subsets of S. With $A - B = A \cap B^c$, the *symmetric difference* is $A \triangle B = (A–B) \cup (B–A)$. Show that:
   a) $(A–B) \cap (B–A) = \phi$.
   b) $A - B$, $A \cap B$, $B - A$ partition S
   c) $A \triangle B = \phi$ iff $A = B$
   d) $A \triangle B = (A \cup B) – (A \cap B)$
   e) $A \cap (B \triangle C) = (A \cap B) \triangle (A \cap C)$.

LESSON 4. THE COUNTING FUNCTION FOR FINITE SETS

In order to rediscover some properties of probability that were inspired by counting, we need to pursue properties of the counting function itself. Let S be a finite set (sample space) with points $s_1, s_2, \cdots, s_m$ when S has size m; the *counting function* begins with $\#(\{s_i\}) = 1$;

$\#(A)$ is the number of elements s_i in subset A.

Obviously (a lemma?), any subset A of S is also finite so that $\#(A)$ is finite. Also, when two subsets C and D have no points in common, then the number of points in both these sets together is just the sum of the numbers in each.

Example: a) Let $S = \{a\ b\ c\ d\ e\}$.
$$\#(\{a\ b\}) = 2 \quad \#(\{c\ d\ e\}) = 3$$
$$\#(\{a\ b\} \cup \{c\ d\ e\}) = \#(\{a\ b\ c\ d\ e\}) = 5$$

$$= 2 + 3 = \#(\{a\ b\}) + \#(\{c\ d\ e\}) .$$

b) Suppose that S is just the points on the page:

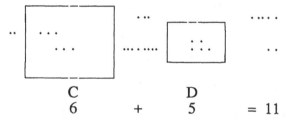

$$
\begin{array}{ccccc}
C & & D & & \\
6 & + & 5 & = & 11
\end{array}
$$

c) The number of animals in a room is the number of females in the room plus the number of males in the room.

At this stage, we cannot prove this result; it must be taken as an axiom which we write as the

First Fundamental Principle of Counting: when two finite (sub)sets have no points in common, the number of points in their union is the sum of the number of points in each.

Formally, we have the

Definition: *Let $\mathscr{F}$ be the collection of all finite subsets of a set S. Then the counting function is the set function # defined on $\mathscr{F}$ with the properties (axioms): $\#(\phi) = 0$; $\#(\{s\}) = 1$;*

$$C \cap D = \phi \text{ implies } \#(C \cup D) = \#(C) + \#(D).$$

Note that the domain of # is a set of sets and that is why # is called a set function; $\#(\{s\}) = 1$ is abbreviated as $\#(s) = 1$.

Example: The *power set* is the collection of all subsets of a given set. The following is a list of the elements of the power set of the set $S = \{a\ b\ c\ d\}$:

$$\phi$$

$$\{a\}\ \{b\}\ \{c\}\ \{d\}$$

$$\{a\ b\}\ \{a\ c\}\ \{a\ d\}\ \{b\ c\}\ \{b\ d\}\ \{c\ d\}$$

$$\{a\ b\ c\}\ \{a\ b\ d\}\ \{a\ c\ d\}\ \{b\ c\ d\}$$

$$\{a\ b\ c\ d\}$$

There is one subset of size 0 and one of size $\#(S)$; there are four subsets of size 1 , four of size 3 and six subsets of size 2.

Exercise 1: Make lists of the power sets of each "S":

$$S_1 = \{0\ 1\} \quad S_2 = \{1\ 2\ 3\ 4\} \quad S_3 = \{a\ e\ i\ o\ u\}\ .$$

Take note of the number of subsets of each size.

The following theorem gives the rule for finding the number of points in the union of two (finite) sets which may not be disjoint. Very likely, you know how to do this already. Incidentally, the Venn diagrams are included for elucidation but are not a part of the logic of the proof.

Theorem: *Let $\mathscr{F}$ be the collection of all finite subsets of a set S. Let A and B be two elements of $\mathscr{F}$. Then,*

$$\#(A \cup B) = \#(A) + \#(B) - \#(A \cap B)\ .$$

Proof: First, $A \cup B = A \cup (A^c \cap B)$ and $A \cap (A^c \cap B) = \phi$

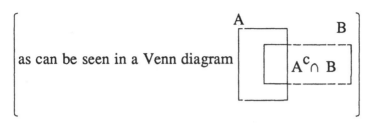

as can be seen in a Venn diagram

Let $C = A$ and $D = A^c \cap B$; then $C \cup D = A \cup B$ and $\#(C \cup D)$ $= \#(A \cup B)$. Now, $C \cap D = \phi$ so by the axiom, $\#(C \cup D) = \#(C)$ $+ \#(D)$ and by substitution,

$$\#(A \cup B) = \#(A) + \#(A^c \cap B) . \tag{1}$$

Second, $B = (A \cap B) \cup (A^c \cap B)$ and $(A \cap B) \cap (A^c \cap B) = \phi.$

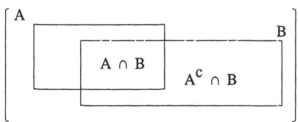

Now let $C = A \cap B$ and $D = A^c \cap B$; then

$$C \cup D = B \text{ and } \#(B) = \#(C \cup D) .$$

As above, $C \cap D = \phi$, $\#(C \cup D) = \#(C) + \#(D)$ and

$$\#(B) = \#(A \cap B) + \#(A^c \cap B) \tag{2}$$

Since all the values of $\#$ are finite, we may subtract equation (2) from equation (1) to get

$$\#(A \cup B) - \#(B) = \#(A) - \#(A \cap B)$$

which is equivalent to the conclusion.

Example: a) Let S be the dots printed on the page with A those inside the rectangle and B those inside the square.

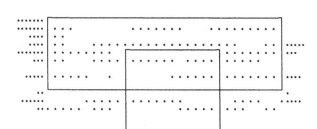

$$\#(A \cup B) = \#(A) + \#(B) - \#(A \cap B) \quad \text{is}$$
$$121 \; = \; 98 \; + \; 39 \; - \; 16$$

b)   It should be noted that when # is interpreted as area, this formula is also valid for the areas of the page enclosed by these figures.

Exercise 2: The following corollary is almost obvious but write out the details as in the theorem.

Corollary: *If $A \subset B$, let $B - A = B \cap A^{c}$. Suppose that A and B are both finite. Then, $\#(B - A) = \#(B) - \#(A)$.*

Exercise 3:   A clerk in a registrar's office reported that 30 students were enrolled in Math 300, 40 students were enrolled in Physics 300 and 10 students were enrolled in both these courses.
    a)   How many students is she talking about?

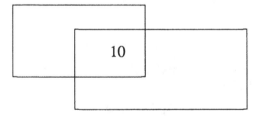

    b)   Later among 84 other student records she found 59 enrolled in Chem 212, 37 enrolled in Math 300 . How many of these students were enrolled in both Math and Chem?
    c)   Can you tell how many students were enrolled in all three courses?

The answer to the last question in the exercise is no! To deal with more than two sets in one conglomeration, we need extensions of these rules. One way to get these is by multiple application of the theorem as in the following.

Examples: a) First we use the associative law for the union of sets: $\#(A_1 \cup A_2 \cup A_3) = \#((A_1 \cup A_2) \cup A_3)$.

Then the theorem may be applied to the right–hand side to get:

$$\#(A_1 \cup A_2 \cup A_3) = \#(A_1 \cup A_2) + \#(A_3) - \#((A_1 \cup A_2) \cap A_3) \ . (3)$$

Next we apply the distributive law for sets to get

$$(A_1 \cup A_2) \cap A_3 = (A_1 \cap A_3) \cup (A_2 \cap A_3) \ ;$$

then we apply the associative law for intersections to get

$$(A_1 \cap A_3) \cap (A_2 \cap A_3) = A_1 \cap A_2 \cap A_3 \ .$$

Using these in the last sharp of (3), yields
$\#((A_1 \cup A_2) \cap A_3)$

$$= \#((A_1 \cap A_3) \cup (A_2 \cap A_3))$$

$$= \#(A_1 \cap A_3) + \#(A_2 \cap A_3) - \#(A_1 \cap A_2 \cap A_3). \ (4)$$

Also, $\#(A_1 \cup A_2) = \#(A_1) + \#(A_2) - \#(A_1 \cap A_2) \ . \ (5)$

Putting (4) and (5) in the right side of (3) yields
$$\#(A_1 \cup A_2 \cup A_3) = \#(A_1) + \#(A_2) + \#(A_3)$$
$$- \#(A_1 \cap A_2) - \#(A_1 \cap A_3) - \#(A_2 \cap A_3)$$
$$+ \#(A_1 \cap A_2 \cap A_3) \ .$$

b)   If S is the dots on the page,

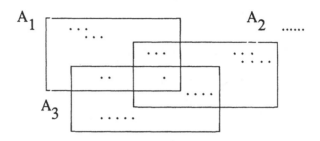

$$\#(A_1 \cup A_2 \cup A_3) = 1 + 2 + 3 + 4 + 5 + 6 + 7 = 28$$

$$\#(A_1) = 1 + 2 + 3 + 6 = 12 \quad \#(A_2) = 1 + 3 + 4 + 7 = 15$$

$$\#(A_3) = 1 + 2 + 4 + 5 = 12 \quad \#(A_1 \cap A_2) = 1 + 3 = 4$$

$$\#(A_1 \cap A_3) = 1 + 2 = 3 \quad \#(A_2 \cap A_3) = 1 + 4 = 5$$

$$\#(A_1 \cap A_2 \cap A_3) = 1$$

and $\quad 28 = 12 + 12 + 15 - 4 - 3 - 5 + 1$

Exercise 4: Make up an example involving four sets. Hint: see the last regular exercise in the previous lesson.

Theorem: *Let S be a set and let # be the counting function on the class $\mathcal{F}$ of finite subsets of S. Let $A_1, A_2, A_3, \cdots$ be elements of $\mathcal{F}$ such that for $i \neq j$, $A_i \cap A_j = \phi$. Then for each positive integer n,*

$$\#(A_1 \cup A_2 \cup \cdots \cup A_n) = \#(A_1) + \#(A_2) + \cdots + \#(A_n) .$$

Proof: Let $\tau(n)$ be the statement "$\# \left[\overset{n}{\underset{i=1}{\cup}} A_i \right] = \overset{n}{\underset{i=1}{\Sigma}} \#(A_i)$" and

let Z be the set of positive integers for which $\tau(n)$ is true. The integer 1 is in Z because $\tau(1)$ is just $\#(A_1) = \#(A_1)$. Next we make the "induction hypothesis" that a positive integer k is in Z :

$$\#(\overset{k}{\underset{i=1}{\cup}} A_i) = \overset{k}{\underset{i=1}{\Sigma}} \#(A_i) . \qquad (6)$$

Now, $\bigcup\limits_{i=1}^{k+1} A_i = (\bigcup\limits_{i=1}^{k} A_i) \cup A_{k+1}$ and (by the lemma you are to

prove below) $(\bigcup\limits_{i=1}^{k} A_i) \cap A_{k+1} = \phi$. Hence, for # of a disjoint

union, (take $C = \bigcup\limits_{i=1}^{k} A_i$ and $D = A_{k+1}$ in the definition),

$$\#(\bigcup\limits_{i=1}^{k+1} A_i) = \#(\bigcup\limits_{i=1}^{k} A_i) + \#(A_{k+1}) .$$

By substitution from the induction hypothesis (6),

$$\#(\bigcup\limits_{i=1}^{k+1} A_i) = \sum\limits_{i=1}^{k} \#(A_i) + \#(A_{k+1}) = \sum\limits_{i=1}^{k+1} \#(A_i) .$$

Thus, assuming $\tau(k)$ is true, we prove that $\tau(k+1)$ is true and the induction is complete.

Did you note the use of the corresponding results from ordinary algebra? For real numbers $a_1, a_2, a_3, \cdots$, and each positive integer $n \geq 2$,

$$\Sigma_{i=1}^{n} a_i = \Sigma_{i=1}^{n-1} a_i + a_n \qquad \Pi_{i=1}^{n} a_i = \Pi_{i=1}^{n-1} a_i \times a_n .$$

"Π" is an abbreviation for product as "Σ" is for the sum.

Exercise 5: Prove by mathematical induction. If $A_1, A_2, A_3, \cdots$ are subsets as in the theorem, then for each positive integer n,

$$\left[\bigcup\limits_{i=1}^{n} A_i \right] \cap A_{n+1} = \phi .$$

A traditional analogy for induction is the dominoe line: the first one is knocked down against the second one which is knocked down against the third one which is $\cdots$.

We should note what happens when S itself is finite, say of size M. Then it is obvious that the largest number of disjoint (nonempty) subsets we can have is M. The theorem is still valid when we recognize that any additional "disjoint" sets A_{M+1}, A_{M+2}, $\cdots$ will have to be ϕ. With this codicil, we have validated the use of the counting function with more than two sets.

Example: In a survey of 116 students taking high–school science, the following enrollments were reported. Physics 52, Chemistry 52, Biology 59, Physics and Chemistry 23, Physics and Biology 24, Chemistry and Biology 25, all three 11. Recalling that three subsets can partition a subspace into eight parts, we use the enrollment figures given and all the properties of the counting function to find the enrollment in each part. With obvious abbreviations, since $P \cap B \cap C \subset C \cap B$,

$$\#(P^C \cap B \cap C) = 25 - 11 = 14 \; ;$$

similarly, $\#(P \cap B^C \cap C) = 23 - 11 = 12$ and

$$\#(P \cap B \cap C^C) = 24 - 11 = 13 \; .$$

At this point, we have the following form.

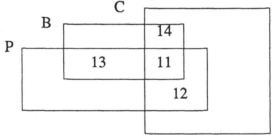

Since $\#(P) = 52$, $\#(P \cap B^C \cap C^C) = 52 - 13 - 11 - 12 = 16$; similarly, $\#(P^C \cap B \cap C^C) = 55 - 13 - 11 - 14 = 17$ and

$$\#(P^C \cap B^C \cap C) = 52 - 12 - 11 - 14 = 15 \; .$$

Altogether now we have

$$11 + 12 + 13 + 14 + 15 + 16 + 17 = 98$$

so there must be $116 - 98 = 18$ students taking none of these sciences.

Exercise 6: In a similar survey, the following tabulations were reported. P—22, B—30, C—25, PBC—5,
BC—14, PB—12, PC—8, None—10 .

Draw a Venn diagram showing the eight regions and label each with its frequency.

Exercise 7: Three inspections are made in the manufacture of a certain nut: Threads, Corners, Smoothness. Suppose that a lot of 1000 nuts is inspected; the following frequencies represent the number of nuts which fail the inspection indicated.

$$\#(T) = 150 \quad \#(C) = 140 \quad \#(S) = 135$$

$$\#(T \cup C) = 210 \quad \#(T \cup S) = 195 \quad \#(C \cup S) = 205$$

$$\#(T \cup C \cup S) = 225 \ .$$

How many nuts failed all three inspections?

LESSON 5. PROBABILITY ON FINITE SAMPLE SPACES

We continue the discussion of experiments with finite sample spaces. The doctor testing salve had a sample space with eight sample points: RRR, RRN, $\cdots$, NNN; the waitress taking egg orders had a sample space with six points: S, OH, OE, SS, B, P. In order to talk about a sample space S in general, it is necessary to have some names for the distinct sample points.

Definition: *When S has size M, label the points* $s_1, s_2, \cdots, s_M$ *(without any ordering being implied by the subscripts). Then each point* s_k, *a one point set* $\{s_k\}$, *is called an elementary event. When the experiment is actually performed and the outcome is* s_k , *we say that the elementary event* $\{s_k\}$ *has occurred.*

In lesson 1, we suggested the consideration of large numbers of repetitions of an experiment. If there are "n" repetitions, then $f_n(\{s_k\})$ or more simply $f_n(s_k)$ denotes the number of times the elementary event $\{s_k\}$ occurred in the first n repetitions.

Example: a) For the egg orders, we have the elementary events $\{S\}$, $\{OH\}$, $\{OE\}$, $\{SS\}$, $\{B\}$, $\{P\}$. (We hope using $\{S\}$ for an event doesn't get scrambled with S as a space.) Two of the final frequencies are $f_{50}(S) = 12$, $f_{50}(OH) = 11$.

b)   For the hospital confinements, the elementary events are the integers $\{1\}$, $\{2\}$, $\{3\}$, $\cdots$, but those observed were
$$\{17\}, \{16\}, \{15\}, \{14\}, \{13\},$$
$$\{12\}, \{11\}, \{10\}, \{9\}, \{8\}, \{6\} .$$
Here we have $f_{50}(17) = 2$, $f_{50}(15) = 4$, $f_{50}(13) = 10$ and so on.

Exercise 1: Find the frequencies $f_{50}(OE)$, $f_{50}(SS)$, $f_{50}(B)$, $f_{50}(P)$ in the example on egg orders in lesson 1.

In applications, the questions of interest may not be specifically about elementary events. For example, the hospital administrator considered lengths of stay under 14 days, that is, the event {13 12 11 10 9 8 7 6 5 4 3 2 1}. In lesson 3, Mary was able to direct her attention to creme centers in the chocolates. In short, we may be concerned with frequencies and relative frequencies of other "compound" events. Of course, any event is made up of the union of elementary events and we have seen some of this before without the verbal descriptions..

Example: a) In the hospital example, the event A that the length of stay was less than 14 days is the union of the elementary events {13}, {12}, {11}, $\cdots$, {2}, {1} .
b)   In some dice games, the events of interest are the sums on the upfaces of six–sided dice:

$$\{Sum\ 3\} = \{12\ 21\} = \{12\} \cup \{21\}$$

$$\{Sum\ 4\} = \{13\} \cup \{22\} \cup \{31\}, \ \text{etc.}$$

Exercise 2: An inspector checks four resistors, recording 1 when the resistor is performing properly and recording 0 otherwise; for example, 0010 would mean that only the third resistor in the line was performing properly. List the 16 quadruples in his sample space. Then list the elementary events in:
a)   the event that exactly three resistors are performing properly;
b)   the event that at least three resistors are performing properly;
c)   the event that exactly one resistor (not saying which one) is performing properly .

The following extends the idea of "occurred".

Definition: *When the experiment is performed and the outcome is s_k and s_k is in the event (subset) A , we say also that the event A has occurred.*

Suppose for the moment that the event A has just two sample points s_1, s_2: $A = \{s_1\} \cup \{s_2\}$. Then $f_n(A)$ denotes the number of occurrences of A in n trials. In the case of the egg orders, for $A = \{S\ OH\} = \{S\} \cup \{OH\}$,

$$f_{50}(A) = 23 = f_{50}(S) + f_{50}(OH).$$

In general, since s_1 and s_2 are distinct points, the number of occurrences of either is just the sum of the numbers of their individual occurrences. More precisely, for the disjoint sets

$$\{s_1\} \text{ and } \{s_2\}, \quad f_n(A) = f_n(\{s_1\} \cup \{s_2\}) = f_n(s_1) + f_n(s_2) .$$

Of course, we have done all this before in ordinary counting. From $A = \{s_1, s_2\}$ and $f_n(A) = f_n(s_1) + f_n(s_2)$, we get

$$f_n(A)/n = f_n(s_1)/n + f_n(s_2)/n .$$

If there is any consistency in our basic assumption about relative frequency (for this experiment), it would follow that

$$P(A) = P(s_1) + P(s_2) .$$

In words, the probability that A *will occur* is the sum of the probabilities that s_1 or s_2 will occur.

It follows that for any event A the probability that A will occur (on the next repetition so to speak) is the sum of the probabilities for each of the elementary events in A. In symbols,

$$P(A) = \sum_A P(s_k) \text{ means the sum } (\Sigma) \text{ of those probabilities } P(s_k)$$

for the elementary events $\{s_k\}$ in the (compound) event A.

Example: a) Recall the sample space for the tossing of a pair of six–sided dice.

  $P(\text{Sum } 5) = P(14 \ 23 \ 32 \ 41) = P(14) + P(23) + P(32) + P(41) .$

b) Suppose given $S = \{1 \ 2 \ 3\}$ and $P(1) = 1/2$, $P(2) = 1/3$,
   $P(3) = 1/6$. Then,
   $P(1 \ 2) = P(\{1 \ 2\}) = P(1) + P(2) = 1/2 + 1/3 + 5/6 $;

   $P(2 \ 3) = P(\{2 \ 3\}) = P(2) + P(3) = 1/3 + 1/6 = 3/6 .$

c) Past experience (the usual source of a large number of repeated trials) suggests that the probabilities for the number of "short–filled" cans in a six–pack of cola (from a certain plant) are

   $P(0) = .980 \quad P(1) = .010 \quad P(2) = .003$
   $P(3) = .003 \quad P(4) = .003 \quad P(5) = .001 \quad P(6) = .000.$

Here $S = \{0 \ 1 \ 2 \ 3 \ 4 \ 5 \ 6\} .$

i) The probability that all cans are "good" is $P(0) = .980.$

ii)   The probability that *not more than one* can is short is

$$P(0\ 1) = P(0) + P(1) = .980 + .010 = .990 .$$

iii)   The probability that *at least one* can is short is

$$P(1\ 2\ 3\ 4\ 5\ 6\ ) = P(1) + P(2) + P(3) + P(4) + P(5) + P(6)$$

$$= .010 + .003 + .003 + .003 + .001 + .000 = .020.$$

Exercise 3: Complete the table for the dice probabilities when each point in $S = \{11\ 12\ \cdots\ 16\ \cdots\ 61\ \cdots\ 66\}$ has probability $1/36$.

Sum	Probability
2	$P(11) = 1/36$
3	$P(12\ 21) = P(12) + P(21) = 2/36$
4	
5	
6	
7	
8	
9	
10	
11	
12	

The properties of relative frequencies also suggest other properties that probability should have:

1) $f_n(S) = n$ since the sample space S contains all possible outcomes of the n repeated trials. Hence, $f_n(S)/n = 1$ and $P(S) = 1$.

2) $f_n(\phi) = 0$ or $f_n(\phi)/n = 0$ since we always have some outcome (with a "physical experimental failure" listed as one of these). Hence, $P(\phi) = 0$.

3) $0 \le f_n(A) \le n$ for all events A since the *frequency* of an event is not negative (certainly) and can't be greater than the number of trials or repetitions. Hence,

$$0 \le f_n(a)/n \le 1 \text{ and } 0 \le P(A) \le 1 .$$

This discussion of what may happen in the physical domain, that is with large numbers of repetitions of an experiment, does not prove the laws of probability in the mathematical domain. Here, we must make an assumption.

Definition: *Let S be a finite set and $\mathcal{F}$ the collection of all its subsets. Then probability is a rule which assigns to each subset A of S (element A in $\mathcal{F}$) a number P(A) with the following*

properties: for each A in $\mathcal{F}$, $P(\phi) = 0 \le P(A) \le 1 = P(S)$

for each A and B in $\mathcal{F}$, with $A \cap B = \phi$,

$$P(A \cup B) = P(A) + P(B) .$$

The triple $[S, \mathcal{F}, P]$ is called a probability space.

We note that probability is another set function. From the last property of the definition, we have

$$P(\{s_1 \ s_2\}) = P(\{s_1\} \cup \{s_2\}) = P(s_1) + P(s_2)$$

since elementary events are disjoint— $\{s_1\} \cap \{s_2\} = \phi$.(This was set up to agree with the physical point of view.) Technically, we now need a mathematical induction to get the general rule

$$P(A) = \sum_A P(s_k)$$

but we can prove the following more general result exactly as with # in the previous lesson. Then, letting $A_k = \{s_k\}$, we get this particular case.

Theorem: *Let $\{A_i\}$ be disjoint subsets of a (finite) probability space. Then for each positive integer n, $P(\bigcup_{i=1}^{n} A_i) = \sum_{i=1}^{n} P(A_i)$.*

Exercise 4: Prove the theorem above by translating the proof for the counting function # .

Exercise 5: For the example of cola cans that may be "short–filled", show that the probability that there is at least one such can is $1 - P(0)$.

Exercise 6 : Consider $S = \{0 \ 1 \ 2 \ 3 \ 4 \ 5 \ 6 \ 7 \ 8 \ 9\}$ with $\mathcal{F}$ its power set and any assignment of probability on $\mathcal{F}$. Letting X be a generic outcome, we have a formal sum for the probability that X is positive:

$$P(X > 0) = 1 - P(0)$$

$= P(1) + P(2) + P(3) + P(4) + P(5) + P(6) + P(7) + P(8) + P(9)$.

Write out similar formal sums for the probabilty that:

X is at most 1;   X is less than 1 ;
X is at least 2;   X is not negative ;
X is not more than 1;   X is at least 9 ;
X is not less than 5;   X is at most 9 ;
X is at least 5;   X is not greater than 6 ;
X is greater than 2 but less than 7 .

Exercise 7: Let (S, $\mathscr{S}$, P) be a finite probability space.

a)   Show that $P(A^c) = 1 - P(A)$.

b)   If $A \subset B$ and both are in $\mathscr{S}$, show that $P(A) \leq P(B)$.

c)   If $S = \{s_1, s_2, ..., s_m\}$, what is the value of

$$\sum_S P(s_k) = P(s_1) + P(s_2) + ... + P(s_m) \ ?$$

Exercise 8: This record has been obtained from the "Returns & Exchanges" counter of a large department store:

Time	(Average) Number of Patrons
9:00——9:59	2
10:00——10:59	3
11:00——11:59	5
12:00——12:59	9
1:00——1:59	4
2:00——2:59	2
3:00——3:59	1
4:00——4:59	3
5:00——5:59	7
6:00——6:59	5

a)   We can use the total number of patrons to calculate various relative frequencies for arrival times (grouped by the hour). Complete the list on the next page and draw a graph as in lesson 1.

b)   What is the proportion of the arrivals:
between 1 and 3 ?   after 5 PM ?
before noon ?   after 12 but before 5 PM ?

Before	Relative frequency or Proportion
9	0/41
10	2/41
11	
12	
13 (for purposes of the graph)	
14	
15	
16	
17	
18	
19	

Theorem: *For events A, B, C in the probability space* $[S, \mathcal{G}, P]$,

$$P(A \cup B) = P(A) + P(B) - P(A \cap B) \; ;$$

$$P(A \cup B \cup C) = P(A) + P(B) + P(C)$$
$$- P(A \cap B) - P(A \cap C) - P(B \cap C)$$
$$+ P(A \cap B \cap C) .$$

Exercise 9: Write out a proof of the last theorem translating an earlier result for # .

Exercise 10: To compare the work of two inspectors (Willing and Able), they are asked to examine a lot which, unknown to them, has 4% defective items. After doing their job, separately, both say that the lot contains 5% defectives. It happens that both correctly identified 2% of the items as defectives. They both called the same 3% of the items defective but only 1% correctly. What percent of the truly defective items did the inspectors miss together? How does their work compare?

Exercise 11: Prove the following

Corollary: *If A and B are events in a probability space with* $P(A) = 1$, *then* $P(A^C \cap B) = 0$, $P(A \cap B) = P(B)$,
$$P(A \cap B^C) = P(B^C) .$$

LESSON *6. ORDERED SELECTIONS

The study of probability was given great impetus when prominent gamblers began asking questions (about their losses) of prominent mathematicians of the 17th and 18th centuries. The latter gentlemen recognized that many card games would have sample spaces with millions of points so that they had to develop formulas for finding the number of points in various subsets without relying on enumeration as we have in the previous lessons. We will consider formulas for ordered selections in this lesson and for unordered selections in lesson 7; lesson 8 will be devoted to the calculation of some classical probabilities.

First we need to clarify the title. "Order" refers to position in a line. When order is "counted" or "considered" or "important" (different writers use different phrases), symbols like abc, acb, cab, cba, bac, bca are judged to be "distinguishable" and each is counted as a selection or arrangement. When order is not considered, these six triplets are judged to be "indistinguishable" and only one (but any one) is counted. Some thought needs to be given to the decision that order does or does not count; for example, when a committee of three is chosen, does it make any difference which member is chosen first?

Here, selection refers to choosing one at a time but this can be "with replacement (repetition)" or "without replacement (repetition)". We have already described *ordered selections*:
a)   *with repetition* in tree diagrams for repeated trials (oh, Doctor!),
b)   *without replacement* in diagrams with destructive sampling (oh, Mary!).
These will now be phrased in this new terminology. It is to be emphasized that the analogy is for the construction of the sample space only; the values of the corresponding probabilities are quite another matter.

Examples: a) To describe a sample space for testing a salve on three patients, the doctor can imagine the possible outcomes R and N as two chips in an urn. In terms of possibilities alone, the outcome for each person corresponds to selecting one of these chips. Since either result on patient one, does not preclude either result on patient two, the chip is *replaced* in the urn and the possibilities for the second outcome also correspond to selecting a chip from the urn. Similarly for patient three (four,···).

b)   Mary is to select two chips (chocolates) from five chips
      (pieces) in an urn (free sample): C_1, C_2, C_3, H_1, H_2 .
      After she selects (eats) the first piece, she may not
      exchange it for another piece but must select the second
      from those that remain.

Exercise 1:   a) Morse code is made up of dots and dashes. Draw
a tree diagram to determine the code words of length three; dots
or dashes may be repeated.
b)   Draw a tree diagram for the selection of two workers from
      a list of five; one worker cannot do two jobs.

      You should recognize that except for the use of different
symbols, the two sample spaces in exercise 1 are the same,
respectively, as those for the Doctor and Mary above. Much of
our interest will be not in the sample points themselves but in
their number, that is, in the sizes of the sample spaces and events.

The Second Fundamental Principle of Counting: if one
experiment, say A, has #(A) possible outcomes, and if for each
of these, a second experiment, say B, has #(B|A) possible
outcomes, then the combined experiment A followed by B has
              #(A) times #(B|A) possible outcomes .
(B|A) may be read as B after A or B given A .

Examples: a) The first fork in Mary's tree has 5 branches and at
each of these, the second fork has 4 branches. There are $5 \cdot 4 =$
20 branches altogether.
b)   This branching is perfectly general: if the first fork has
      #(A) branches and at each of these, the second fork has
      #(B|A) branches, then the tree has #(A)·#(B|A) branches
      altogether.
c)   When A is a toss of a coin, #(A) = 2. The second toss B
      also has two outcomes so #(B|A) = 2. Therefore the
      sample space for two tosses has #(A)·#(B|A) = $2 \cdot 2 = 4$
      outcomes.
d)   Some club has thirty members. Experiment A is to choose
      a president and experiment B is to choose a secretary.
              #(A) = 30 and #(B|A) = 29 so there are
                  #(A)·#(B|A) = $30 \cdot 29 = 870$
      such executive committees.
e)   Suppose the club just mentioned decides that it also needs a
      treasurer. Now we think of A as selecting the president and
      secretary and B as selecting the treasurer from those that

remain. Then #(A) = 870 and #(B|A) = 28 so there are
$$870 \cdot 28 = 24360$$
possible executive committees of President, Secretary and Treasurer.

f)   In fact, this last example illustrates how the second fundamental principle can be carried over to use with more than two experiments. Let the selection numbers be symbolized by #(P) for selecting the president, #(P|S) for selecting the secretary after the president and #(T|P ∩ S) for selecting the treasurer after the president and secretary.

Then, #(P)·#(S|P)·#(T|P ∩ S) = 30·29·28 = 24360 .

In many instances, it is better to think the problem through with the principle in mind rather than to worry about the symbolization; we need symbols to state rules but they should not be allowed to confuse the issue. Formally, we would prove the extension started in part f) above by mathematical induction but we leave that for the reader.

Examples: a) How many ways can one dress a mannequin using one of five hats, one of 7 dresses and one pair of 10 pairs of shoes? 5·7·10 = 350

b)   In how many ways could the birthdays of five people occur in a 365 day year? There are no restrictions so selection is

with replacement: 365·365·365·365·365

c)   How many of the quintuplets of days in b) would indicate that the five people had different birthdays? This restriction implies selection without replacemnt:
$$365 \cdot 364 \cdot 363 \cdot 362 \cdot 361$$

d)   How many license plates can be made using 3 Roman letters followed by three digits from {1 2 3 4 5 6 7 8 9} ?

There are no other restrictions: 26·26·26·9·9·9

e)   How many of the license plates in d) will end in an even digit? The restriction is on the last digit only—
$$26 \cdot 26 \cdot 26 \cdot 9 \cdot 9 \cdot 4$$

Exercise 2: a) How many five digit numbers can be formed using any of the digits in d) above?

b)   How many of the numbers in a) are made up of five distinct digits?

c)   How many of the numbers in a) are odd?

Other names for ordered selections are ordered samples and permutations. One must always specify whether the selection is with or without replacement (repetition). The following theorem gives us the general formulas for the number of permutations.

Theorem: *Let Ω be a set of distinguishable objects say $a_1, a_2, \cdots, a_n$. The number of different (distinguishable) ordered samples of size r from Ω is:*

   a)   *n^r if sampling is with replacement between each single selection; r can be any positive integer;*
   b)   *$n(n - 1)(n - 2) \cdots (n - r + 1)$ if sampling is without replacement between each single selection; r must be less than or equal to n.*

Proof: In case a), each single selection is an experiment A with #(A) = n   which is repeated r times so that by the Second Fundamental Principle of Counting, the number is

$$n \cdot n \cdot n \cdots n = n^r.$$

In case b), the *first* selection A has n possible outcomes;

after A, the *second* selection B has only n − 1 possible outcomes.

After A and B, the *third* selection C has only n − 2 possible outcomes; then n − 3 remain for the *fourth* selection, n − 4 for the *fifth* selection, etc. After the (r −1)st selection, there remain

$$n - (r - 1)$$

for the rth selection. But n − (r − 1) = n − r + 1. Hence the experiment A B C $\cdots$ has (the product)
$$n(n - 1)(n - 2) \cdots (n - r + 1)$$
possible outcomes.

The product $n(n - 1)(n - 2) \cdots (n - r + 1)$ is also symbolized by $_nP_r$ or $(n)_r$ or P^n_r or P(r,n) but we shall use only the last for "the number of *permutations* (without replacement) of size r from n distinct objects". In particular, the number of permutations of size r = n (that is, of all n objects at once) is
$$P(n,n) = n(n - 1)(n - 2) \cdots (n - n + 1)$$
$$= n(n - 1)(n - 2) \cdots 3 \cdot 2 \cdot 1$$

which is also written as n! , read n *factorial*.

Examples: a) $3! = 3 \cdot 2 \cdot 1 = 6$ is the number of permutations of three distinct letters as listed earlier.

b) $6 \cdot 5 \cdot 4 = 6 \cdot 5 \cdot 4 \cdot 3 \cdot 2 \cdot 1 / 3 \cdot 2 \cdot 1 = 6! / 3!$

c) $9 \cdot 8 \cdot 7 \cdot 6! = 9 \cdot 8 \cdot 7 \cdot 6 \cdot 5 \cdot 4 \cdot 3 \cdot 2 \cdot 1 = 9!$

d) $7! / 4! = 7 \cdot 6 \cdot 5 \cdot 4 \cdot 3 \cdot 2 \cdot 1 / 4 \cdot 3 \cdot 2 \ 1 = 7 \cdot 6 \cdot 5$

For $r \leq n$, we can write

$$P(r,n) = n(n - 1)(n - 2) \cdots (n - r + 1)$$

$$= n(n - 1)(n - 2) \cdots (n - r + 1)(n - r)!/(n - r)!$$

$$= \frac{n(n - 1)(n - 2) \cdots (n - r + 1)(n - r)(n - r - 1) \cdots 3 \cdot 2 \cdot 1}{(n - r)!}$$

$$= n!/(n - r)!$$

Examples: a) $P(4,7) = 7 \cdot 6 \cdot 5 \ 4 = 7 \cdot 6 \cdot 5 \cdot 4 \cdot 3! / 3!$

b) $P(3,7) = 7!/(7 - 3)! = 7!/4!$

$$= 7 \cdot 6 \cdot 5 \cdot 4 \cdot 3 \cdot 2 \cdot 1/4 \cdot 3 \cdot 2 \cdot 1 = 7 \cdot 6 \cdot 5.$$

c) $P(3,9) = 9! /(9 - 3)! = 9! / 6! = 9 \cdot 8 \cdot 7 \cdot 6!/6! = 9 \cdot 8 \cdot 7.$

d) $P(10,10) = 10! /(10 - 10)! = 10! / 0! = 10! = 3628800.$

Note that we must take $0!$ to be 1 for otherwise this formula would be inconsistent—we know that the number of permutations of all 10 things is $10!$

Exercise 3:   Evaluate $4!$; $7!$; $6! / 3!$; $P(3,6)$; $P(2,7)$; $P(5,7)$

Exercise 4:   a) How many itineraries does a candidate have for visiting 6 major cities?

b)   In how many ways can five different books be arranged on a shelf?

c)   There are $5! = 5 \cdot 4 \cdot 3 \cdot 2 \cdot 1 = 120$ permutations of the digits 1,2,3,4,5 at once. In how many of these are the digits 1 and 2 next to each other?

d)   Johnny has a choice of three flavors of icecream or two kinds of pie but not both. How many choices of dessert does he have? If Johnny can have both an icecream and a pie or only one, how many choices does he have?

e)   If there are three pieces of pie, two dips of icecream and in how many ways can Johnny pile his plate?

Exercise 5:   A box contains two good and two defective fuses. Two fuses are to be selected, one at a time without replacement.

   a)   In how many ways can such a selection occur?

   b)   In how many of these ways will both defective fuses be selected?

Exercise 6:   A case of 24 cans of cola contains 3 which are defective, say underfilled. Six are to be selected one at a time without replacement.

  a)   In how many ways can this selection occur?

  b)   In how many of these ways will none of the defective cans be selected?

  c)   In how many of these ways will at least one defective can be selected?

LESSON *7. UNORDERED SELECTIONS

In this lesson, we continue "selections from an urn" but without distinguishing the order of selection. The classical example is again from gambling—in many card games, the order in which the dealer passes you the cards does not effect the value of the hand. Here we have sampling without replacement as well as without regard to order. In collecting baseball pictures (in packs of chewing gum in the old days), matchbooks, stuffed dolls, etc., usually order is not important but one takes account of repetitions—replications.

First we consider selection of distinct objects without replacement and without regard to order; equivalently, from the urn with size n, we select r in one "handful". Such selections are called *combinations*. Note that a combination is really a subset of the population in the urn so we are going to be talking about the number of subsets of size r.

Example: Select three letters from {a b c d e} without regard to order and without replacement. With a little care, we can actually write down all possible subsets of size three:

$$\{a\ b\ c\}, \{a\ b\ d\}, \{a\ b\ e\}, \{a\ c\ d\}, \{a\ c\ e\},$$

$$\{a\ d\ e\}, \{b\ c\ d\}, \{b\ c\ e\}, \{b\ d\ e\}, \{c\ d\ e\}.$$

Each of these combinations can be arranged into $3! = 6$ ordered samples without replacement; {a b c} yields abc acb bca bac cab cba (lesson 6). If A is the experiment to select a subset of size three and B is the experiment to order the subset selected, then the combined experiment A followed by B is the same as selecting an ordered sample (permutation) of size three. Since $\#(A \cap B) = \#(A) \cdot \#(B\,|\,A)$, $P(3,5) = \#(A) \cdot 3!$ and hence

$$\#(A) = P(3,5)/3! = 5!/(5-3)! \cdot 3! = 5!/2! \cdot 3!$$

$$= 5 \cdot 4 \cdot 3 \cdot 2 \cdot 1/2! \cdot 3 \cdot 2 \cdot 1 = 5 \cdot 4/2 \cdot 1 = 10\ .$$

The ten subsets were listed at the beginning of this example.

We write $C(3,5) = P(3,5)/3!$; $C(3,5)$ is the number of combinations of size three from 5 distinct objects. Other symbols used are $_5C_3 = 5_{(3)}/3! = \begin{bmatrix} 5 \\ 3 \end{bmatrix}$ with the last being the most common. The same kind of reasoning as in the example above will yield a proof of the

Theorem: *If Ω is a set with n distinguishable objects and $\begin{bmatrix} n \\ r \end{bmatrix}$ is the number of (unordered samples without replacement) subsets of size $r \le n$, then $\begin{bmatrix} n \\ r \end{bmatrix} = n!/r!(n - r)!$.*

Examples: a) $\begin{bmatrix} 5 \\ 3 \end{bmatrix} = \dfrac{5!}{3!\ 2!} = \dfrac{5 \cdot 4 \cdot 3 \cdot 2 \cdot 1}{3 \cdot 2 \cdot 1 \cdot 2 \cdot 1} = \dfrac{5 \cdot 4 \cdot 3}{3 \cdot 2 \cdot 1}$

$$= \frac{60}{6} = 10$$

b) $\begin{bmatrix} 5 \\ 2 \end{bmatrix} = \dfrac{5!}{2!\ 3!} = \dfrac{5!}{2 \cdot 1 \cdot 3 \cdot 2 \cdot 1} = \dfrac{5 \cdot 4}{2 \cdot 1} = 10$

c) $\begin{bmatrix} 10 \\ 3 \end{bmatrix} = \dfrac{10!}{3!\ 7!} = \dfrac{10 \cdot 9 \cdot 8 \cdot 7 \cdot 6 \cdot 5 \cdot 4 \cdot 3 \cdot 2 \cdot 1}{3 \cdot 2 \cdot 1 \cdot 7 \cdot 6 \cdot 5 \cdot 4 \cdot 3 \cdot 2 \cdot 1}$

$$= \frac{10 \cdot 9 \cdot 8}{3 \cdot 2 \cdot 1} = 120$$

d) $\begin{bmatrix} 10 \\ 7 \end{bmatrix} = \dfrac{10!}{7!\ 3!} = \dfrac{10!}{3!\ 7!} = \begin{bmatrix} 10 \\ 3 \end{bmatrix} = 120$

e) $\begin{bmatrix} 11 \\ 6 \end{bmatrix} = \dfrac{11!}{6!\ 5!} = \dfrac{11 \cdot 10 \cdot 9 \cdot 8 \cdot 7 \cdot 6 \cdot 5!}{6 \cdot 5 \cdot 4 \cdot 3 \cdot 2 \cdot 1 \cdot 5!}$

$$= \frac{11 \cdot 10 \cdot 9 \cdot 8 \cdot 7 \cdot 6}{6 \cdot 5 \cdot 4 \cdot 3 \cdot 2 \cdot 1} = 462$$

Exercise 1: a) Evaluate $\begin{bmatrix} 12 \\ 2 \end{bmatrix}$; $\begin{bmatrix} 7 \\ 4 \end{bmatrix}$; $\begin{bmatrix} 10 \\ 6 \end{bmatrix}$; $\begin{bmatrix} 9 \\ 9 \end{bmatrix}$.

b)   How many committees of three can be formed in a club of 10 members? Of 30 members?

c)   How many juries of size 12 could be selected from a panel of 75 ?

Example: Counting the number of juries (committees) can be more involved.

a)   Select 12 out of 75 : $\begin{bmatrix} 75 \\ 12 \end{bmatrix}$

b)   First select 12 out of 75, then select a chairman : $\begin{bmatrix} 75 \\ 12 \end{bmatrix} \cdot 12$

c)   First select a chairman, then 11 out of 74 :

$\begin{bmatrix} 75 \\ 1 \end{bmatrix} \cdot \begin{bmatrix} 74 \\ 11 \end{bmatrix}$ which is not the same as b).

d) First select 12 out of 75, then select 2 alternates out

of 63: $\begin{bmatrix} 75 \\ 12 \end{bmatrix} \cdot \begin{bmatrix} 63 \\ 2 \end{bmatrix}$.

e) Since the alternates are not used in the same way as the regular jurors, the number in d) is not $\begin{bmatrix} 75 \\ 14 \end{bmatrix}$ either numerically or conceptually.

The following examples concern a standard deck of 52 cards in four *suits*: red hearts, red diamonds, black clubs, black spades, each with 13 *denominations*: A(ce) 2 3 4 5 6 7 8 9 10 J(ack) Q(ueen) K(ing). We consider individual *hands* dealt without replacement and without regard for order.

Examples: a) The number of "poker" hands is the number of combinations of size 5 :

$$\begin{bmatrix} 52 \\ 5 \end{bmatrix} = \frac{52!}{5! \ 47!} = \frac{52 \cdot 51 \cdot 50 \cdot 49 \cdot 48}{5 \cdot 4 \cdot 3 \cdot 2 \cdot 1} = 2,598,960 .$$

b) The number of poker hands containing three kings and two jacks is found by looking at a sequence of "experiments".

A — select all kings $\begin{bmatrix} 4 \\ 4 \end{bmatrix} = 1$ way

B — select three of these for the hand $\begin{bmatrix} 4 \\ 3 \end{bmatrix}$ ways

C — select all jacks $\begin{bmatrix} 4 \\ 4 \end{bmatrix} = 1$ way

D — select two of these for the hand $\begin{bmatrix} 4 \\ 2 \end{bmatrix}$ ways

Hence (with the extended second fundamental principle of counting) there are

$$1 \cdot \begin{bmatrix} 4 \\ 3 \end{bmatrix} \cdot 1 \cdot \begin{bmatrix} 4 \\ 2 \end{bmatrix} = \begin{bmatrix} 4 \\ 3 \end{bmatrix} \cdot \begin{bmatrix} 4 \\ 2 \end{bmatrix} = 4 \cdot 6 = 24$$

hands containing three kings and two jacks.

c) How many poker hands contain three cards of one denomination and two cards of a second denomination?

Select one denomination for the triple in $\begin{bmatrix} 13 \\ 1 \end{bmatrix}$ ways;

select three of these four cards in $\begin{bmatrix} 4 \\ 3 \end{bmatrix}$ ways.

Select a second (different) denomination for the double in $\begin{bmatrix} 12 \\ 1 \end{bmatrix}$ ways; select two of these four cards in $\begin{bmatrix} 4 \\ 2 \end{bmatrix}$ ways.

The number of such hands (called a full house) is

$$\begin{bmatrix} 13 \\ 1 \end{bmatrix} \cdot \begin{bmatrix} 4 \\ 3 \end{bmatrix} \cdot \begin{bmatrix} 12 \\ 1 \end{bmatrix} \cdot \begin{bmatrix} 4 \\ 2 \end{bmatrix} = \frac{13}{1} \frac{4 \cdot 3 \cdot 2}{3 \cdot 2 \cdot 1} \frac{12}{1} \frac{4 \cdot 3}{2 \cdot 1}$$

$$= 13 \cdot 4 \cdot 12 \cdot 6 = 3744$$

d)   How many hands contain exactly two pair (and of course a non–matching fifth card)?

Select two denominations for the pairs (the order between the pairs does not count) : $\begin{bmatrix} 13 \\ 2 \end{bmatrix}$

Select two of four cards in one denomination : $\begin{bmatrix} 4 \\ 2 \end{bmatrix}$;

select two of four cards in the second denomination : $\begin{bmatrix} 4 \\ 2 \end{bmatrix}$.

Select one denomination for the fifth card : $\begin{bmatrix} 11 \\ 1 \end{bmatrix}$;

select one of those four cards : $\begin{bmatrix} 4 \\ 1 \end{bmatrix}$.

The number of such hands is

$$\begin{bmatrix} 13 \\ 2 \end{bmatrix} \cdot \begin{bmatrix} 4 \\ 2 \end{bmatrix} \cdot \begin{bmatrix} 4 \\ 2 \end{bmatrix} \cdot \begin{bmatrix} 11 \\ 1 \end{bmatrix} \cdot \begin{bmatrix} 4 \\ 1 \end{bmatrix} = 78 \cdot 6 \cdot 6 \cdot 11 \cdot 4 = 123{,}552$$

e)   Note the three different kinds of hands herein. In b) only KKKJJ is of interest and in c), JJJKK is also allowed. But in d), neither of these is allowed; we must have KKJJ2, JJKK5, etc.

Exercise 2:   Find the number of poker hands:
a)   Royal flush—A, K, Q, J, 10 all in one suit
b)   One pair—like 2, 3, 4, 5, 5
c)   Containg 0 aces   d) Containing 2 aces
e)   Three of a kind—like 2, 2, 2, 4, K .

Example: Unknown to the plumber, his box of quarter inch washers contains 5 slightly undersize, 42 satisfactory, and 3 slightly oversize. He selects six washers in one bunch (without replacement, without regard to order).
a)   In how many ways could he obtain 6 satisfactory

washers? $\begin{bmatrix} 42 \\ 6 \end{bmatrix}$

b) In how many ways could he obtain 5 satisfactory washers and 1 undersize washer? $\begin{bmatrix} 42 \\ 5 \end{bmatrix} \cdot \begin{bmatrix} 5 \\ 1 \end{bmatrix}$

c) In how many ways could he obtain 5 satisfactory and 1 not satisfactory? We interpret "one not satisfactory" to mean one either too fat or too thin. We wind up with two mutually exclusive (disjoint) events :

$$\{5 \text{ satisfactory } \& 1 \text{ undersize}\} \text{ and}$$

$$\{5 \text{ satisfactory } \& 1 \text{ oversize}\}$$

Now we can apply the first fundamental principle of counting; that is, we add the number of ways he might select "5 satisfactory and 1 undersize" to the number of ways he might select "5 satisfactory and 1 oversize" : $\begin{bmatrix} 42 \\ 5 \end{bmatrix} \cdot \begin{bmatrix} 5 \\ 1 \end{bmatrix} + \begin{bmatrix} 42 \\ 5 \end{bmatrix} \cdot \begin{bmatrix} 3 \\ 1 \end{bmatrix}$.

Exercise 3: It may be convenient here and in remaining exercises to leave your answers in combinatorial form as in this last example. In a supermarket, a tray of unlabeled cans happens to have (from our superman vision) 3 cans of lima beans, 7 cans of cut beans and 8 cans of baked beans. A customer is going to select a combination of four cans. How many ways could she get:

a) exactly one can of each kind of beans ?
b) four cans of cut beans?
c) two cans of lima beans and two cans of baked beans?
d) at least one can of each kind of beans?

We turn now to selection one at a time without regard to order but with replacement; strictly speaking these are no longer subsets and so no longer combinations. As indicated above, this is the "collector's" problem which we will illustrate by thinking of the "free toy", one of six, in each box of "Goody" cereal.

The company machine packages one of the six different toys in each box of cereal; we think of the box "selecting" the toy. If you bought eight boxes, you might get all six toys; the diagram following represents toy 1 selected by (in) two boxes, toy 2 selected by (in) two boxes, toys 3, 4, 5, 6 in one box each.

1	2	3	4	5	6
BB	BB	B	B	B	B

But you might have missed one of the toys if the selection is

1	2	3	4	5	6
BBB	BB		B	B	B

or

1	2	3	4	5	6
BB	BB	BB	B	B	

and so on.

We can get all the different patterns with eight boxes and 6 toys by arranging the 8 B's and 5 bars | in the middle; note that two of the bars are fixed on the ends. The number of such patterns is the number of ways 13 objects can be divided into one group of 8 and one group of 5; equivalently, the number of such patterns is the number of ways 8 things (corresponding to the boxes) can be selected from 13 things (leaving 5 to correspond to the bars). This number is $\begin{bmatrix} 13 \\ 8 \end{bmatrix}$.

Another way to think of this is to consider the placing of eight B's in six cells. We need $6 + 1 = 7$ bars to form the six cells but only $7 - 2 = 5$ of these can be interchanged with the B's ; $8 + 5 = 13$ things are being arranged. Note that these are *permutations of non–distinct objects*. In general, that is for "n" cells (toys), we would have $n - 1$ bars to be arranged with "r" B's. The number of distinguishable arrangements is then

$$\begin{bmatrix} n - 1 + r \\ r \end{bmatrix} .$$

Or, the number of unordered selections of size r with replacements from n distinct objects is $\begin{bmatrix} n - 1 + r \\ r \end{bmatrix}$. (Unfortunately, some writers call these combinations; we do not.)

Exercise 4: Complete the following chart summarizing the formulas for permutations and combinations. There are n distinguishable objects and r are to be selected. The number of distinguishable selections is:

	Ordered	Unordered
with replacements		
without replacements		

It should be pointed out that arrangements in a circle obey other laws. For example, the number of ways that 4 people can sit at a bridge table is $6 = 4!/4$ because the following are not distinguishable "within the table":

$$\begin{matrix} & 1 & & 2 & & 3 & & 4 \\ 4 & & 2 & 1 & & 3 & 2 & & 4 & 3 & & 1 \\ & 3 & & & 4 & & & 1 & & & 2 \end{matrix}$$

Exercise 5: A "bridge" hand is a combination of size 13 from the standard deck. There are $\begin{bmatrix} 52 \\ 13 \end{bmatrix}$ individual hands; this is:

$$\frac{52 \cdot 51 \cdot 50 \cdot 49 \cdot 48 \cdot 47 \cdot 46 \cdot 45 \cdot 44 \cdot 43 \cdot 42 \cdot 41 \cdot 40}{13 \cdot 12 \cdot 11 \cdot 10 \cdot 9 \cdot 8 \cdot 7 \cdot 6 \cdot 5 \cdot 4 \cdot 3 \cdot 2 \cdot 1}$$
$$= 635,013,559,600 \ .$$

    a)   How many of these hands contain no aces?
    b)   How many of these hands contain 6 hearts and 7 clubs?
    c)   How many of these hands are all red cards?
    d)   How many of these hands are all one suit?

INE–Exercises: We have deliberately chosen a mixture of "perms and combs".
1.   A grocery firm codes its stock with 5 of the 26 Roman letters. How many items can be coded if
    a)   no letter can be used twice in one code;
    b)   "O" does not end a code;
    c)   "O" does not begin a code?
2.   A committee of two algebraists, one analyst, and three applied mathematicians is to be formed in a department of 8 algebraists, 6 analysts, and 10 applied mathematicians.
    a)   How many such committees are possible?
    b)   How many such committees are possible if one of the applied mathematicians will not serve with one of the algebraists?
3.   How many different arrangements (linear) of 4 different algebra books, 6 different calculus books and 5 different statistics books can be made if
    a)   there are no restrictions;
    b)   books on each subject are to be together?
4.   How many kinds of salad can be made using one or more of Boston lettuce, Romaine lettuce, iceberg lettuce, watercress, escarole, endive, dandelion, spinach?
5.   Let $0 \le k \le n$ be integers. The following two relations for the combinatorial can be proved by brute force manipulation of the symbols. Instead, give a "verbal argument" by concentrating on the meanings of the two

sides of the equalities: $\begin{bmatrix} n \\ k \end{bmatrix} = \begin{bmatrix} n \\ n-k \end{bmatrix}$;

(Pascal's Triangle): $\begin{bmatrix} n \\ k \end{bmatrix} = \begin{bmatrix} n-1 \\ k \end{bmatrix} + \begin{bmatrix} n-1 \\ k-1 \end{bmatrix}$.

6.   Ten equally skilled men apply for two jobs in a "team". How many paired comparisons for "compatibility" must the staff psychologist make for these men?

7.   Dominoes can be thought of as unordered pairs of integers: (0,0), (0,1), (0,2), $\cdots$, (1,1), (1,2), $\cdots$, (n,n) . How many such dominoes are there?

LESSON 8. SOME UNIFORM PROBABILITY SPACES

First we recall the basic properties of a probability function.

Let $S = \{s_1, s_2, \cdots s_m\}$ be a finite set; a probability function P assigns a value $P(\{s_k\}) = P(s_k) \geq 0$ to each elementary event $\{s_k\}$; for any event A of S,

$$P(A) = \sum_A P(s_k),$$

where $\sum_A$ indicates the sum of the probabilities $P(s_k)$ for those s_k which are in the subset A. In particular,

$$P(S) = P(s_1) + P(s_2) + \cdots + P(s_m) = 1.$$

A probability on a finite set is called *uniform* if each elementary event has the same probability; since there are m of these events, each elementary event has probability $1/m$. Then,

$$P(\{s_1\} \cup \{s_2\}) = P(s_1, s_2) = 1/m + 1/m = 2/m;$$

if A has (any) three elementary events,

$$P(A) = 1/m + 1/m + 1/m = 3/m;$$

if A has #(A) elementary events,

$$P(A) \text{ is } 1/m + 1/m + \cdots \text{ #(A) times so}$$

$$P(A) = \#(A)/m \text{ which is } \#(A)/\#(S) .$$

It is just such an assumption of a uniform probability distribution which is made in many games. The same principle is expressed in some elementary books as "The probability is the number of outcomes favorable to A divided by the number of possible outcomes." Other modes are "Select a point at random" or "The outcomes are equally likely." Many writers begin probability discussion with such uniformity; for example, with a well made six–sided die,(no weights, homogeneous material, etc.) the lack of any physical peculiarities suggests that no "upface" outcome should occur more often than any other. One winds up with a uniform distribution on {1 2 3 4 5 6}. In real life, uniformity is difficult to establish and very likely false (at least microscopically); otherwise, we would have little need to study all the distributions which lie ahead. It may be of interest to note that a respectable mathematician (d'Alembert) was mislead into assigning probability 1/3 to each of the points "two heads, one head, two tails" in the experiment of tossing two coins.

Examples: a) We have already worked with the uniform distribution for tossing a pair of six–sided dice; each elementary event $\{11\}$, $\{12\}$, $\cdots$, $\{66\}$ has probability 1/36. Mathematically this must be an assumption; physically or empirically, the collective experience of humans tossing well made dice seems to be in agreement.

b)   Following a good shuffle (and an honest dealer), it is not unreasonable to assume that all single poker hands are equally likely; ie., each has probability $1/\begin{bmatrix} 52 \\ 5 \end{bmatrix}$ of occurring. (Here the repeated trials are: shuffle, deal 5, record, replace; shuffle, deal 5, record, replace; $\cdots$). The probability that an individual hand is a full house is

$$\text{\# (Full houses)} \div \text{\#(Individual hands)} =$$

$$\begin{bmatrix} 13 \\ 1 \end{bmatrix} \cdot \begin{bmatrix} 4 \\ 3 \end{bmatrix} \cdot \begin{bmatrix} 12 \\ 1 \end{bmatrix} \cdot \begin{bmatrix} 4 \\ 2 \end{bmatrix} \div \begin{bmatrix} 52 \\ 5 \end{bmatrix} = 3744 \div 2598960 .$$

The probability that an individual is dealt two pair is

$$\begin{bmatrix} 13 \\ 2 \end{bmatrix} \cdot \begin{bmatrix} 4 \\ 2 \end{bmatrix} \cdot \begin{bmatrix} 4 \\ 2 \end{bmatrix} \cdot \begin{bmatrix} 11 \\ 1 \end{bmatrix} \cdot \begin{bmatrix} 4 \\ 1 \end{bmatrix} \div \begin{bmatrix} 52 \\ 5 \end{bmatrix}$$

$$= 123552 \div 2598960 .$$

Exercise 1:   Continuing the example for poker, find the probabilities for:
   a)   Royal flush—10 J Q K A in one suit;
   b)   Straight flush—any five cards in sequence in a single suit from A 2 3 4 5 to 10 J Q K A ;
   c)   Three of a kind—three cards of one denomination, one card of a second denomination and one card of a third denomination.

Example: Let us consider the birth months of six people. Counting orders and repetitions, there are 12^6 possible arrangements of these birth months. Assume that all of these are equally likely, that is, the distribution is uniform. (This is not quite true in the United States.) Then,
   a)   the probability that these six people have birthdays in different months is $12 \cdot 11 \cdot 10 \cdot 9 \cdot 8 \cdot 7 \div 12^6$;

b)   the probability that at least two of these people have birthdays in the same month is 1 – the probability that the birth months are all different,   that is,

$$1 - P(6,12) \div 12^6.$$

Exercise 2: Assume that there are 365 days in a year and that the distribution of birthdays is uniform (again not quite true). Find the probability that:

a)   Five people have the same birthday;
b)   Five people all have different birthdays;
c)   In a group of seven people, at least two have the same birthday.

Under the assumption of a uniform distribution for birthdays, the probability that r (independent—no twins–···) people have different birthdays is

$$p_r = 365 \cdot 364 \cdot 363 \cdot \ldots \cdot (365 - (r - 1)) \div 365^r$$

$$= (1 - 1/365)(1 - 2/365) \cdots (1 - (r - 1)/365) .$$

The probability that that in a group of r people at least two have the same birthday is $1 - p_r$ and this is increasing in r:

r	$1 - p_r$
2	.002740
5	.027136
10	.116948
20	.411438
21	.443688
22	.475695
23	.507297
24	.538344
25	.568700
30	.706316
40	.891232
50	.970374
100	.9999997

If every time that you are in a group of at least 23 people, you are able to make a bet that at least two people in the room have the same birthday, then the relative frequency interpretation of probability tells you that you will win at least 50.7297% of those bets (in the long run).

Exercise 3:   Show that $1 - p_r$ is increasing in r. Hint: consider p_r/p_{r+1} .

Exercise 4:   Recall the plumber's box of washers with 5 undersize, 42 satisfactory and 3 oversize. The plumber is to select 6 in one bunch so we assume that every half dozen has the same chance of being selected namely $1 \div \begin{bmatrix} 50 \\ 6 \end{bmatrix}$. Find the probability that among those selected:

a)   all six washers are satisfactory;
b)   5 washers are satisfactory and 1 is over size;
c)   only half of the washers are satisfactory.

Note that if the plumber withdrew the washers "one at a time without replacement" as a permutation, the calculations would be more complicated.

Example; Let S be the set of triples denoting the sex of children in families with three single births :

$$S = \{bbb \ bbg \ bgb \ gbb \ bgg \ gbg \ ggb \ ggg\}$$

Assume a uniform distribution on S (again not quite true although for the U.S., the ratio of boys to girls is 105/205 = which is the repeating decimal .51219̇) Let A be the event that the family selected has at least one girl and let B be the event that the family has children of both sexes. Then

$$P(A) = P(\{bbg \ bgb \ gbb \ ggb \ gbg \ bgg \ ggg\}) = 7/8;$$
$$P(B) = 6/8 \ ; \ P(A \cup B) = P(A) = 7/8;$$
$$P(A \cap B) = P(B) = 6/8.$$

Exercise 5:   Continue with the example above.

a)   List the points in B, A ∪ B, A ∩ B to verify the probabilities given.
b)   Let C be the event that the family selected has at most one girl. Find P(C) and P(B ∩ C).
c)   Do you see any relation among the values of P(B), P(C), and P(B ∩ C)?

Example: Consider n seats in a row (like a lunch counter). If k < n seats are selected "at random", what is the probability that they are all next to each other? There are $n - k + 1$ such k–tuples. (Write out some cases: n = 5 , k = 2; n = 6, k = 3; etc.) Hence the probability is $(n - k + 1)/\begin{bmatrix} n \\ k \end{bmatrix}$.

Exercise 6: a) Is $(n - k + 1)/\begin{bmatrix} n \\ k \end{bmatrix}$ the probability that when k people sit at a lunch counter (at random), they will all be side by side? Why?
b) What is the corresponding probability (all side by side) if the n seats are in a circle? (Try some cases first!)

Exercise 7: If a committee of three is chosen "at random" from your class of 29 students and 1 instructor, what is the probability that you and the instructor will be on the committee? Both will not be on it?

Exercise 8: A closet contains 10 distinct pairs of shoes. If six shoes are withdrawn "at random", what is the probability that there will be no pair among them?

INE–Exercises:
1.   (Chevalier de Méré, 1658) Which of the following events is more probable: at least one ace (1) in one roll of four fair six–sided dice or at least one double ace (11) in 24 rolls of two such dice?
2.   Find the probability that in a sample of n digits from this one set {0 1 2 3 4 5 6 7 8 9} there are no repetitions. Note cases!
3.   Consider the sample space {a b c d}. Determine the probability function P such that $P(\{a\}) = P(\{b\ d\})$,
$P(\{b\ c\}) = P(\{a\ c\ d\})$, $P(\{c\ d\}) = P\{(a\})$.
4.   In the year past, Ms Sims kept a record of her mode of getting to work: drove her own car 38 days; rode a bus 26 days; rode in a friend's car 52 days; walked 104 days. Assuming the same general conditions for this year as last, what odds would you give that tomorrow Ms Sims:
     a) rides to work? b) walks or drives her own car to work?
     c) does not take the bus?
5.   A college registrar lists the enrollment by college and sex:

	Science	Arts	Business	Education
Male	353	308	412	228
Female	329	350	219	321

     Let S A B E M F denote the event that a student is a member of the corresponding college/sex. If one student is to be "selected at random", find the probability that the student will be from Science, say P(S). Similarly, find P(A ∪ B), P(E ∪ M), P(E ∩ M), P(S ∪ A ∪ F), P(F ∩ E), P(E ∩ (F ∪ S)).

6.   If the digits 1 2 3 4 5 6 are arranged in a row "at random", what is the probability that the sequence is:
a) 1 3 5 2 4 6? b) odd odd odd even even even?
c) odd even odd even odd even?

7.   If schools A and B each have three runners in a six man race, what is a probability that the finishing sequence is:
a) AAABBB ? b) ABABAB ? What are your assumptions? Compare with problem 6.

8.   Four "experts" each rank three drinks, say A B C, best to worst, say 1 2 3 . Suppose that unknown to them, the drinks are all the same so that the ranks are really being assigned "at random". What is the probability that A is rated best (with four ranks 1)?

9.   In the form   ABBAABBABBBAAAB   there are 8 *runs*: like letters separated by other letters. If a letters A and b , letters B are arranged in a row "at random", show that the probability that there are r runs is:

$$2 \binom{a-1}{k-1} \binom{b-1}{k-1} \div \binom{a+b}{a} \quad \text{for } r = 2k \geq 2$$

$$\left[\binom{a-1}{k-1} \binom{b-1}{k} + \binom{a-1}{k} \binom{b-1}{k-1} \right] \div \binom{a+b}{a}$$

for r = 2k–1 ≥ 1 .

10.   a) Toss a fair coin n times. Show that the probability of getting exactly k heads is $\binom{n}{k} \div 2^n$.

b)   For k = 0(1)n, let B_k be the event "getting exactly k heads in n tosses". Verify that $\{B_0, B_1, \cdots, B_n\}$ is a partition of the sample space.

c)   Use the properties of probability to show that

$$\sum_{k=0}^{n} \binom{n}{k} = 2^n.$$

LESSON 9. CONDITIONAL PROBABILITY/INDEPENDENCE

We will motivate the definition of conditional probability by looking at a particular case wherein we can count. As before, we let #(A) denote the number of elements in the set A.

Example: The following records were obtained from a health spa—38 males were overweight, 62 females were overweight, and so on.

	MALES	FEMALES
Overweight	38	62
Average weight	27	41
Underweight	6	15

With obvious abbreviations, we have

$$\#(O \cap M) = 38 \qquad \#(O \cap F) = 62$$
$$\#(A \cap M) = 27 \qquad \#(A \cap F) = 41$$
$$\#(U \cap M) = 6 \qquad \#(U \cap F) = 15$$

Using the first fundamental principle of counting, we get

$$\#(M) = \#(O \cap M) + \#(A \cap M) + \#(U \cap M)$$
$$= 38 + 27 + 6 = 71.$$

Similarly, $\#(F) = 62 + 41 + 15 = 118$; $\#(O) = 100$; $\#(A) = 68$; $\#(U) = 21$. If we let the sample space S be the set of these records from the spa, then $\#(S) = 189$.

a) Consider the selection "at random" of an individual record from this set; at random is to mean that every record has the same chance of being chosen: 1/189 . The probability that the record is that of an overweight male is

$$P(O \cap M) = \#(O \cap M) \div \#(S) = 38/189.$$

The probability that the record is that of a male is

$$P(M) = \#(M) \div \#(S) = 71/189.$$

b) Suppose now that we restrict the selection to "males", that is, the records of males. Then at random means that every "male" has the same probability of being chosen: 1/71 . The probability that the "man selected" will be overweight is

$$P(O|M) = \#(O \cap M) \div \#(M) = 38/71.$$

This is the probability of "O" *when* "M" is known; the bar | on the left may be read "given" or "restricted to".

c)   If we restrict our selection to overweight people, the probability that the selected person is male is

$$P(M|O) = \#(M \cap O) \div \#(O) = 38/100 .$$

Exercise 1:   Continuing with the example above, find the probability that

a)   when an average weight person is selected, the result is a male;

b)   when a female is selected, she is not overweight.

Example: The following relationships exist for the probabilities computed (above) for the health spa.

d) $P(O|M) = \#(O \cap M) \div \#(M) = 38/71$

$$= \frac{38}{189} \div \frac{71}{189} = P(O \cap M) \div P(M) .$$

e) $P(M|O) = \#(O \cap M) \div \#(O) = 38/100$

$$= \frac{38}{189} \div \frac{100}{189} = P(M \cap O) \div P(O) .$$

The relative frequency like $\#(A \cap B) \div \#(B)$ used in these examples is more properly called a *conditional probability* — a calculation on the condition that (when) the event B has already been observed. A clue in the English description of these cases is the use of the subjunctive "if" or "when".

Definition: *For the probability function P, events A and B with $P(B) > 0$, $P(A \cap B) \div P(B)$ is called the conditional probability of A given B. We write $P(A|B) = P(A \cap B) \div P(B)$, equivalently, $P(A \cap B) = P(A|B) \cdot P(B)$. Similarly, $P(B|A) = P(A \cap B) \div P(A)$, equivalently, $P(A \cap B) = P(B|A) \cdot P(A)$.*

Example: Consider the selection with equal probabilities (1/4) of a family with two children: S = {bb bg gb gg} . Let A be the event that both children are boys and let B be the event that at least one child is a boy.

$$P(A) = 1/4 ; P(B) = 3/4 ; P(A \cap B) = 1/4 ;$$

$$P(A|B) = P(A \cap B) \div P(B) = (1/4) \div (3/4) = 1/3.$$

If it is known that the family selected has at least one boy, then the probability that the family has two boys is 1/3 ; 1/3 of the families with a boy actually have two boys.

Example: Two cards are drawn one at a time without replacement from a standard deck.

a) Given that the first card is red, what is the probability that the second card is also red? We assume that every pair of cards drawn in this way has the same probability of appearing—$1/52 \cdot 51$. There are $26 \cdot 25$ such red,red pairs. If A is the event that the first card is red and B is the event that the second card is red, then $P(A \cap B) = (26 \cdot 25) \div (52 \cdot 51)$. Also, $P(A) = 26 \div 52$ so that the conditional probability of getting a red card on the second draw given that the first draw was red is
$$P(B \mid A) = P(A \cap B) \div P(A) =$$

$$(26 \cdot 25/52 \cdot 51) \div (26/52) = 25/51$$

as you should have guessed.

b) Consider the same question as in a) but with replacement (and shuffling) between the first and second draws. Then the probability that any particular pair appears is $1/52 \cdot 52$ .Here
$$\#(A \cap B) = 26 \cdot 26 \text{ so}$$

$$P(A \cap B) = 26 \cdot 26 \div 52 \cdot 52 = 1/4.$$

With $P(A) = 26/52$ we get

$$P(B \mid A) = P(A \cap B) \div P(A) = (1/4) \div (1/2) = 1/2.$$

Of course, there are still 26 red cards for the second drawing; this is a case of repeated trials.

Exercise 2: Let A be the event that the family selected has children of both sexes and let B be the event that the family selected has at most one girl. Assume a uniform distribution for S in each case. Find $P(A \mid B)$:

a) among families with two children so S has 4 points;
b) among families with three children; S has 8 points;
c) among families with four children; S has 16 points.

We have already dealt with independent trials—each repetition starts anew so that the outcome on the next trial is in no way affected by the previous outcomes. Events will be independent if the occurrence of one does not affect the probabilities for the occurrence of the other. More precisely,

Definition: *For a given sample space S and probability P , events A and B are stochastically independent, briefly independent, if $P(A|B) = P(A)$ or $P(B|A) = P(B)$.*

Combining this with the multiplication in the first definition of this lesson, we get

$$P(A \cap B) = P(A|B) \cdot P(B) = P(A) \cdot P(B)$$
$$= P(B|A) \cdot P(A) = P(B) \cdot P(A) .$$

That is, two events are independent iff the probability of the *joint* occurrence $P(A \cap B)$ is the product of the probabilities of their individual occurrences $P(A) \cdot P(B)$.

Examples: a) Refer to exercise 2. You should have found:

Children	P(A)	P(B)	P(B ∩ A)
2	2/4	3/4	1
3	6/8	4/8	3/8
4	14/16	5/16	4/16

In the case of three children, we have $P(B|A) = P(B)$ but this equality does not hold in the other cases. Hence A and B are independent events among families with three children but not among families with two or four children. (The ambitious student can show that "three" is the only case in which this independence occurs.) Also in the case with three children,

$$P(A|B) = P(A \cap B) \div P(B) = (3/8) \div (4/8) = 3/4 = P(A) .$$

b) One card is drawn from a standard deck. A is the event that the card is a deuce (2) and B is the event that the card is a heart. Then

$$P(A) = 4/52 ; P(B) = 13/52 ; P(A \cap B) = 1/52 .$$
$$P(A|B) = P(A \cap B) \div P(B)$$
$$= (1/52) \div (13/52) = 1/13 = P(A);$$
$$P(B|A) = P(A \cap B) \div P(A)$$
$$= (1/52) \div (4/52) = 1/4 = P(B).$$

Again, A and B are independent events, as you should expect.

Obviously, "independent trials" have outcomes which are "independent events". but otherwise, *independence* is not easy to "see". Fortunately, a great deal of statistics can be done with the assumption of independence and a great many physical experiments seem to tolerate this assumption in their mathematical models. It will be convenient to have additional facts about independent events.

Theorem: *If for a given sample space S and probability P, A and B are independent events, then A and B^C are also independent events; so are A^C and B; so are A^C and B^C .*

Partial proof: $A = (A \cap B) \cup (A \cap B^C)$ as in

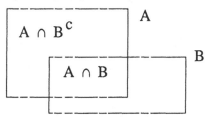

Since $A \cap B$ and $A \cap B^C$ are disjoint,

$$P(A) = P(A \cap B) + P(A \cap B^C)$$

and so by subtraction, $P(A \cap B^C) = P(A) - P(A \cap B)$. By independence, $P(A \cap B) = P(A) \cdot P(B)$ so by substitution,

$$P(A \cap B^C) = P(A) - P(A) \cdot P(B).$$

By factoring, $P(A \cap B^C) = P(A) \cdot (1 - P(B)) = P(A) \cdot P(B^C)$.

Exercise 3: Write out the proofs for the other cases of the theorem.

Example: Let the sample space S consist of the three–tuples
$$(1\ 1\ 1\ ),\ (1\ 0\ 0),\ (0\ 1\ 0),\ (0\ 0\ 1);$$
let the probability be uniform. Let A be the event that the first component is 1; let B be the event that the second component is 1; let C be the event that the third component is 1. Then

$$P(A) = P(\{(1\ 1\ 1)\} \cup \{(1\ 0\ 0\ )\}) = 1/2 .$$

Similarly, $P(B) = 1/2$ and $P(C) = 1/2$.

$$P(A \cap B) = P(\{(111)\}) = 1/4;$$

similarly, $P(A \cap C) = 1/4$ and $P(B \cap C) = 1/4$. It is easy to check (do it!) that A and B are independent; similarly, A and C are independent and B and C are independent. However,

$$P(A \cap B \cap C) = P(\{(1\ 1\ 1)\}) = 1/4$$
$$\neq 1/8 = P(A) \cdot P(B) \cdot P(C).$$

We say that A, B, C are pairwise independent but not mutually independent. The existence of such cases makes the following definition necessary.

Definition: *A collection of events* $\{A_1, A_2, \cdots\}$ *in a probability space are mutually stochastically independent iff they are independent in pairs, in triples, in quadruples,* $\cdots$:

$$P(A_i \cap A_j) = P(A_I) \cdot P(A_J) \text{ for } i \neq j;$$

$$P(A_i \cap A_j \cap A_k) = P(A_i) \cdot P(A_j) \cdot P(A_k) \text{ for } i \neq j \neq k;$$

$$P(A_i \cap A_j \cap A_k \cap A_m) = P(A_i) \cdot P(A_j) \cdot P(A_k) \cdot P(A_m)$$

for $i \neq j \neq k \neq m;$ *etc.*

Exercise 4: Probability in $[S, \mathscr{S}\!P]$ is uniform;
$$S = \{0\ 1\ 2\ 3\ 4\ 5\ 6\ 7\}.$$
Let A = $\{0\ 1\ 6\ 7\}$, B = $\{0\ 3\ 4\ 7\}$, C = $\{0\ 2\ 5\ 7\}$. Show that A B   C   are independent in pairs but not independent in triples. (They are pairwise independent but not mutually independent.)

Example: Assume that treatments of individual patients can be viewed as independent trials. Past experience has shown that a certain salve will relieve the itching in 70% of the patients, say $P(R) = .70$. We are going to do some calculations involving three patients.

  a)   The probability that all three patients are relieved is

P({R on first} $\cap$ {R on second} $\cap$ {R on third})

= P({R on first}) $\cdot$ P({R on second}) $\cdot$ P({R on third})

by independence of the events first, second, third. This is usually shortened to

$$P(RRR) = P(R) \cdot P(R) \cdot P(R);$$

then with $P(R) = .70$, we get $P(RRR) = (.70)^3$.

  b)   The probability that none of the three are relieved is
P({N on first} $\cap$ {N on second} $\cap$ {N on third})

$$= P(N) \cdot P(N) \cdot P(N) = (.30)^3 .$$

Here we have used the complement $P(N) = 1 - P(R)$ as well as the product rule for independent events.

  c)   In each of the following events, exactly one patient is relieved: {RNN} {NRN} {NNR} . As these are disjoint, the probability that exactly one patient (not saying which one) is relieved is the sum of the probabilities for these three events:

$$P(RNN) + P(NRN) + P(NNR)$$
$$= P(R) \cdot P(N) \cdot P(N) + P(N) \cdot P(R) \cdot P(N) + P(N) \cdot P(N) \cdot P(R)$$
$$= (.70)(.30)(.30) + (.30)(.70)(.30) + (.30)(.30)(.70)$$
$$= 3(.70)(.30)^2 .$$

d)   Similarly, the probability that exactly two patients are relieved is: $P(RRN) + P(RNR) + P(NRR)$

$$= (.70)(.70)(.30) + (.70)(.30)(.70) + (.30)(.70)(.70)$$

$$= 3(.70)^2(.30) .$$

In summary we have:

Number of Patients Relieved	Probability	
3	$(.70)^3$	= .343
2	$3(.70)^2(.30)^1$	= .441
1	$3(.70)^1(.30)^2$	= .189
0	$(.30)^3$	= .027

Note that the sum of the values in the right hand column is 1.

Exercise 5: Construct a table like that above for a doctor treating 5 patients with the assumption of independent single outcomes and $P(R) = .70$.

Exercise 6: Assume that cashiers work independently and that past experience has shown that the probability that a cash drawer is "short" by more than five cents is 1/10 . (A smaller figure would be more realistic but complicates the arithmetic.) Construct a probability table similar to those above for the number of cashiers in 5 who are "short". Note that you have already done half the thinking for this problem in exercise 5.

Mathematically, $P(A \cap B \cap C) = P(A)P(B)P(C)$ does not imply $P(A \cap B) = P(A)P(B)$ for the first of these equalities holds when $P(C) = 0$ no matter what the status of the second equality. On the other hand, when the first equality holds and $P(C) = 1$, then the three events are mutually independent.

Exercise 7:   Prove the following "oddball" results in $[S, \mathscr{S}P]$.
  a) The last statement in the previous paragraph.
  b) If A and B are mutually independent and mutually exclusive, then P(A) and/or P(B) equals 0.
  c) For any event A, A and ϕ are independent.
  d) For any event A, A and S are independent.

INE–Exercises:
1. Suppose that the probability of success on any one trial is .10 .How many independent trials must be made in order to guarantee that the probability of at least one success among those trials is more than .5? Hint: derive $1 - (.90)^n > .5$; then find n by (arithmetical) trial.
2. Suppose that a box of 50 items contains 6 defectives.
  a) If the defectives are located by selecting one at a time at random and testing, what is the probability that the last (sixth) defective is located on the tenth inspection?
  b) In the same vein, suppose that the last defective was obtained on the eighth inspection; what is the probability that the first item selected was defective?
  c) On the other hand, if ten items are withdrawn "in one bunch", what is the probability that this subset contains all six defectives?
3. The following is a sketch of irrigation ditches from a dam to a farm with electrically controlled gates at the points a b c d e.

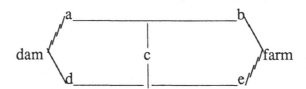

The gates operate independently with the same probability of failing to open, say θ (a Greek symbol). What is the probability that water will flow to the farm at a given signal from the damsite? Hint: convince yourself that you need to consider only a & b open, a & c & e open, d & e open, d & c & b open.
4. Denote the recessive gene by a and the dominant gene by A. Suppose it is known that one parent is aa and the other parent is either AA or Aa. "At random", the progeny gets one gene from each parent.

    a)   If it is observed that the progeny is aa, what is the probability that the second parent was Aa?

    b)   If it is observed that the progeny is Aa, what is the probability that the other parent was Aa?

5.   Let the sample space be $S = \{a\ b\ c\ d\}$ with probabilities $\sqrt{2}/2 - 1/4,\ 1/4,\ 3/4 - \sqrt{2}/2,\ 1/4$, respectively. Let $A = \{a\ c\}, B = \{b\ c\}, C = \{c\ d\}$.

    a)   Show that $P(A \cap B \cap C) = P(A) \cdot P(B) \cdot P(C)$.

    b)   Is the following true of false?
$$P(A \cap B) = P(A) \cdot P(B).$$

6.   Let $S = \{a\ b\ c\ d\}$ with uniform probability $1/4$. Let $A = \{a,b\}, B = \{a,c\}, C = \{a,d\}$.

    a)   Show that A, B, C are pairwise independent.

    b)   Are A, B, C mutually independent?

7.   An urn contains g green balls and r red balls. Two balls are to be drawn at random without replacement.

    a)   Find the probability that the second ball is green knowing that the first ball was green.

    b)   Find the probability that both balls are green.

8.   What is the probability of getting three diamond cards when drawing 3 cards without replacement from a standard deck?

LESSON *10. BAYES' RULE

Here it is convenient to begin with an

Example: Among men in a certain age group, incidence of lung cancer is said to be 2 in 100. A certain test will show a positive reaction in 99% of such men who actually have lung cancer but also in .1% of such men who do not have lung cancer. In what proportion of men in this age group does the test show a positive reaction? If the test shows a positive reaction in such a man, what is the probability that he has lung cancer?

a) We begin the analysis by setting up an appropriate sample space. Let C be the event that the man has cancer; then C^c is the event that he doesn't have cancer. Let T be the event that the test shows a positive reaction; then T^c is the event that the test does not show a positive reaction . For each man, a sample space is

$$\{ \{C \cap T\} \quad \{C^c \cap T\} \quad \{C \cap T^c\} \quad \{C^c \cap T^c\} \}$$

b) We now translate each phrase of the problem into a probability statement: the first sentence talks about all men in the group so we get

$$P(C) = .02 \text{ and } P(C^c) = 1 - P(C) = .98;$$

the first part of the second sentence talks about men with cancer so that $P(T|C) = .99$;

the second part of the second sentence talks about men without cancer so that $P(T|C^c) = .001$.

The first question is to find $P(T)$; the second question is to find $P(C|T)$.

c) From the decomposition of T :

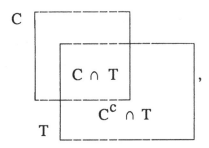

$$P(T) = P(C \cap T) + P(C^c \cap T) .$$

Then, $P(C \cap T) = P(T \cap C)$
$$= P(T|C) \cdot P(C)$$
$$= (.99)(.02) = .0198;$$
$$P(C^c \cap T) = P(T|C^c) \cdot P(C^c)$$
$$= (01)(.98) = .00098.$$
Hence, $P(T) = .0198 + .00098 = .0278$.

d) $P(C|T) = P(C \cap T) \div P(T) = .0198 \div .02078 \approx .95.$
(The $\approx$ indicates a numerical approximation.)

The next example will use different notation but the basic elements are the same: there are two criteria for classification of each point of a sample space S so that $S = C \cup C^c = T \cup T^c$

$$= (C \cap T) \cup (C \cap T^c) \cup (C^c \cap T) \cup (C^c \cap T^c). \qquad (*)$$

Some probabilities or conditional probabilities are given and others are to be found.

Exercise 1: a) Make a Venn diagram showing all the sets in (*).
b)   Formally reduce the second line of (*) to   S ; mention specifically each rule from the algebra of sets that you use.

Example: A company has two plants which manufacture its circuit breakers. Because of various conditions, the plants do not function alike. Plant I makes 40% of the circuit breakers of which .3% are defective. Plant II makes 60% of the circuit breakers with .1% defective. A single circuit breaker is to be selected. Let A be the event that the circuit breaker was manufactured in plant I and B be the event that it was manufactured in plant II; let D be the event that it is defective. Then,
$$P(A) = .4 , P(B) = .6 , P(D|A) = .003 , P(D|B) = .001 .$$
From this information, we get

$$P(D \cap A) = P(D \mid A) \cdot P(A) = (.003) \cdot (.4) = .0012$$
$$P(D \cap B) = P(D \mid B) \cdot P(B) = (.001) \cdot (.6) = .0006 \ .$$

In this formualtion, $A \cup B$ is the entire sample space and of course $A \cap B = \phi$. Thus $D = (D \cap A) \cup (D \cap B)$ as in

	A	B
D	$D \cap A$	$D \cap B$
D^c	$D^c \cap A$	$D^c \cap B$

Since $D \cap A$ and $D \cap B$ are disjoint,

$$P(D) = P(D \cap A) + P(D \cap B) = .0012 + .0006 = .0018.$$

That is, about .18% of the population of circuit breakers is defective. It is now simple to compute other conditional probabilities:

$$P(A \mid D) = P(A \cap D) \div P(D) = .0012 \div 0018 \approx .67$$
$$P(B \mid D) = P(B \cap D) \div P(D) = .0016 \div .0018 \approx .33$$

Thus when (if) a defective item happens to have been selected, there is a 67% chance that it came from plant I and a 33% chance that it came from plant II.

When we write out the general formulas used in the discussion of the previous examples, we obtain:

Bayes' rule: If $A_1, \cdots, A_n$ and D are events in a sample space S such that $A_1, \cdots, A_n$ are disjoint, $A_1 \cup \cdots \cup A_n = S$,

$\quad$ $P(A_1), P(A_2), \cdots P(A_n)$, and $P(D)$ are positive,

then,

$$P(D) = P(D \mid A_1) \cdot P(A_1) + \cdots + P(D \mid A_n) \cdot P(A_n)$$

and

$$P(A_i \mid D) = P(A_i \cap D) \div P(D) = P(D \mid A_i) \cdot P(A_i) \div P(D)$$

for each i.

Exercise 2: Using the notation $A_1 \cup A_2 \cup \cdots \cup A_n = \cup_{i=1}^{n} A_i$ and properties of P , derive the two formulas in Bayes' rule.

What we are really doing here is building a new sample space and a new probability function. The events $A_1, A_2, \cdots, A_n$ are a partition of the sample space S. Let the derived sample space $\tilde{S}$ have these A's as elementary events and let $\tilde{\mathcal{F}}$ be the collection of all subsets of $\tilde{S}$. The null set is still the null set ϕ. The union of two "elementary events" $\{A_1, A_2\}$ in $\tilde{\mathcal{F}}$ is $A_1 \cup A_2$ relative to S ; similarly, $\{A_1, A_2, A_3\}$ is $A_1 \cup A_2 \cup A_3$ and so on. The set D is fixed in S but is not made up of elementary events in $\tilde{S}$; P(D) is positive. Let Q be a function on $\tilde{\mathcal{F}}$ with

$$Q(A_i) = P(A_i \cap D) \div P(D) .$$

Then as you are asked to show in the next exercise, Q has all the properties of a probability function on $\tilde{\mathcal{F}}$. We call the triple $[\tilde{S}, \tilde{\mathcal{F}}, Q]$ a derived probability space. It is in this sense that the conditional probability $P(A_i | D) = Q(A_i)$ is truly a probability.

Exercise 3: Show that the triple defined above makes up a finite probability space.

Exercise 4: Insurance claims in a certain office are distributed as follows: 60% auto with 5% of these for amounts over $5000; 10% life with 90% of these for amounts over $5000; 30% medical with 40% of these for over $5000.
    a) What percent of the claims in this office are for amounts over $5000?
    b) If a claim for over $5000 is made, what is the probability that it is for a life policy? Hint: draw a Venn diagram to illustrate the partitions.

Exercise 5: A patient is to be treated simultaneously with three drugs A B C which do not interact. The past history has been;

P(Response to A) = .2; P(Cure | Response to A) = .2;

P(Response to B) = .4; P(Cure | Response to B) = .5;

P(Response to C) = .4; P(Cure | Response to C) = .3 .

      a)   What assumptions are hidden in the simple phrase that the drugs do not interact?

      b)   What is the probability that a patient receiving all three drugs will be cured?

      c)   If the patient is cured, what is the probability that the cure was due to drug B?

Exercise 6: An electric device has three components, say A B C; failure of one component may or may not shut off the device. Previous experience (relative frequencies from the past) indicates

that $P(A \text{ fails}) = .2$; $P(B \text{ fails}) = .4$; $P(C \text{ fails}) = .4$;

$P(\text{more than one fails}) = 0$; $P(\text{shutoff} \mid A \text{ fails}) = .05$;

$P(\text{shutoff} \mid B \text{ fails}) = .03$; $P(\text{shutoff} \mid C \text{ fails}) = .02$.

Find the most likely cause of the present shutoff.

Exercise 7: A certain test is 98% accurate on those who have this type of cancer and 95% accurate on those who do not have this type. If one–quarter percent of the people under test have this type of cancer, what is the probability that this person has this type of cancer when her test "says" she does?

LESSON 11. RANDOM VARIABLES

So far, we have concentrated on the probability function defined on subsets of a sample space S. In most applications, there is another kind of function defined on S itself which is equally important. For a change, we start with a

Definition: *Let S be a finite sample space and let $\mathcal{F}$ be its power set (the set of all subsets). Let X be a real valued function on S. Then X is a (real) random variable for $[S, \mathcal{F}]$.*

The importance of including $\mathcal{F}$ as part of the definition will be brought out in Part II. Of course, this kind of thing was done in a different form in lessons 8 and 9.

Example: For the case of exercise 6 lesson 9, let X be the number of cashiers who are "short". Then,

$$p_X(0) = P(X = 0) = \qquad (9/10)^5$$

$$p_X(1) = P(X = 1) = \quad 5 \cdot (1/10) \cdot (9/10)^4$$

$$p_X(2) = P(X = 2) = \quad 10 \cdot (1/10)^2 \cdot (9/10)^3$$

$$p_X(3) = P(X = 3) = \quad 10 \cdot (1/10)^3 \cdot (9/10)^2$$

$$p_X(4) = P(X = 4) = \quad 5 \cdot (1/10)^4 \cdot (9/10)$$

$$p_X(5) = P(X = 5) = \qquad (1/10)^5$$

Exercise 1: Make a table of probabilities for exercise 5 lesson 9.

Inspired by such examples, we make the following

Definition: *Let X be a real random variable for the triple $[S, \mathcal{F}, P]$. Then for each real number x , the probability density function (PDF) of X has value*

$$p_X(x) = P(X = x) = P(X^{-1}(x))$$

$$= P(\{s \text{ in } S : X(s) = x\}) = \sum_{s \in X^{-1}(x)} P(\{s\}) \,.$$

(The cumbersome summation notation means "The probability that $X = x$ is the sum of the probabilities $P(\{s\})$ over all s in S for which $X(s) = x$".)

Of course, if x is not a value of X, that is, if x is not in $X(S)$, then $X^{-1}(x) = \phi$ and $p_X(x) = P(\phi) = 0$. Since these do not contribute to the probabilities for X, they are often left unmentioned.

At present the set S is finite, so the range of the RV X will also be finite; for simplicity we say that X is a random variable with the *induced* sample space $\tilde{S} = X(S)$. If the distinct values of X are $x_1, x_2, \cdots, x_d$, then $\tilde{S} = \{x_1, x_2, \cdots, x_d\}$ and these *induced elementary events* have probabilities $p_X(x_i) = P(X = x_i)$. The induced $\tilde{\mathcal{F}}$ is then the power set of $\tilde{S}$ and the new probability space is the triple $[\tilde{S}, \tilde{\mathcal{F}}, p_X]$. Indeed, in most applications, this is our starting point; we have some knowledge of a numerical outcome X and we examine properties of various p_X which might be associated with the experiment. Often, what is of the greatest interest is a "tail" probability: $P(X \leq y)$ and/or $P(X > z)$.

Example: Suppose that a certain concrete product is sold in bags that contain $X = 1, 2, 3, 4$ cubic feet. Suppose that past history has shown "demand" as $P(X = 1) = .4$, $P(X = 2) = .3$, $P(X = 3) = .2$, $P(X = 4) = .1$.

a)   Then $P(X \leq 1) = P(X = 1) = .4;$

$P(X \leq 2)\quad = P(\{X = 1\} \cup \{X = 2\})$

$\qquad\qquad = P(X = 1) + P(X = 2) = .4 + .3 = .7;$

$P(X \leq 2.5) = P(X \leq 2) = .7$ because there are no values of X between 2 and 2.5;

$P(X \leq 3)\quad = P(X = 1) + P(X = 2) + P(X = 3)$

$\qquad\qquad = .4 + .3 + .2 = .9;$

$P(X \leq 3.999) = P(X \leq 3) = .9;$

$P(X \leq 4) = P(X = 1) + P(X = 2) + P(X = 3) + P(X = 4)$

$\qquad\qquad = .4 + .3 + .2 + .1 = 1.0.$

b)   These "left tail" probabilities can be given in the
following compact form and/or graph.

$$
\begin{array}{llll}
P(X \le y) & = & 0 & \text{when} & y < 1 \\
& = & .4 & \text{when} & 1 \le y < 2 \\
& = & .7 & \text{when} & 2 \le y < 3 \\
& = & .9 & \text{when} & 3 \le y < 4 \\
& = & 1.0 & \text{when} & 4 \le y
\end{array}
$$

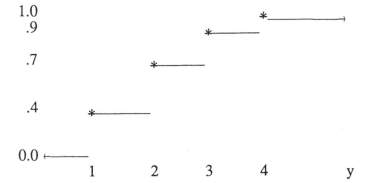

The * emphasizes that the value of the function at 1 2 3 4 is
on the upper branch of this *step function*.

Example: Let the probability in $[S, \mathscr{P}, P]$ be uniform;
$S = \{a\ b\ c \cdots x\ y\ z\}$, the 26 Latin letters. Let $X(\alpha) = 1$ if α is a
vowel $\{a\ e\ i\ o\ u\}$ and $X(\alpha) = 2$ if α is a consonant. Then,
$$P(X = 1) = 5/26 \text{ and } P(X = 2) = 21/26.$$

$$
\begin{array}{lll}
P(X \le x) & = 0 & \text{if } x < 1 \\
& = 5/26 & \text{if } 1 \le x < 2 \\
& = 1 & \text{if } 2 \le x.
\end{array}
$$

This is just like an unbalanced coin. The graph is:

```
     1                                    *—————————→



  5/26                    *————————————


     0      ←————————————
                          1              2
```

Exercise 2:   a) Find $P(X \leq y)$ when X is the upface on a single roll of a fair six–sided die; graph this function.
b)   Find $P(X \leq y)$ when X is the sum of the upfaces on two independent rolls of a fair six–sided die and graph this function.

This discussion leads to the following

Definition: *Let X be a real RV with $X(S) = \{x_1, x_2, \cdots, x_d\}$ and probability density function p_X . Then for each real number y, the cumulative distribution function F has value*

$$F_X(y) = P(X \leq y)) = \sum_{x_i \leq y} p_X(x_i) \,.$$

As above, the notation means the sum of the probabilities $p_X(x_i)$ for those elementary events $\{x_i\}$ which have value at most y. If the interval of real numbers less than or equal to y is represented by $(-\infty, y]$, then this set can be represented by $X^{-1}((-\infty, y]) = \{x_i \in \tilde{S} : x_i \leq y\}$. Note that we are using the induced sample space which, as noted earlier, is all we may have anyway. "Cumulative distribution function" may be shortened to "distribution function" and most often we shall use CDF.

Exercise 3:   Find the CDF for the example involving cashiers and sketch the function.

If the distinct values of X are $\{x_1, x_2, \cdots, x_d\}$ as in the last definition above, $X^{-1}(x_i) \cap X^{-1}(x_j) = \phi$ since a function cannot have two values: $X(s) = x_i \neq x_j$. Then,

$$p_X(x_1) + p_X(x_2) + \cdots + p_X(x_d)$$
$$= P(\{x_1, x_2, \cdots, x_d\}) = P(X(S)) = P(S) = 1 \,.$$

Thus we see a PDF characterized as

a function p_X on $X(S) = \{x_1, x_2, \cdots, x_d\}$

with $p_X(x_i) \geq 0$ and $\sum_{i=1}^{d} p_X(x_i) = 1$.

This is emphasized in examples like the following.

Example: Let the RV Y (for a change) have PDF $P(Y = i) =$ $p_Y(i) = ci$ for c fixed and $i = 1(1)6$. Then

$$1 = \sum_{i=1}^{6} p_Y(i) = \sum_{i=1}^{6} ci$$

$$= c1 + c2 + c3 + c4 + c5 + c6 = c21$$

which implies that $c = 1/21$. Therefore, $p_Y(i) = i/21$, $i = 1(1)6$, and, of course, zero elsewhere.

Exercise 4: Find the value of the constant c which will make p_Y a proper PDF on the range indicated. Then find and graph its CDF.

a) $p_Y(y) = c/y$ for $y = 1(1)6$.

b) $p_Y(y) = (c+y)/c$ for $y = 2(2)8$ (y takes values 2 thru 8 in steps of 2).

Exercise 5 Five out of a group of 100 typists have not passed a certain speed test. Ten of these 100 typists are to be selected "at random" for transfer (in a large company). Let X be the number of the five "failures" selected for transfer. Make a table of the PDF and CDF of X.

Exercise 6: Say that 20% of the income tax returns are in error. Suppose that six returns are to be selected "at random" from the large number available so that these are essentially independent Bernoulli trials. Let X be the number of defective returns among the six selected. Find the PDF and CDF of X.

Exercise 7: Let A and B be independent events of $[S, \mathcal{S}, P]$ with $P(A) = a$ and $P(B) = b$. For any susbset C, let $I_C(s) = 1$ if $s \in C$ and equal 0 otherwise. Determine the CDF of $X = I_A - I_B$. Determine the CDF of $Y = I_B \cdot I_A$.

Exercise 8: Let X have range $x_1 < x_2 < x_3 \cdots$. Show that the CDF F is non–decreasing (monotonically increasing): for real numbers $b < c$, $F(b) \leq F(c)$. Hint: show that

$$\{x_i \le b\} \subset \{x_i \le c\}.$$

Given a CDF as a certain kind of step–function, we can discover other probabilities.

Example: Let the RV X have CDF

$$
\begin{aligned}
F(x) \;\; &= 0 && \text{for } x < 2 \\
&= 1/3 && \text{for } 2 \le x < 4 \\
&= 5/6 && \text{for } 4 \le x < 6 \\
&= 1 && \text{for } 6 \le x \;.
\end{aligned}
$$

Since there are no distinct values for F other than at the jump points 2, 4, 6, we know that other values of X must have probability zero and we can ignore them here. Then:

$$P(X = 2) = P(X \le 2) = F(2) = 1/3;$$

$$P(X = 4) = P(X \le 4) - P(X \le 2)$$

just as $\{2,4\} - \{2\} = \{4\}$ so that

$$P(X = 4) \;\;\; = F(4) - F(3) = 5/6 - 1/3 = 1/2;$$

$$P(X = 6) \;\;\; = P(X \le 6) - P(X \le 4) = F(6) - F(4)$$

$$= 1 - 5/6 = 1/6.$$

Exercise 9: Let the RV Y have CDF
$$F(y) = j(j+1)/42 \;\; \text{for } j \le y < j + 1, \, j = 1(1)6.$$
Find the (essential) range and probabilities for Y. Hint: assume automatically that $F(y) = 0$ for $y < 1$ and $F(y) = 1$ for $y \ge 7$.

We are not limited to obtaining individual values. For the distribution function F,

$$F(b) - F(a) = P(X \le b) - P(X \le a) = P(a < x \le b)$$

$$\text{just as } \{(-\infty,b]\} - \{(-\infty,a]\} = (a,b].$$

In summation notation, this is

$$F(b) - F(a) = \sum_{a < x_i \le b} p_X(x_i) \; .$$

We emphasis that this is the sum of the probabilities for values in the range of X strictly greater than a but less than or equal to b . Note also that $P(X > b) = 1 - P(X \le b)$.

Example: Let $P(Y = i) = p_Y(i) = i/21$ for $i = 1(1)6$,

$\qquad\qquad p_Y(y) = 0$ elsewhere.

$P(1 < Y \leq 5) = P(Y \leq 5) - P(Y \leq 1) = 15/21 - 1/21 = 14/21.$

Note that $\quad P(1 < Y < 6) \qquad = P(1 < Y \leq 5)$.

$\qquad\qquad\quad P(1 < Y \leq 3) \qquad = 6/21 - 1/21 = 5/21$.

$\qquad\qquad\quad P(1 \leq Y < 5) \qquad = P(0 < Y \leq 4)$

$\qquad\qquad\qquad\qquad\qquad\quad = P(Y \leq 4) = F(4) = 10/21$.

$\qquad\quad P(Y > 4) \quad = 1 - P(Y \leq 4) = 11/21$.

Exercise 10 : a) Find the constant c so that $p_X(x) = cx$ is a probability distribution on $X(S) = \{1\,2\,3\,4\,5\,6\,7\,8\,9\}$.

b)   Show that $P(X \leq x) = F_X(x) = x(x+1)/90$ for $x \, \varepsilon \, X(S)$.

c)   Evaluate $F_X(3)$, $P(2 < X \leq 7)$, $F_X(6) - F_X(2)$, $P(X > 3)$, $P(3 \leq X \leq 6)$, $P(3 < X < 6)$.

LESSON 12. EXPECTATION

"Expectation" in probability/statistics is a generalization of the *simple arithmetic mean* SAM.

Example: a) 4, 3, 8, 9 has SAM

$$\frac{4 + 3 + 8 + 9}{4} = 6 .$$

b)   4, 3, 8, 9, 3, 3, 8, 9, 3, 4, 8, 3 has SAM

$$\frac{4(2) + 3(5) + 8(3) + 9(2)}{2 + 5 + 3 + 2}$$

or

$$4 \left[\frac{2}{12} \right] + 3 \left[\frac{5}{12} \right] + 8 \left[\frac{3}{12} \right] + 9 \left[\frac{2}{12} \right] = 65/12 .$$

Definition: *The numbers $x_1, x_2, \cdots, x_n$ have SAM*

$\bar{x} = \sum\limits_{j=1}^{n} x_i/n.$ *If these numbers are grouped so that*

the distinct values are $y_1, y_2, \cdots, y_d$

with corresponding frequencies $f_1, f_2, \cdots, f_d$,

then the SAM is $\sum\limits_{j=1}^{d} y_j \cdot f_j \Big/ \sum\limits_{j=1}^{d} f_j = \sum\limits_{j=1}^{d} y_j(f_j/n) .$

Exercise 1:   Make a table of distinct values with the corresponding frequencies and find the SAM of our expenditures (tax included) for (Tuesday Lunch) burritoes over the last year:

3.29 6.21 4.14 5.14 5.29 6.83 5.29 6.21 5.50 6.57 5.50

6.37 4.64 5.86 5.50 6.53 6.21 4.10 5.50 4.64 5.64 5.84

6.76 6.96 6.73 4.28 6.55 5.61 5.67 5.73 4.64 5.06 5.53

5.36 6.83 5.37 6.21 5.29 7.57 5.73 4.14 5.52 6.26 4.14

4.96 6.83 5.68 4.14 5.52 5.52 6.83 5.73

Suppose that y_1, y_2, $\cdots$, y_d in the definition are the distinct values of some RV X. Suppose that the x_1, x_2, $\cdots$, x_n represent results of repeated (trials) observations of X. Then in the relative frequency interpretation of probability, $f_j/n = \#(x_i = y_j)/n$ will "converge" to $P(X = y_j)$ as the number of trials n tends to ∞. Again, this is not a mathematical theorem but it does lead to a:

Definition: *Let X be a real RV with range $S = \{y_1, y_2, \cdots, y_d\}$ and probability $P(X = y_j) = p_X(y_j)$. Then the mean or expected value of X is* $E[X] = \sum_{j=1}^{d} y_j \cdot p_X(y_j)$. *This is also called the mean of the distribution of X. If h is a real valued function on S, then the mean of h(X) is* $E[h(X)] = \sum_{j=1}^{d} h(y_j) \cdot p_X(y_j)$.

Example: Suppose the random variable X has

$$p_X(1) = P(X = 1) = .4;$$
$$p_X(2) = P(X = 2) = .3;$$
$$p_X(3) = .2; \quad p_X(4) = .1 .$$

a) $E[X] = \sum_{j=1}^{4} y_j p_X(y_j)$

$$= 1(.4) + 2(.3) + 3(.2) + 4(.1) = 2.0 .$$

b)   For $h(x) = x^2$, $E[h(X)] = \sum_{j=1}^{4} h(y_j) p_X(y_j)$

$$= 1^2(.4) + 2^2(.3) + 3^2(.2) + 4^2(.1) = 5.0 .$$

We also have $E[h(X)] = E[X^2]$.

c)   For $h(x) = (x - 2)^2$,

$$E[h(X)] = E[(X - 2)^2] = \sum_{j=1}^{4} (y_j - 2)^2 p_X(y_j)$$

$$= (1-2)^2(.4) + (2-2)^2(.3) + (3-2)^2(.2)$$
$$+ (4-2)^2(.1) = 1.0.$$

Exercise 2:   a) Rephrase exercise 5 of lesson 9 in terms of a random variable X ; compute E[X] and E[X^2] .
b)   Do the same for exercise 6 of lesson 9.

A traditional symbol for E[X] is the Greek letter   μ (pronounced mu by most American statisticians) or when more than one RV is in the problem   E[X] = μ_X , E[Y] = μ_Y, etc.

After "*the* mean μ", the next most used "average" is E[(X – μ)^2] = σ^2 (or σ_X^2), the Greek sigma squared being another traditional symbol; this "mean" is called the *variance* of X and abbreviated Var X. (In physics, with density p_X, the variance is the moment of inertia about the center of gravity which is the mean.) We also call E[X] the first *moment*, E[X^2] the second moment, etc.

Although the distributions in the following example are artificial, they are designed to convey some sense of the "variation" in "variance". Each *probability histogram* is drawn with contiguous bars of equal width and heights proportional to the relative frequencies.

Example: a) The random variable A has a uniform distribution on its range S_A = {–3, –2, –1, 0, 1, 2, 3} ; the corresponding probability histogram is:

The mean μ_A = E[X] = (–3)(1/7) + (–2)(1/7) + (–1)(1/7)
$$+ (0)(1/7) + (1)(1/7) + (2)(1/7) + (3)(1/7) = 0 .$$

The variance σ_A^2 = E[(X–μ)^2] reduces to E[X^2]

$$= (-3)^2(1/7) + (-2)^2(1/7) + (-1)^2(1/7)$$
$$+ 0^2(1/7) + 1^2(1/7) + 2^2(1/7) + 3^2(1/7)$$
$$= 28/7.$$

b)   The RV B has distribution given by

$p_B(-3) = 2/7$, $p_B(-1) = p_B(0) = p_B(1) = 1/7$, $p_B(3) = 2/7$;

the corresponding histogram is:

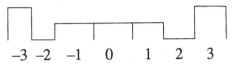

$$-3 \; -2 \quad -1 \quad 0 \quad 1 \quad 2 \quad 3$$

It is easy to see that $E[B] = 0$; then $\sigma_B^2 = E[B^2]$

$$= (-3)^2(2/7) + (-1)^2(1/7) + 0(1/7) + 1^2(1/7) + 3^2(2/7)$$

$$= 38/7.$$

c)   The RV C has $p_C(-3) = 3/7$, $p_C(0) = 1/7$, $p_C(3) = 3/7$ with histogram:

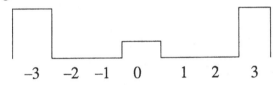

$$-3 \qquad -2 \quad -1 \quad 0 \quad 1 \quad 2 \quad 3$$

$E[C] = 0$ and $\sigma_C^2 = 9(3/7) + 0(1/7) + 9(3/7) = 54/7$.

d)   Let D have $p_D(-2) = p_D(2) = 3/7$, $p_D(0) = 1/7$.

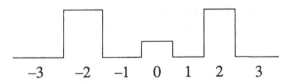

$$-3 \qquad -2 \quad -1 \quad 0 \quad 1 \quad 2 \quad 3$$

$\mu_D = 0$ and $\sigma_D^2 = 24/7$.

e)   The RV G has $p_G(-4) = 1/8$, $p_G(-2) = 2/8$, $p_G(1) = 2/8$, $p_G(2) = 3/8$.

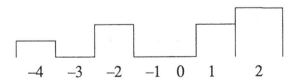

$$-4 \quad -3 \quad -2 \quad -1 \; 0 \quad 1 \qquad 2$$

$E[G] = (-4)(1/8) + (-2)(2/8) + (1)(2/8) + 2(3/8) = 0$.

Var $G = 16/8 + 8/8 + 2/8 + 12/8 = 48/8$.

Look at the results for A, B, C, D. These distributions are all symmetric about zero; formally,

$$P(X = 3) = P(X = -3), P(X = 2) = P(X = -2),$$
$$P(X = 1) = P(X = -1).$$

Graphically, D is "more compact" than B which is "more compact" than C and $\sigma_A^2 < \sigma_B^2 < \sigma_C^2$. Although A has the smaller variance, it is not clear that A is "more compact" than D. Also, the distributions with smaller variances do not have necessarily the smaller spread; the variables A, B, C have maximum − minimum = 3 − (−3) = 6. Then along comes G with mean 0, variance 48/8 in between 38/7 and 54/7 and max − min = 2 − (−4) = 6. Although a comparison of variances does not put the distributions in a nice order, this numerical characteristic has other properties which make it useful in statistics later on.

Exercise 3: Compute the variances; draw the corresponding histograms.
 a) $p_X(i) = P(X = i) = 1/5$ for i=1(1)5.
 b) $p_X(j) = j/15$ for j=1(1)5.
 c) $p_Z(z) = (15-z)/60$ for z=1(1)5.

Exercise 4: First find the constant c which will make the distribution a proper probability; then find the mean and variance of X. Draw probability histograms.
 a) $p_X(1) = p_X(5) = 5c; p_X(2) = p_X(4) = 4c;$
 $p_X(3) = 3c$.

 b) $p_X(1) = p_X(5) = 3c; p_X(2) = p_X(4) = 4c;$
 $p_X(3) = 5c.$

 c) $p_X(4) = p_X(3) = 3c; p_X(5) = p_x(2) = 4c; p_X(1) = 5c.$

In a later lesson, we will prove a general version of the following theorem; even in that fancy form, we are just extending the algebra for sums:

$$\sum_{i=1}^{n}(a_i + b_i) = \sum_{i=1}^{n}a_i + \sum_{i=1}^{n}b_i \; ; \; \sum_{i=1}^{n}ca_i = c\sum_{i=1}^{n}a_i \; .$$

Of course, for now, our spaces are finite and our functions have finite ranges.

Theorem: *Let X be a real RV for [S, $\mathcal{F}$, P]. Let h_1, h_2, h_3, $\cdots$ be real valued functions defined on X(S). Let c be a real number. Then, a) $E[(h_1 + h_2)(X)] = E[h_1(X) + h_2(X)]$*

$$= E[h_1(X)] + E[h_2(X)]$$

and by induction, $E[\sum_{i=1}^{n} h_i(X)] = \sum_{i=1}^{n} E[h_i(X)]$ for a positive integer n.

b) $E[cX] = cE[X]$.

(The expected value of this sum of random variables is the sum of their expected values; a "constant" can be brought outside the expectation sign.)

Perhaps we should note that here we are dealing with composite functions. First, the random variable X is a function on S and h is a function on X(S); then

$$Y(s) = h_1(X(s)) \text{ and } Z(s) = h_2(X(s))$$

are functions on S. When X(s) is labeled x_j , the sum of these two functions on S has several representations:

$$(Y + Z)(s) = Y(s) + Z(s)$$

$$(h_1 + h_2)(X)(s))) = (h_1 + h_2)(X(S))$$

$$(h_1 + h_2)(x_j) = h_1(x_j) + h_2(x_j).$$

For these finite–space cases, it is really "obvious" that functions of random variables are random variables and all the algebra of real valued functions carries over.

Example: a) For $h_1(x) = x$ and $h_2(x) = x^2$,

$$E[h_1(X) + h_2(X)] = E[X + X^2] = E[X] + E[X^2] .$$

b)   In particular, when X(S) = { 1, 2, 3, 4, 5, 6} (say for the upfaces of the toss of a die),

$$E[X + X^2] = \sum_{x=1}^{6} (x + x^2)p(x)$$

$$= \sum_{x=1}^{6} xp(x) + \sum_{x=1}^{6} x^2 p(x) \ .$$

c)   Continuing with the die,

$$E[2X] = \sum_{x=1}^{6} (2x)p(x)$$

$$= 2p(1) + 2 \cdot 2p(2) + 2 \cdot 3p(3) + 2 \cdot 4p(4) + 2 \cdot 5p(5) + 2 \cdot 6p(6)$$

$$= 2[1p(1) + 2p(2) + 3p(3) + 4p(4) + 5p(5) + 6p(6)] = 2 \cdot E[X] \ .$$

Exercise 5: Obviously, the following theorem will be useful in calculations. State the properties of ordinary algebra, summation (Σ), and probability which justify each of the equalities labeled $1,2,\cdots,8$ in the proof.

Theorem: *Let the real random variable X have sample space*

$$\{x_1, x_2, \cdots, x_d\} \text{ and density } p_X. \text{ Then,}$$

$$\sigma_X^2 = Var\ X = E[X^2] - (E[X])^2 \ .$$

Proof: Let $\mu = E[X] = \sum\limits_{j=1}^{d} x_j p_X(x_j)$; then,

$$\sigma^2 = E[(X - \mu)^2]$$

$$\overset{1}{=} \sum_{j=1}^{d} (x_j - \mu)^2 p_X(x_j)$$

$$\overset{2}{=} \sum_{j=1}^{d} (x_j^2 - 2x_j\mu + \mu^2)p_X(x_j)$$

$$\overset{3}{=} \sum_{j=1}^{d} \{ (x_j^2 p_X(x_j) - 2\mu x_j p_X(x_j) + \mu^2 p_X(x_j) \}$$

$$\overset{4}{=} \sum_{j=1}^{d} x_j^2 p_X(x_j) - 2\mu\sum_{j=1}^{d} x_j p_X(x_j) + \mu^2\sum_{j=1}^{d} p_X(x_j)$$

$$\overset{5}{=} E[X^2] - 2\mu E[X] + \mu^2$$

$$\overset{6}{=} E[X^2] - 2\mu^2 + \mu^2$$

$$\overset{7}{=} E[X^2] - \mu^2 \overset{8}{=} E[X^2] - (E[X])^2 .$$

Exercise 6: Find $E[X]$ and $E[X^2]$ when

$$p_X(x) = \begin{bmatrix} 6 \\ x \end{bmatrix}\begin{bmatrix} 7 \\ 9-x \end{bmatrix} \div \begin{bmatrix} 13 \\ 9 \end{bmatrix} \quad \text{for } x = 0(1)6 .$$

Exercies 7: Let $E[X] = \mu$ and Var $X = \sigma^2$.
  a)   What is $E[(X-\mu)]$?
  b)   What is the variance of the RV $Y = (X - \mu)/\sigma$?
  c)   What is the mean and variance of $Z = \alpha X + \beta$ when α and β are constants?
        Hint: work out sums as in the last theorem.

Exercise 8: Let X be a random variable with a uniform distribution on $\{1, 2, 3, \cdots, n\}$, ie, the set of positive integers from 1 to (a fixed but unknown) n. Find the mean and variance of this distribution. Hint: First prove by induction:

$$\sum_{j=1}^{n} j = n(n + 1)/2 \text{ and } \sum_{j=1}^{n} j^2 = n(n + 1)(2n + 1)/6 .$$

INE–Exercises:
  1.   Sketch the probability histogram and CDF for the RV X in exercise 2 at the end of lesson 11. Find the mean and variance of X.
  2.   In a certain group, the probability that a man will live 20 years from date is .82 and the probability that his wife will live 20 years from date is .89 . Assuming that their demises are independent events (true?), what is the probability that 20 years from date, both, neither, at least one of a man and his wife will be alive? What is the expected number of these two who will be alive?

3. A poker hand may contain 0, 1, 2, 3, 4 kings. Find the mean and variance of the number of kings in a poker hand.

4. Let A, B C be events in a sample space with $P(A) = .1$, $P(B) = .4$, $P(A \cap B) = .1$, and $P(C) = .3$; assume that $A \cap C = \phi = B \cap C$. Define the RV X as $I_A + I_B - I_C$. Find the CDF of X and the mean and variance of the distribution.

5. Assign a uniform probability distribution to the sample space $S = \{1\ 2\ 3\ 4\ 5\ 6\ 7\ 8\ 9\ 10\ 11\ 12\}$; let $X(s) = s$ be the associated RV. Let $A = \{X$ is divisible by 2$\}$ and $B = \{X$ is divisible by 3$\}$.

    a)   Compute the variance of X.

    b)   Are A and B independent events?

6. The probability space is $[S, \mathscr{S}, P]$; indicator functions are $I_A(s) = 1$ if $s \in A$, $I_A(s) = 0$ otherwise, etc. Show that:

    a) $E[I_A] = P(A)$;

    b) $P(A \cap B) = P(A) \cdot P(B)$ iff

$$P(I_A = i, I_B = j) = P(I_A = i) \cdot P(I_B = j)$$

    for all $i, j \in \{0, 1\}$ iff $E[I_A \cdot I_B] = E[I_A] \cdot E[I_B]$.

LESSON 13. A HYPERGEOMETRIC DISTRIBUTION

There are many instances wherein you use samples to help you make decisions, though not in the formal ways we shall be developing in this course. For example, the authors examine a box of strawberries in the supermarket and, seeing at most one or two berries with "spots", buy the box. The other way around, a nut broker examines a handful of nuts from a large sack of unshelled pecans and, on the basis of size, color, fullness, and taste, decides what price to pay. First we need the following.

Definition: *The collection of objects which are under study in an experiment is called a population. A sample is a part of the population.*

In ordinary language, we might refer to the population of students at this university this semester. But a statistician would be more specific referring to the population of the gradepoints of these students or the population of their social preferences or the population of their working hours or even all three. "Population" is not restricted to people; other examples are: a population of the lifetimes of certain electrical components; a population of patients' reactions to a medicine; a population of egg orders at a lunch counter, etc. Corresponding samples would be the gradepoints or the lifetimes or the reactions or $\cdots$ that we *actually* observe. This is not a sample space which consists of the *possible* values that we could observe.

Consider the following simplified model involving the pecans. Each pecan is either "defective" or "non–defective". There are twelve nuts in (a handful) the sample. If there are no defective pecans in the sample, the broker will buy the sack; otherwise, he will "reject" (not buy) the sack. What is the probability that he will buy a 25 pound sack?

Obviously, we cannot answer this question unless we know the numbers of pecans, defective and non–defective, in the sack and how the sample has been drawn.

Definition: *For a population of size N , let a sample space S be all samples of size n, drawing without replacements and without regard to order. Assign a uniform distribution to S so that each sample of size n has the same probability (of being selected). Then each "point" in S is said to be a simple random sample from the population.*

The anxious reader may turn to lesson 14 for a discussion of the physical techniques of performing such sampling. You should recognize the definition of a simple random sample as a general form of dealing a hand of cards.

Example: There are ten pairs of vowels from {a e i o u}: ae, ai, ao, au, ei, eo, eu, io, iu, ou. These pairs could be the points of an S.

Exercise 1: Continuing with the vowels in the example above, list all simple random samples of each size: n = 1, 2, 3, 4, 5 .

Exercise 2: a) Consider hands of size 5 from an ordinary straight deck, each with probability $1/\binom{52}{5}$. Let X be the number of red cards in a hand. Write the formula for
$$P(X = x) = p_X(x) \text{ for } x = 0, 1, 2, 3, 4, 5.$$
b)   Consider hands of size 13 from an ordinary straight deck each with probability $1/\binom{52}{13}$. Let X be the number of aces in a hand. Find $P(X = x) = p_X(x)$ for x = 0, 1, 2, 3, 4.

In the case of the pecans, suppose that there are D defective pecans among the 2500 in a sack. We assume that shaking the trees, hulling the nuts, and sacking them has thoroughly mixed the good and the bad so that the handful may be regarded as a random sample. If X is the number of defective pecans in the sample, then
$$P(X = 0) = \binom{D}{0}\binom{2500-D}{12-0}/\binom{2500}{12}.$$
Here is a general result.

Theorem: *Let a population consist of D "defective" and N-D "non-defective" items. For a random sample of size n < N, let X be the number of defective items in the sample. Then the probability density of X is*

$$P(X = x) = h_X(x; N,D,n) = \binom{D}{x}\binom{N-D}{n-x}/\binom{N}{n} \quad \text{for } x = 0(1)m$$

where m is the minimum of D and n; we take $\binom{a}{b} = 0$ whenever b > a.

Notes: (1) "m is the minimum of D and n" simply says that the number of defectives in the sample cannot be greater than the number of defectives in the population or the number of items in the sample; with the symbol for min, "∧", $m = D \wedge n$. (2) The name comes from the use of these terms as coefficients in the hypergeometric function in analysis.

Proof of the theorem: This is just an application of counting:

$\begin{bmatrix} D \\ x \end{bmatrix}$ is the number of ways of choosing x of the D defective items;

$\begin{bmatrix} N-D \\ n-x \end{bmatrix}$ is the number of ways of choosing the non–defective items;

$\begin{bmatrix} N \\ n \end{bmatrix}$ is the numer of random samples of size n.

To return to the nuts, we realize that we cannot know D at this stage of buying and selling. For no one would examine all N = 2500 nuts and, even if they did, the process (shelling, tasting) would render them unsaleable. The most that we can do is to examine (calculate) the probabilities $P(X = 0)$ for various values of D:

D	5	10	20	25	50	100	125	200	300	500
P	.976	.953	.908	.886	.784	.612	.540	.367	.215	.068

Under this particular rule (the broker buys the lot if there are no defective items in the sample), at least 90% of the sacks will be sold when $D \leq 20$. The broker will buy less than 37% of the sacks when $D \geq 200$. We will continue the formalization of this problem in lesson 15.

Exercise 3: For $h_X(x; D,N,n) = \begin{bmatrix} D \\ x \end{bmatrix} \begin{bmatrix} 10-D \\ n-x \end{bmatrix} / \begin{bmatrix} 10 \\ n \end{bmatrix}$, compute $P(X \geq 1)$ for each pair of n = 2,3,4,5,6 and D = 2,4,6.

Today most of us have access to a computer of some sort to do this arithmetic but "in the old days", we needed tables of the hypergeometric.

For each triple (N, D, n) we get a different distribution. This phrase should remind you of "function". Here, the domain is the collection of these triples of integers with
$$0 \leq D \leq N \text{ and } 1 \leq n \leq N ;$$
the range is the collection of functions

$$\{p_X(0) \ p_X(1) \ p_X(2) \ \cdots \ p_X(m)\} \ .$$

The triples index the family of distributions.

This is the same kind of indexing implied by the simple equation $y = bx$ in the (x,y) plane: for each real number b, we get one line thru the origin with slope b. In many problems, the domain of the parameters is implied by some notation.

Definition: *An index of a family (collection) of lines, sets, functions, $\cdots$ is called a parameter.*

As you may have noted during this discussion, a "parameter" may have "components" (N, D, n) ; rather than introduce more terminology, we simply call the components parameters. To emphasize the dependence on its parameters, we may symbolize the probability as

$$P(X = x \mid N, D, n) = p_X(x ; N, D, n)$$

The bar or semicolon is read "given".

Exercise 4: What are the parameter values (index sets) for A,B,C, "assumed" in the (real) relations:
a) $y = A \sin(x/B)$
b) $y = \exp(Ax)$
c) $y = x/(x - A)^{1/2}$
d) $x^2 + y^2 = C$
e) $y = A(x - B) \geq 0$
f) $y = e^{x-B}/B \geq 0$?

Returning once more to a seller–buyer format, we note that the seller is more likely to have some idea of the "proportion of defectives" in the product rather than their actual number. Similarly, the buyer will focus on some proportion that he can tolerate (without losing his shirt). Then we treat the hypergeometric in a "reparamterized" form:

$$p_X(x; N, \theta, n) = \begin{bmatrix} N\theta \\ x \end{bmatrix} \begin{bmatrix} N-N\theta \\ n-x \end{bmatrix} / \begin{bmatrix} N \\ n \end{bmatrix}$$

for $\theta = D/N$ in the set $\{0, 1/N, 2/N, ..., N/N\}$.

Exercise 5: Sketch $R(\theta) = P(X \geq c) = \sum_{x=c}^{n} p_X(x; N, \theta, n)$ for

θ = 0, 2/20, 4/20, 6/20, 10/20, 15/20, 20/20 .

    a) n = 5 c = 2 b) n = 8 c = 2 c) n = 10 c = 5 .

In the next lesson (and later), we will need the mean or expected value of the hypergeometric random variable X :

if you believe that for n > D, $P(X = x) = \begin{bmatrix} D \\ x \end{bmatrix} \begin{bmatrix} N-D \\ n-x \end{bmatrix} / \begin{bmatrix} N \\ n \end{bmatrix}$,

then you must believe that $\displaystyle\sum_{x=0}^{D} \begin{bmatrix} D \\ x \end{bmatrix} \begin{bmatrix} N-D \\ n-x \end{bmatrix} = \begin{bmatrix} N \\ n \end{bmatrix}$.   (*)

Now $E[X] = \displaystyle\sum_{x=0}^{D} x \begin{bmatrix} D \\ x \end{bmatrix} \begin{bmatrix} N-D \\ n-x \end{bmatrix} / \begin{bmatrix} N \\ n \end{bmatrix}$ so that

$$\begin{bmatrix} N \\ n \end{bmatrix} E[X] = \sum_{x=0}^{D} x \begin{bmatrix} D \\ x \end{bmatrix} \begin{bmatrix} N-D \\ n-x \end{bmatrix} = \sum_{x=1}^{D} x \cdot \frac{D!}{x!\,(D-x)!} \begin{bmatrix} N-D \\ n-x \end{bmatrix}$$

since, when x = 0, x! ≠ 0 and the first term disappears.

Then when x ≥ 1, x! = x(x–1)! and the last sum simplifies to

$$\sum_{x=1}^{D} \frac{D!}{(x-1)!(D-x)!} \begin{bmatrix} N-D \\ n-x \end{bmatrix}.$$

Now let x–1 = y or x = y+1 . Then, as in integration by substitution, the summand and the limits change:

$$\begin{bmatrix} N \\ n \end{bmatrix} E[X] = \sum_{y=0}^{D-1} \frac{D!}{y!\,(D-1-y)!} \begin{bmatrix} N-D \\ n-1-y \end{bmatrix}$$

$$= D \sum_{y=0}^{D-1} \begin{bmatrix} D-1 \\ y \end{bmatrix} \begin{bmatrix} N-D \\ n-1-y \end{bmatrix} \quad \text{using } D! = D(D-1)!$$

Let N–1 = M and D–1 = d so that N–D = N–1–d = M–d. Then the last summation above can be written as

$$\sum_{y=0}^{d} \begin{bmatrix} d \\ y \end{bmatrix} \begin{bmatrix} M-d \\ n-1-y \end{bmatrix}.$$

This sum has the same form as the left– hand side of (*) so

$$\binom{N}{n} E[X] = D \binom{M}{n-1} = D \binom{N-1}{n-1}$$

from which we get $E[X] = D \binom{N-1}{n-1} / \binom{N}{n} = nD/N$.

Exercise 6: a) Using "tricks" like those in the above proof, show

that for $n \le D$, $\sum_{x=0}^{n} x \binom{D}{x} \binom{N-D}{n-x} = D \binom{N-1}{n-1}$ whence, again,

$E[X] = nD/N$.

b)   Show that for $n > D$ in this hypergeometric

$$E[X(X-1)] = n(D/N)(1-D/N)(N-n)/(N-1) +$$

$$(nD/N)^2 - nD/N.$$

c)   What is the variance of the hypergeometric distribution?

Hint: $E[X^2] = E[X(X-1)] + E[X]$.

Exercise 7: Formalize the results on the mean and variance of this hypergeometric distribution as a theorem.

Exercise 8: Consider a lake with no in/out streams for fish. One day a warden "selects" 1000 fish, tags them and returns them to the lake. After a reasonable period of time, he "selects" 1000 fish again and finds that 400 of these had his tags. How many fish are in the lake?

Hint: one "solution" is to find the N which maximizes the

probability $\begin{bmatrix} 1000 \\ 400 \end{bmatrix} \begin{bmatrix} N - 1000 \\ 600 \end{bmatrix} \div \begin{bmatrix} N \\ 1000 \end{bmatrix}$; we select N

which maximizes the probability of observing what in fact we did observe!

LESSON *14. SAMPLING AND SIMULATION

In the previous lesson, the nut broker's sample was assumed to be "a random one", that is, a selection with uniform probability on the $\begin{bmatrix} 2500 \\ 12 \end{bmatrix}$ possible sample points. Presumably, after shaking the nuts from the trees, removing the hulls, and putting them in 25 pound sacks, we have gotten them pretty well mixed up. Moreover, the broker trusts that the sack has not been "salted" with lots of good nuts on top; etc. The essence of "shake the urn, withdraw the chip" mentioned in earlier sections seems to be there.

But the physical process of mixing the chips is not always well done. The result of the 1969 draft lottery has now become classical. There were two bowls, each with chips numbered 1–366, supposedly well–mixed. The capsules were drawn one–by–one to establish draft order (with one bowl) and birthday order (from the second bowl). Look at one summary of the results:

Office of the Director, Selective Service System, Wash., DC

	Draft # 1–183	Draft # 184–366
Jan thru Jun births	73	109
July thru Dec births	110	74

If the sampling were really random (uniform), all four numbers should be close to $366(1/4) = 91.5$. Obviously, something is amiss; the Jan thru Jun numbers are almost opposite to those for July thru Dec. Further details will be discussed (much) later but the point has been made.

Today it is relatively easy to get access to a computer with a random number generator (or even to program one's own, e.g., Yakowitz, 1977). However, there is still a catch. Because of the finite number of significant digits in a machine, there is bound to be some recycling after (depending on the machine) 2^{15} or 10^9 integers have appeared "at random". Indeed, there is really more to it than that and we have to be content with "pseudo–random numbers".

This is more obviously true for tables of random numbers; in fact, most of these were generated some years ago by (old fashioned) computers. The general ideas of sampling can be used in either mode but here we use the table following. In a large table, the digits would occur with more or less equal frequencies approximating a uniform distribution on 0 thru 10; moreover, pairs of these digits approximate a uniform distribution on 00 thru 99, triples on 000 thru 999, etc.

Example : To simulate successive tosses of a fair coin, we read the integers (left to right, then down the lines) and declare 0 2 4 6 8 to be head and 1 3 5 7 9 to be tail.

> Thus,   4 2 1 3 0 5 7 0 7 8 4 0 9 9
> yields H H T T H T T H T H H H T T

For the next experiment, we continue where we left off here.

Example : To simulate successive tosses of an unfair coin, say with P(H) = 6/10 P(T) = 4/10, we declare 0 1 2 3 4 5 to be head and 6 7 8 9 to be tail.

> Thus,   1 0 2 8 0 0 7 5 2 3 8 5 6 9 1 8 8 5 4
> yields H H H T H H T H H H T H T T H T T H H

Exercise 1: Simulate some tosses of a three sided coin with P(H) = 4/10 P(O) = 3/10 P(T) = 3/10 .

Example : a) To simulate tossing of a fair six–sided die, we can read 1 2 3 4 5 6 and simply skip over 0 7 8 9 .

> 　　　　　2 9 7 3 7 1 2 6 5 5 1 4 2 0 9 4 8
> yields   2 　 3 　1 2 6 5 5 1 4 2 0 9 4 8

b)   For two tosses of a six–sided die (or one toss of a pair of such dice), we look at pairs of digits, skipping those with 0 7 8 9 .

> 　　　　 46 28 93 22 19 55 52 96 78 46 66 36
> yields   46 　　　 22 　　 55 52 　　　 46 66 36

Exercise 2: Let X be a random variable defined on the digits by X(0) = X(1) = 1, X(2) = 2, X(3) = 3, X(4) = 4, X(5) = 5, X(6) = 6, X(7) = X(8) = X(9) = 0 . Impose a uniform distribution on the ten digits.
>   a)   Find the conditional probability distribution of X when X > 0.
>   Hint: P(X = 1 | X > 0) = P(X = 1)/P(X > 0), etc.
>   b)   Describe this in terms of a "loaded" six–sided die.
>   c)   Simulate some tosses.

Example : For the draft lottery, we try to pick out one sequence from 366! sequences. We read triples, discard any duplicate triples and any representing numbers above 366. We could use two sequences as with the two bowls but this isn't really necessary since the days are naturally ordered.

388 459 462 675 128 348 244 793 436 111 · · ·
yields   128 348 244   111 · · ·

Example : To pick a pair of vowels, we should use a first sequence to order the pairs (removing our prior prejudices) and the next digits to select the pair: 5 1 4 0 (4) 6 (0) 7 2 8

(8 2 1 8 6 7 7 6 8 4 0 2 8 7 4 7 5 4) 3 (4 8 4 1 6 3 3) 9 · · ·

Discarding duplicates in parentheses yields

ae–5 ai–1 ao–4 au–0 ei–6 eo–7 eu–2 io–8 iu–3 ou–9

The next digit is 6 so our sample of size one (pair) is ei ; the next five pairs (sampling with replacement) would be eu from 2, au from 0, ao from 4, ae from 5, ao from 4.

Example : For some non–uniform distributions, we can wind up using longer spans of digits. Suppose we want to simulate a sample of size ten for a loaded die:

$$P(1) = 6/21 \approx .285714 \qquad P(2) = 5/21 \approx .238095$$

$$P(3) = 4/21 \approx .190476 \qquad P(4) = 3/21 \approx .142857$$

$$P(5) = 2/21 \approx .095238 \qquad P(6) = 1/21 \approx .047619$$

Since all this is "pseudo" anyway, we might be content to use the approximations even though they add up to only .999999. Taking the random digits in spans of six, we declare "values":

          1 to 285714 to be 1
     285715 to 523809 to be 2
     523810 to 714285 to be 3
     714286 to 857142 to be 4
     857143 to 952380 to be 5
     952381 to 999999 to be 6 .

The next ten sextuplets of digits are:

197690 037375 057106 668997 955485
084123 570964 150417 633170 922158

Exercise 3: Translate the observations in the last example to the die.

Exercise 4: Make up a scheme to simulate observations of X where

$$p(x) = \begin{bmatrix} 4 \\ x \end{bmatrix} \begin{bmatrix} 6 \\ 3-x \end{bmatrix} \div \begin{bmatrix} 10 \\ 3 \end{bmatrix} \text{ for } x = 0(1)\ 3 \ .$$

Hint: $1/30 \approx .033333$.

Since we are limited as well in the physical precision of our observations, all experiments have only a finite number of outcomes (microscopically). Even though we work with other types of mathematical distributions, their simulation will still entail the techniques illustrated herein. We can use more decimal digits and let the computer "read" the table for us but these are differences of degree not kind.

Some Random Digits
(Generated by STSC–APL)

```
4 2 1 3 0 5 7 0 7 8 4 0 9 9 1 0 2 8 0 0 7 5 2 3 8 5 6
9 1 8 8 5 4 2 9 7 3 7 1 2 6 5 5 1 4 2 0 9 4 8 4 6 2 8
9 3 2 2 1 9 5 5 5 2 9 6 7 8 4 6 6 6 3 6 3 8 8 4 5 9 4
6 2 6 7 5 1 2 8 3 4 8 2 4 4 7 9 3 4 3 6 1 1 1 5 1 4 0
4 6 0 7 2 8 8 2 1 8 6 7 7 6 8 4 0 2 8 7 4 7 5 4 3 4 8
4 1 6 3 3 9 6 2 0 4 5 4 1 9 7 6 9 0 0 3 7 3 7 5 0 5 7
1 0 6 6 6 8 9 9 7 9 5 5 4 8 5 0 8 4 1 2 3 5 7 0 9 6 4
1 5 0 4 1 7 6 3 3 1 7 0 9 2 2 1 5 8 9 7 0 2 9 9 9 8 4
1 7 5 9 0 0 5 5 9 9 8 1 0 8 8 9 5 0 0 9 0 4 4 6 1 5 0
2 7 7 3 8 7 3 7 9 2 1 9 6 5 3 2 2 4 7 6 8 4 4 0 9 2 9
1 5 6 7 3 5 3 7 5 4 4 9 0 2 3 8 9 1 3 6 2 0 1 6 9 6 9
4 6 8 1 4 5 0 2 9 3 9 6 9 6 5 7 8 0 1 8 7 7 7 0 8 9 4
4 3 0 9 5 1 6 9 5 2 6 4 1 0 4 6 7 7 0 1 6 2 4 2 0 1 1
4 8 5 5 1 7 5 3 9 9 5 2 4 4 2 2 5 0 1 0 7 9 0 7 4 7 8
6 2 0 9 9 5 5 2 3 5 9 9 6 1 1 3 9 7 6 9 7 1 9 4 0 4 5
7 8 1 1 6 8 5 6 9 0 9 1 9 5 2 7 5 8 8 6 7 6 3 5 6 3 8
4 9 2 9 1 4 8 9 3 5 5 8 9 6 2 9 5 1 2 8 1 1 8 6 6 8 6
3 8 6 1 0 6 6 4 5 0 3 8 3 6 5 0 7 2 4 1 0 8 9 5 0 6 4
5 8 4 6 2 4 2 9 2 9 4 1 9 8 2 0 7 0 9 6 5 1 1 8 5 6 5
5 1 9 6 2 3 3 8 2 1 7 6 2 8 0 1 1 3 7 7 6 5 3 1 2 1 2
6 4 8 5 3 4 6 8 7 8 2 2 1 5 0 9 7 1 1 4 7 2 4 0 9 1 2
4 7 2 2 2 3 2 9 7 6 5 3 9 1 1 7 6 0 7 1 8 6 3 3 2 3 0
9 7 3 7 8 8 1 3 9 3 3 0 5 7 8 4 8 6 4 3 3 2 9 6 7 1 7
2 6 8 3 6 1 3 1 2 1 2 7 0 4 9 2 1 6 0 8 1 1 8 9 0 0 8
1 8 2 3 5 6 0 9 8 4 0 1 9 8 6 8 2 7 5 2 2 0 4 4 9 0 5
```

```
3 8 7 6 3 6 6 6 2 0 9 2 6 5 3 1 4 1 5 5 4 8 0 3 1 2 8
9 8 9 8 7 8 4 6 6 8 4 1 4 9 4 3 5 2 8 9 1 4 6 2 4 7 0
3 1 8 2 7 4 7 7 5 6 7 0 6 3 4 4 5 3 8 1 3 3 8 0 4 7 2
9 9 9 7 3 7 1 6 5 2 2 5 5 7 6 9 7 5 1 1 5 2 5 3 6 0 8
8 3 1 7 6 4 5 5 4 6 9 6 2 0 6 3 0 1 6 4 5 7 3 3 3 0 5
4 1 3 6 5 8 5 9 0 4 1 9 5 6 1 4 0 4 9 4 8 9 5 1 8 1 6
6 9 4 1 1 2 9 5 8 0 9 9 4 2 4 4 8 1 3 7 8 6 7 6 3 1 6
9 1 0 1 1 1 6 6 5 3 4 3 0 8 4 1 6 9 8 6 2 1 0 1 1 3 5
9 2 0 0 8 8 9 5 1 4 8 9 5 5 4 7 3 1 5 3 8 9 0 3 4 8 8
2 7 8 4 7 4 5 5 0 4 7 2 3 1 7 6 6 3 1 1 4 5 1 9 4 0 1
5 9 1 3 3 3 1 3 3 6 6 0 3 0 5 4 2 2 4 6 3 0 9 1 3 9 3
1 9 1 0 9 7 2 1 2 4 2 2 5 8 8 6 5 7 9 8 7 2 9 2 3 2 7
9 7 4 6 4 1 1 8 5 9 8 2 6 9 5 8 0 1 9 7 8 0 5 3 9 1 2
5 8 5 0 7 7 7 0 8 0 9 2 4 6 7 9 6 9 1 8 2 4 9 7 3 1 9
2 5 3 2 8 2 6 3 5 4 6 6 0 8 3 5 9 7 7 7 4 4 2 7 2 6 4
2 3 3 6 9 7 9 0 9 0 4 1 7 0 6 9 1 5 0 5 3 4 3 4 5 9 6
5 0 2 3 7 5 5 2 1 1 3 8 2 1 4 6 5 0 4 3 0 4 1 2 4 3 9
8 3 4 8 9 3 5 5 1 5 4 7 0 2 1 3 3 2 1 8 2 8 7 9 2 2 5
7 5 4 3 8 4 0 9 9 5 9 4 5 8 4
```

LESSON 15. TESTING SIMPLE HYPOTHESES

This lesson contains some general ideas which will be used later but here the only example for which details are discussed is that of a hypergeometric distribution. Although the following will require some elaboration, it is still useful to begin with the technical meaning of the words in the title.

Definition: *A (statistical) hypothesis is an assumption about a family of probability distributions (associated with some experiment). When the hypothesis determines exactly one distribution, the hypothesis is said to be simple; otherwise the hypothesis is composite.*

Example : A trivial but easy case— When we assume that a (six–sided) die is fair, we assign probabilities 1/6 to each upface and there is only one distribution; the hypothesis is simple. When we assume only that the die is not fair, without assigning probability values to each of the upfaces, then the distribution is not determined exactly and the hypothesis is composite.

Exercise 1: Rewrite the notion of a choice between a fair and an unfair (two–sided) coin in this phraseology. Include simple and composite cases.

Definition: *A test of a hypothesis is a procedure for deciding whether or not to reject the hypothesis.*

One version of this technique can be abstracted from the discussion of lesson 13. The producer believes that the proportion of defective items in the lot (population), say $\theta = D/N$, is no more than $\theta_o = D_o/N$. The consumer believes that the proportion of defectives in the lot may be as high as $\theta_a = D_a/N > \theta_o$. These two agree that the consumer will decide whether or not to buy the lot after looking at a sample. One procedure is to reject the lot when the observed number X of defectives in the sample is greater than 0 .

Given (N,D,n) we can calculate the hypergeometric probabilities. We have one distribution if the producer is correctso that $D = N\theta_o$; we have another distribution if the consumer is correct so that $D = N\theta_a$. The corresponding simple hypotheses are;

H_o: the RV X has a hypergeometric distribution

with known parameters (N, n, θ_o);

H_a: the RV X has a hypergeometric distribution

with known parameters (N, n, θ_a).

H_o is called the null hypothesis and often indicates that there is no (null) difference from the "past". H_a is called the alternative hypothesis and usually indicates a new (alternative) condition for the "future", generated by a change in the manufacturing process, a new medicine, a new training program, even a new doubt, $\cdots$. Often, the simple hypotheses are shortened to

$$H_o: \theta = \theta_o \text{ vs } H_a: \theta = \theta_a \text{ or even } \theta = \theta_o \text{ vs } \theta = \theta_a$$

but we should not forget that a distribution is being assumed as well.

If the consumer were not able to zero in on a value θ_a, the alternative might be only $\theta > \theta_o$. But we cannot calculate probabilities knowing only that θ is bigger; we need exact values. Thus in this case, the alternative would not determine exactly one probability distribution; the hypothesis $H_a: \theta > \theta_o$ is not simple but composite. Similarly, one has occasion to consider a composite null hypothesis like $H_o: \theta \leq \theta_o$,but only later.

Exercise 2: Describe some other dichotomous populations in engineering, psychology, $\cdots$ where there may be a difference of opinion regarding the proportions of yes/no, good/bad, for/against, $\cdots$

The spirit of a test is that when the observed value of X is x , we must decide which of H_o , H_a is more likely to be the true (correct) description of the distribution. You should have noticed the little hedge (more likely) in this last statement. This is caused by the necessity of dealing with samples; for as suggested in lesson 13, the population may be so large that every item cannot be examined or examination itself may destroy the item. There are also cases where a lot is so small that very few items can be examined; often, too, the examination is very expensive. (How many rockets are sent to Mars?) Moreover, if the whole lot could be examined without measurement errors and the like, there would be no need to interject probability or statistics; more precisely, there would be no sampling and no random variable X.

This dependence on samples also means that no matter what procedure we use to reject one of H_o, H_a, we may be wrong. For example, the handful examined may be the only defective pecans in the sack, yet the broker would not purchase the remaining *good* nuts.

Definition: *Rejecting H_o when H_o is true is called a Type I error. Rejecting H_a when H_a is true is a Type II error.*

Since we consider only H_o vs H_a, we can rephrase the type II error as not rejecting H_o when H_o is false. Much as in jury trials wherein the jury may (I) convict an innocent man or (II) let a guilty man go free, we cannot avoid these errors. So (as in our legal system), we try to minimize the frequency of these errors. That is, we look for a scheme which will minimize the probabilities.

Definition: *(Here alpha, α, and beta, β, are traditional symbols.) The size of type I error is*

$$P(Type\ I) = P(Reject\ H_o | H_o\ true) = \alpha\ .$$

The size of type II error is

$$P(Type\ II) = P(Do\ not\ reject\ H_o | H_o\ false) = \beta\ .$$

Example: In our example of the pecans, the lot (sack of N = 2500 pecans) is rejected when the number X of defectives in the sample (handful of n = 12 pecans) is greater than or equal to one: $X \geq 1$. Suppose we consider

$$H_o : \theta = .01\ or\ D = 25\ and\ H_a : \theta = .05\ or\ D = 125\ .$$

Then, (via lesson 13)

$$\alpha = 1 - P(X = 0\ |\ D = 25) = 1 - .886$$

$$\beta = P(Do\ not\ reject\ H_o\ |\ Ho\ is\ false)$$

$$= P(X < 1\ |\ H_a)$$

$$= P(X = 0\ |\ D = 125) = .540\ .$$

Exercise 3: Continuing with the last example, find α and β when

$H_o : \theta = .02$ and $H_a : \theta = .04$.

We need some inspiration to devise a reasonable test; historically, this has come thru the expected values of X. If H_o were true, θ_o would be D_o/N and $E[X]$ would be nD_o/N ; if H_a were true, θ_a would be D_a/N and $E[X]$ would be nD_a/N . Since $\theta_a > \theta_o$, it follows that $D_a > D_o$. As our intuition suggests, we expect more defectives in the sample when H_a is true then when H_o is true. Hence, some reverse logic suggests the test: reject H_o when the observed value of X is too big, say, $X \geq r$. (When somebody gets "too many heads" while tossing a coin, you begin to get suspicious!)

Definition: *Let a random variable X have a sample space* $S = C \cup C^C$. *If a test of* H_o *vs* H_a *is to reject* H_o *when the observed value of X is in C , then C is called the critical region of the test.*

For our example, the errors have sizes:

$$\alpha = P(X \; \varepsilon \; C | H_o) = P(X \geq r | \theta = \theta_o)$$

$$= P(\text{Reject } H_o | H_o \text{ is true});$$

$$\beta = P(C \; \varepsilon \; C^C | H_a) = P(X < r | \theta = \theta_a)$$

$$= P(\text{Do not reject } H_o | H_o \text{ is false}).$$

The probabilities in the example and exercises are to be determined from the tables at the end of this lesson. The heading HYPERA is followed by the parameters N D n ; then,

column 1 contains some values of X ;

column 2 is $h(x; N, D, n) = \begin{bmatrix} D \\ x \end{bmatrix} \begin{bmatrix} N-D \\ n-x \end{bmatrix} \div \begin{bmatrix} N \\ n \end{bmatrix}$;

column 3 is $P(X \leq x) = \sum_{k=0}^{x} h(k; N, D, n)$;

column 4 is $P(X \geq x) = \sum_{k=x}^{m} h(k; N, D, n)$.

In applications, fewer decimal places may be retained in the calculations, but here we want to get a clear picture for several sets of parameter values.

Example: X is hypergeometric with $N = 100, n = 12$.
$\qquad$ $H_o : D = 10$ or $\theta = .1$ $\qquad$ $H_a : D = 20$ or $\theta = .2$

Test 1—the critical region is $C_1 = \{X \geq 1\}$

$\qquad$ $\alpha = P(X \geq 1 \mid D = 10) = .739$ from HYPERA 100 10 12

$\qquad$ $\beta = P(X < 1 \mid D = 20) = .057$ from HYPERA 100 20 12 .

Test 2—the critical region is $C_3 = \{X \geq 2\}$

$\qquad$ $\alpha = P(X \geq 2 \mid D = 10) = .343$ from HYPERA 100 10 12

$\qquad$ $\beta = P(X < 2 \mid D = 20) = .257$.

Test 3— $C_3 = \{X \geq 3\}$

$\qquad$ $\alpha = P(X \geq 3 \mid D = 10) = .098$

$\qquad$ $\beta = P(X < 3 \mid D = 20) = .555$.

Exercise 4: Let X be hypergeometric with $N = 100, n = 15$.

$\qquad$ Take $H_o : \theta = .2$ or $D = 20$ vs $H_a : \theta = .3$ or $D = 30$.

Find α and β when

$\qquad$ a) $C = \{X \geq 4\}$, b) $C = \{X \geq 5\}$, c) $C = \{X \geq 6\}$.

These results show that when the sample size n remains the same, the sizes of type I and type II errors, α and β, are inversely related. In general, for a critical region C, $\alpha = P(C \mid H_o)$; then α can be reduced only by making C "smaller". But when C is reduced, C^c is enlarged so that $\beta = P(C^c \mid H_a)$ must increase (at least not decrease). Thus for a given N, n, θ_o, θ_a , we cannot minimize α and β at the same time.

$\qquad$ We summarize this lesson as follows:

1) $\qquad$ a random sample of size n is drawn from a (dichotomous) population of size N with θ the proportion of "defectives" ;

2) for $H_o: \theta = \theta_o$ vs $H_a: \theta = \theta_a > \theta_o$, we reject H_o when the observed value of the hypergeometric X is in the critical region $\{X \geq r\}$;

3) given N, n, r, θ_o, θ_a , $\alpha = P(X \geq r \mid \theta_o)$ and $\beta = P(X < r \mid \theta_a)$ are determined.

In the next lesson, we examine changes in n and r required by a given α and β .

Exercise 5: Let N = 100 and n = 15. For $\theta = .4$ vs $\theta = .5$, determine α and β when the critical region is:
   a) $\{X \geq 7\}$ b) $\{X \geq 8)$ c) $\{X \geq 9\}$ d) $\{X \geq 10\}$.

Exercise 6: Let N = 100 and n = 15. For $\theta = .5$ vs $\theta = .3$, determine α and β when the critical region is:
   a) $\{X \leq 7\}$ b) $\{X \leq 6\}$ c) $\{X \leq 5\}$ d) $\{X \leq 4)$.

Table HYPERA N D n

This table (generated by STSC–APL) contains values of hypergeometric distributions; each row lists value x, probability for x, cumulative probability of at most x, and cumulative probability of at least x . The param– eters are population size N, number of defectives D and sample size n. For example, in the first table,
   $P(X = 3 \mid N = 100, D = 10, n = 12) = .080682221045$

   $P(X \leq 3 \mid 100, 10, 12) = .982581094248$

   $P(X \geq 3 \mid 100, 10, 12) = .098101126797$

x	HYPERA	100 10 12	
0	.260750268922	.260750268922	1.000000000000
1	.396076357857	.656826626779	.739249731078
2	.245072246424	.901898873203	.343173373221
3	.080682221045	.982581094248	.098101126797
4	.015496890018	.998077984266	.017418905752
5	.001792411376	.999870395641	.001922015734
6	.000124473012	.999994868653	.000129604359
7	.000005020760	.999999889414	.000005131347
8	.000000109464	.999999998878	.000000110586
9	.000000001118	.999999999996	.000000001122
10	.000000000004	1.000000000000	.000000000004

x	HYPERA	100 20 12	
0	.057354756035	.057354756035	1.000000000000
1	.199494803600	.256849559635	.942645243965
2	.297817242517	.554666802153	.743150440365
3	.251676542972	.806343345125	.445333197847
4	.133703163454	.940046508579	.193656654875
5	.046887684718	.986934193297	.059953491421
6	.011088303818	.998022497116	.013065806703
7	.001774128611	.999796625727	.001977502884
8	.000189668355	.999986294081	.000203374273
9	.000013137202	.999999431284	.000013705919
10	.000000555805	.999999987088	.000000568716
11	.000000012792	.999999999880	.000000012912
12	.000000000120	1.000000000000	.000000000120

x	HYPERA	100 30 12	
0	.010128218629	.010128218629	1.000000000000
1	.061799300112	.071927518741	.989871781371
2	.164283139463	.236210658204	.928072481259
3	.251362180599	.487572838803	.763789341796
4	.246294394700	.733867233503	.512427161197
5	.162632489199	.896499722702	.266132766497
6	.074116368776	.970616091478	.103500277298
7	.023456609019	.994072700497	.029383908522
8	.005108920525	.999181621022	.005927299503
9	.000745580441	.999927201463	.000818378978
10	.000069075835	.999996277298	.000072798537
11	.000003640360	.999999917659	.000003722702
12	.000000082341	1.000000000000	.000000082341

x	HYPERA	100 40 12	
0	.001332188501	.001332188501	1.000000000000
1	.013050009801	.014382198301	.998667811499
2	.055984542046	.070366740347	.985617801699
3	.139046575016	.209413315363	.929633259653
4	.222608218655	.432021534018	.790586684637
5	.241928931972	.673950465990	.567978465982
6	.182940087448	.856890553438	.326049534010
7	.096934487895	.953825041333	.143109446562
8	.035701318086	.989526359419	.046174958667
9	.008907931218	.998434290637	.010473640581
10	.001428340695	.999862631333	.001565709363
11	.000132049987	.999994681320	.000137368667
12	.000005318680	1.000000000000	.000005318680

x	HYPERA	100 50 12	
0	.000115572371	.000115572371	1.000000000000
1	.001778036472	.001893608843	.999884427629
2	.011979520731	.013873129574	.998106391157
3	.046749349195	.060622478769	.986126870426
4	.117708182795	.178330661564	.939377521231
5	.201472610551	.379803272115	.821669338436
6	.240393455771	.620196727885	.620196727885
7	.201472610551	.821669338436	.379803272115
8	.117708182795	.939377521231	.178330661564
9	.046749349195	.986126870426	.060622478769
10	.011979520731	.998106391157	.013873129574
11	.001778036472	.999884427629	.001893608843
12	.000115572371	1.000000000000	.000115572371

x	HYPERA	100 10 15	
0	.180768730941	.180768730941	1.000000000000
1	.356780390015	.537549120956	.819231269059
2	.291911228194	.829460349150	.462450879044
3	.129738323642	.959198672792	.170539650850
4	.034487402487	.993686075279	.040801327208
5	.005690421410	.999376496689	.006313924721
6	.000585434301	.999961930991	.000623503311
7	.000036717134	.999998648125	.000038069009
8	.000001327125	.999999975250	.000001351875
9	.000000024576	.999999999827	.000000024750
10	.000000000173	1.000000000000	.000000000173

x	HYPERA	100 20 15	
0	.026193691710	.026193691710	1.000000000000
1	.119062235044	.145255926754	.973806308290
2	.236347421804	.381603348557	.854744073246
3	.271104395598	.652707744156	.618396651443
4	.200381509790	.853089253946	.347292255844
5	.100763273494	.953852527440	.146910746054
6	.035480025878	.989332553318	.046147472560
7	.008870006470	.998202559788	.010667446682
8	.001579590193	.999782149981	.001797440212
9	.000199227592	.999981377573	.000217850019
10	.000017532028	.999998909601	.000018622427
11	.000001048566	.999999958168	.000001090399
12	.000000040853	.999999999021	.000000041832
13	.000000000967	.999999999988	.000000000979
14	.000000000012	1.000000000000	.000000000012
15	.000000000000	1.000000000000	.000000000000

x	HYPERA	100 30 15	
0	.002847892342	.002847892342	1.000000000000
1	.022884849176	.025732741518	.997152107658
2	.081502182152	.107234923670	.974267258482
3	.170498817836	.277733741506	.892765076330
4	.234074648216	.511808389722	.722266258494
5	.223151164632	.734959554354	.488191610278
6	.152425658902	.887385213256	.265040445646
7	.075861618255	.963246831512	.112614786744
8	.027695511427	.990942342938	.036753168488
9	.007404702708	.998347045646	.009057657062
10	.001435373140	.999782418786	.001652954354
11	.000197709799	.999980128585	.000217581214
12	.000018688986	.999998817571	.000019871415
13	.000001141635	.999999959206	.000001182429
14	.000000040182	.999999999388	.000000040794
15	.000000000612	1.000000000000	.000000000612

x	HYPERA	100 40 15	
0	.000209972409	.000209972409	1.000000000000
1	.002738770556	.002948742965	.999790027591
2	.015908177909	.018856920874	.997051257035
3	.054573888106	.073430808980	.981143079126
4	.123626562852	.197057371832	.926569191020
5	.195824475557	.392881847390	.802942628168
6	.223982243285	.616864090674	.607118152610
7	.188292764959	.805156855634	.383135909326
8	.117238891390	.922395747023	.194843144366
9	.054036032245	.976431779269	.077604252977
10	.018274003632	.994705782901	.023568220731
11	.004449838547	.999155621448	.005294217099
12	.000754650982	.999910272429	.000844378552
13	.000084072523	.999994344952	.000089727571
14	.000005496267	.999999841219	.000005655048
15	.000000158781	1.000000000000	.000000158781

x	HYPERA	100 50 15	
0	.000008884673	.000008884673	1.000000000000
1	.000185097362	.000193982036	.999991115327
2	.001715902575	.001909884610	.999806017964
3	.009392308830	.011302193441	.998090115390
4	.033956808849	.045259002290	.988697806559
5	.085910726387	.131169728677	.954740997710
6	.157153767781	.288323496458	.868830271323
7	.211676503542	.500000000000	.711676503542

8	.211676503542	.711676503542	.500000000000
9	.157153767781	.868830271323	.288323496458
10	.085910726387	.954740997710	.131169728677
11	.033956808849	.988697806559	.045259002290
12	.009392308830	.998090115390	.011302193441
13	.001715902575	.999806017964	.001909884610
14	.000185097362	.999991115327	.000193982036
15	.000008884673	1.000000000000	.000008884673

x	HYPERA	2500 25 12	
0	.886147803068	.886147803068	1.000000000000
1	.107891372127	.994039175195	.113852196932
2	.005777550150	.999816725345	.005960824805
3	.000179621051	.999996346396	.000183274655
4	.000003604071	.999999950466	.000003653604
5	.000000049067	.999999999533	.000000049534
6	.000000000464	.999999999997	.000000000467
7	.000000000003	1.000000000000	.000000000003
8	.000000000000	1.000000000000	.000000000000

x	HYPERA	2500 50 12	
0	.784292737816	.784292737816	1.000000000000
1	.192937942882	.977230680698	.215707262184
2	.021310153937	.998540834635	.022769319302
3	.001396814678	.999937649314	.001459165365
4	.000060488596	.999998137910	.000062350686
5	.000001822333	.999999960244	.000001862090
6	.000000039146	.999999999389	.000000039756
7	.000000000604	.999999999993	.000000000611
8	.000000000007	1.000000000000	.000000000007
9	.000000000000	1.000000000000	.000000000000

x	HYPERA	2500 100 12	
0	.612034032102	.612034032102	1.000000000000
1	.307426052123	.919460084225	.387965967898
2	.070039115222	.989499199447	.080539915775
3	.009568985490	.999068184937	.010500800553
4	.000873089918	.999941274854	.000931815063
5	.000056041208	.999997316063	.000058725146
6	.000002594500	.999999910563	.000002683937
7	.000000087283	.999999997846	.000000089437
8	.000000002117	.999999999963	.000000002154
9	.000000000036	1.000000000000	.000000000037
10	.000000000000	1.000000000000	.000000000000

x	HYPERA	2500 200	12
0	.366820617145	.366820617145	1.000000000000
1	.384608772891	.751429390036	.633179382855
2	.183822839270	.935252229306	.248570609964
3	.052956383203	.988208612509	.064747770694
4	.010241237720	.998449850229	.011791387491
5	.001400633297	.999850483526	.001550149771
6	.000138903258	.999989386784	.000149516474
7	.000010064326	.999999451110	.000010613216
8	.000000528750	.999999979860	.000000548890
9	.000000019643	.999999999503	.000000020140
10	.000000000490	.999999999993	.000000000497
11	.000000000007	1.000000000000	.000000000007

x	HYPERA	2500 300	12
0	.214893591967	.214893591967	1.000000000000
1	.353411115158	.568304707125	.785106408033
2	.265381086245	.833685793370	.431695292875
3	.120315782293	.954001575663	.166314206630
4	.036679298137	.990680873800	.045998424337
5	.007921256543	.998602130343	.009319126200
6	.001242585442	.999844715785	.001397869657
7	.000142656734	.999987372519	.000155284215
8	.000011896181	.999999268700	.000012627481
9	.000000702713	.999999971413	.000000731300
10	.000000027910	.999999999323	.000000028587
11	.000000000669	.999999999993	.000000000677
12	.000000000007	1.000000000000	.000000000007

x	HYPERA	2500 500	12
0	.068265861109	.068265861109	1.000000000000
1	.205930199424	.274196060534	.931734138891
2	.284007754935	.558203815468	.725803939466
3	.236792000599	.794995816067	.441796184532
4	.132928039493	.927923855560	.205004183933
5	.052931104938	.980854960498	.072076144440
6	.015329846089	.996184806587	.019145039502
7	.003253681619	.999438488206	.003815193413
8	.000502274874	.999940763080	.000561511794
9	.000054997883	.999995760963	.000059236920
10	.000004054649	.999999815612	.000004239037
11	.000000180707	.999999996318	.000000184388
12	.000000003682	1.000000000000	.000000003682

LESSON *16. AN ACCEPTANCE SAMPLING PLAN

First we rewrite the essential ideas of the previous lesson in a different terminology. The lot (population of size N) consists of objects from a (manufacturing) process and are either defective or non–defective; the proportion of defectives in the lot is $\theta = D/N$ but D is unknown. The process has been "well– tuned" and the producer believes that $\theta \leq \theta_o$, a known number. If that is true, the producer considers this a "good lot" and expects to sell it. On the other hand, the consumer is unwilling to buy a "bad lot" wherein the proportion of defectives is $\theta \geq \theta_a$, a known number greater than θ_o. In order to decide whether or not to buy the lot, the consumer insists on examining a sample (of size n); the consumer and the producer agree that the consumer can refuse the lot when the number X of defectives in the sample is too big, say $X \geq c$.

The *producer's risk* is

P(Consumer refuses the lot | lot is good)

and the *consumer's risk* is

P(Consumer buys a lot | lot is bad).

Each wants their own risk to be as small as possible. On the basis of their previous experience (both passed this course!), they establish upper bounds, α and β, for these risks; some common values of α, β are .001, .005, .01, .025, .05, .10. The problem now is to find (n,c), the sampling plan, such that

$$P(X \geq c \,|\, \theta \leq \theta_o) \leq \alpha \quad \text{and} \quad P(X < c \,|\, \theta \geq \theta_a) \leq \beta \qquad (*)$$

where, of course, X is hypergeometric and the population size N is fixed.

Exercise 1:   Show that the hypergeometric density is a non–decreasing function of θ or D   when x is "large". Specifically, show that for $x \geq n(D + 1)/(N + 1)$,

$$h(x \,;\, N, n, D) \leq h(x \,;\, N, n, D{+}1) .$$

Hint: start with the inequality on h and simplify.

In the problem at hand, we want the event $\{X \geq c\}$ to have "small" probability when H_o is true; obviously then, c will be "large" and as a corollary to the exercise, it follows that

$P(X \geq c \mid \theta = D/N)$ will be non–decreasing in θ or D . We conclude that the probability of the event $\{X \geq c\}$ when $\theta < \theta_o$ is at most equal to the probability when $\theta = \theta_o$; formally,

$$P(X \geq c \mid \theta < \theta_o) \leq P(X \geq c \mid \theta = \theta_o) .$$

Similarly,

$$P(X < c \mid \theta > \theta_a) \leq P(X < c \mid \theta = \theta_a) .$$

This simplifies the inequalities (*) to

$$P(X \geq c \mid \theta = \theta_o) \leq \alpha \ \text{ and } \ P(X < c \mid \theta = \theta_a) \leq \beta. \quad (**)$$

Inconveniently, we cannot eliminate the remaining inequalities. Since the distribution function inceases only by "jumps", adding values at each new x, for example,

$$P(X \geq 10) = P(X = 10) + P(X \geq 11)) ,$$

we may not be able to find exact probabilities such as

$$P(X \geq c \mid \theta = \theta_o) = .01 \ \text{ or } \ P(X < c \mid \theta = \theta_a) = .05.$$

Exercise 2: Look at the tables in the previous section for
$$P(X \geq c \cdots) \text{ and } P(X < c \cdots);$$
note that none of these values is exactly .01 or .05 .

Now we have reduced the problem to this: given N, the producer's values θ_o and α, the consumer's values θ_a and β, we are to find the sampling plan (n, c). To do this, we must solve the inequalities (**); that is, find n and c for which the inequalities are true. Of course, we want the smallest n since that will reduce the costs in sampling. However, the solution must be obtained by arithmetical trials. Today, we can use a computer but some tables and graphs for the solution can be found in hand–books of statistics and quality control.

Exercise 3: Discuss the parallel problem of testing $H_o : \theta = \theta_o$ vs $H_a : \theta = \theta_a$ where θ_o , θ_a are given with $\theta_a < \theta_o$. Write out the pair of inequalities needed to determine n and c given N, $\theta_o, \theta_a, \alpha, \beta$.

In the tables at the end of this lesson, each column 1 is

$c = 0(1)m$; each column 2 is $P(X \geq c \,|\, \theta = \theta_o)$; each column 3 is $P(X \leq c \,|\, \theta = \theta_a)$. The notation n HYPER N D_o D_a represents sample size n, the population size N, the null value D_o, the alternative value D_a. Given N, D_o, α, D_a, β, we search for n and c such that $P(X \geq c \,|\, D_o) \leq \alpha$ and $P(X \leq c-1 \,|\, D_a) \leq \beta$. (The tables are printed in the order in which the writers tried them out.)

Example: N = 500, D_o = 25, D_a = 100. Suppose that α = .10 and β = .10 . None of the values for n = 10 or n = 20 have

$P(X \geq c \,|\, D = 25) \leq .10$ and $P(X \leq c-1 \,|\, D = 100) \leq .10$.

For n = 40, we find $P(X \geq 5 \,|\, D = 25) = .0408 \leq .10$

$P(X \leq 4 \,|\, D = 100) = .0677 \leq .10$.

This is a solution and we could quit here but, as noted above, we really want the smallest possible sample size.

For n = 35, $P(X \geq 4 \,|\, D = 25) = .0884 \leq .10$,

$P(X \leq 3 \,|\, D = 100) = .0542 \leq .10$.

For n = 33, $P(X \geq 4 \,|\, D = 25) = .0740 \leq .10$,

$P(X \leq 3 \,|\, D = 100) = .0934 \leq .10$.

For n = 31, $P(X \geq 4 \,|\, D = 25) = .0610 \leq .10$,

$P(X \leq 3 \,|\, D = 100) = .0996 \leq .10$.

For n = 30, $P(X \geq 4 \,|\, D = 25) = .0550 \leq .10$,

$P(X \leq 3 \,|\, D = 100) = .1152 > .10$.

Apparently, we have struck bottom. We conclude that we should take a sample of size n = 31 and reject the lot when the observed number of defectives in the sample is at least 4. Then, at least 93% of "good lots" will be sold and at most 10% of the "bad lots" will be purchased.

Exercise 4: Suppose that the procedure in the example is carried out for several lots with the following observed values of X ; which lots are rejected and which are sold?
                6 4 6 2 2 1 7 5 2 3 5 8 1 1

Exercise 5: Continuing with the example for N = 500,

find the smallest n and the value c such that

$$P(X \geq c \mid D = 25) \leq .05 \quad \text{and} \quad P(X \leq c-1 \mid D = 100) \leq .05.$$

Example: In some cases, we may not find a solution. For $N = 10$, $D_o = 2$, the value of α must be at least .0222:

n	$P(X \geq 2 \mid \cdots)$	$P(X \geq 1 \mid \cdots)$
2	.0222	.3778
3	.0667	.5333
4	.1333	.6667
5	.2222	.7778
6	.3333	.8667
7	.4667	.9333
8	.6222	.9778
9	.8000	1.0000
10	1.0000	1.0000

Exercise 6: You will need a calculator for these.
  a)   Find the sampling plan for
        $N = 100$, $D_o = 10$, $D_a = 50$.
  b)   Find the sampling plan for
        $N = 100$, $D_o = 10$, $D_a = 20$.

In the last exercise, note that $D_a = 20$ is "closer" to $D_o = 10$ than $D_a = 50$. To distinguish "20 from 10", we need larger samples than to distinguish "50 from 10". In general terms, the finer the distinction to be made, the more information to be used. We summarize these two lessons as follows:

I. The *assumptions*
    We will be able to take a random sample of size n from a finite population of size N with observations classified as defective (proportion θ) or non–defective. The random variable X giving the number of defectives in the sample is hypergeometric.

II. The *hypotheses*

$$H_o : \theta = \theta_o \text{ (or } \theta \leq \theta_o) \quad H_a : \theta = \theta_a > \theta_o \text{ (or } \theta > \theta_o)$$

III. The *test*
    Given α and β, we find n and c so that

$$P(X \geq c \mid \theta = \theta_o) \leq \alpha \quad \text{and} \quad P(X < c \mid \theta = \theta_a) \leq \beta.$$

We reject H_o when the observed x is in the critical region

$$\{X \geq c\}.$$

IV. The *data*

We now send our assistant (sic!) to get the random sample from the specified population.

V. The *analysis*

We do whatever is necessary to determine the number of defectives in the sample. (This procedure is simple enough now but as problems go beyond the hypergeometric, more calculations become necessary and we may wind up using some statistical package.)

VI. The *conclusion*

We reject H_o or we do not reject H_o.

Remember that rejection of H_o does not mean that H_o is false but only that if H_o were true, the sample results we actually have would be a "rare event". Conversely, not rejecting H_o does not mean that H_o is true but only that if it were true, the sample results we have are not a "rare event". Rather than "You can prove *anything* with statistics", we find "You *prove* nothing with statistics." More discussion of this topic can be found in Varderman, 1986.

Exercise 7: Consider the output of one canning process. The lot size is 2500 cans; a sample consists of 24 cans randomly selected at different times of the work day. The null hypothesis is that the proportion of defective cans is at most .01; the alternative is that the proportion is .05. When should the lot be rejected (the process shut down and inspected for malfunctions)?

Table n HYPER   N, D_o, D_a

This table (generated by STSC–APL) contains cumulative values of hypergeometric distributions. The columns are, in order, value x, probability at least x for $D = D_o$, probability at most x for $D = D_a$. For example, $P(X \geq 3 | n=10, N=500, D_o=25) = .0105$

$$P(X \leq 3 | n=10, N=500, D_a=100) = .8813$$

10	HYPER	500, 25, 100		20	HYPER	500, 25, 100	
x				x			
0	1.0000	.1050		0	1.0000	.0105	
1	.4041	.3734		1	.6488	.0653	
2	.0845	.6784		2	.2637	.2004	
3	.0105	.8813		3	.0716	.4079	
4	.0008	.9687		4	.0137	.6306	
5	.0000	.9942		5	.0019	.8083	
6	.0000	.9992		6	.0002	.9176	
7	.0000	.9999		7	.0000	.9707	
8	.0000	1.0000		8	.0000	.9914	
9	.0000	1.0000		9	.0000	.9979	
10	.0000	1.0000		10	.0000	.9996	

40	HYPER	500, 25, 100		35	HYPER	500, 25, 100	
x				x			
0	1.0000	.0001		0	1.0000	.0003	
1	.8822	.0011		1	.8445	.0031	
2	.6119	.0062		2	.5359	.0161	
3	.3225	.0240		3	.2511	.0542	
4	.1300	.0677		4	.0884	.1343	
5	.0408	.1505		5	.0240	.2631	
6	.0101	.2759		6	.0051	.4280	
7	.0020	.4319		7	.0009	.6006	
8	.0003	.5945		8	.0001	.7513	
9	.0000	.7387		9	.0000	.8624	
10	.0000	.8486		10	.0000	.9326	
11	.0000	.9213		11	.0000	.9707	
12	.0000	.9633		12	.0000	.9887	
13	.0000	.9847		13	.0000	.9962	
14	.0000	.9943		14	.0000	.9988	
15	.0000	.9981		15	.0000	.9997	
16	.0000	.9994		16	.0000	.9999	
17	.0000	.9998		17	.0000	1.0000	
18	.0000	1.0000		18	.0000	1.0000	

33	HYPER	500, 25, 100		31	HYPER	500, 25, 100	
x				x			
0	1.0000	.0005		0	1.0000	.0008	
1	.8264	.0048		1	.8062	.0073	
2	.5030	.0233		2	.4688	.0333	
3	.2234	.0739		3	.1965	.0996	
4	.0740	.1731		4	.0610	.2203	
5	.0188	.3215		5	.0144	.3875	
6	.0037	.4980		6	.0027	.5711	

7	.0006	.6691	7	.0004	.7350
8	.0001	.8070	8	.0000	.8564
9	.0000	.9007	9	.0000	.9318
10	.0000	.9550	10	.0000	.9717
11	.0000	.9821	11	.0000	.9897
12	.0000	.9937	12	.0000	.9967
13	.0000	.9980	13	.0000	.9991
14	.0000	.9995	14	.0000	.9998
15	.0000	.9999	15	.0000	1.0000
16	.0000	1.0000	16	.0000	1.0000

30	HYPER	500, 25, 100	45	HYPER	500, 25, 100
x			x		
0	1.0000	.0010	0	1.0000	.0000
1	.7954	.0090	1	.9110	.0003
2	.4512	.0397	2	.6787	.0023
3	.1833	.1152	3	.3948	.0101
4	.0550	.2472	4	.1787	.0321
5	.0125	.4230	5	.0636	.0802
6	.0022	.6081	6	.0181	.1645
7	.0003	.7664	7	.0042	.2866
8	.0000	.8783	8	.0008	.4351
9	.0000	.9447	9	.0001	.5894
10	.0000	.9781	10	.0000	.7280
11	.0000	.9924	11	.0000	.8363
12	.0000	.9977	12	.0000	.9108
13	.0000	.9994	13	.0000	.9560
14	.0000	.9999	14	.0000	.9804
15	.0000	1.0000	15	.0000	.9921
16	.0000	1.0000	16	.0000	.9971
17	.0000	1.0000	17	.0000	.9990
18	.0000	1.0000	18	.0000	.9997
19	.0000	1.0000	19	.0000	.9999
20	.0000	1.0000	20	.0000	1.0000

50	HYPER	500, 25, 100	47	HYPER	500, 25, 100
x			x		
0	1.0000	.0000	0	1.0000	.0000
1	.9330	.0001	1	.9205	.0002
2	.7364	.0008	2	.7029	.0015
3	.4657	.0040	3	.4234	.0070
4	.2329	.0144	4	.1998	.0234
5	.0927	.0401	5	.0745	.0612
6	.0296	.0915	6	.0222	.1312
7	.0077	.1767	7	.0054	.2382

8	.0016	.2956	8	.0011	.3760
9	.0003	.4378	9	.0002	.5278
10	.0000	.5851	10	.0000	.6724
11	.0000	.7187	11	.0000	.7927
12	.0000	.8255	12	.0000	.8807
13	.0000	.9011	13	.0000	.9376
14	.0000	.9489	14	.0000	.9704
15	.0000	.9759	15	.0000	.9872
16	.0000	.9897	16	.0000	.9950
17	.0000	.9960	17	.0000	.9982
18	.0000	.9986	18	.0000	.9994
19	.0000	.9995	19	.0000	.9998
20	.0000	.9999	20	.0000	1.0000
21	.0000	1.0000	21	.0000	1.0000

48 HYPER	500, 25,	100	49 HYPER	500, 25,	100
x			x		
0	1.0000	.0000	0	1.0000	.0000
1	.9249	.0002	1	.9291	.0001
2	.7144	.0013	2	.7256	.0010
3	.4376	.0059	3	.4517	.0049
4	.2106	.0199	4	.2217	.0169
5	.0803	.0533	5	.0864	.0463
6	.0245	.1166	6	.0270	.1035
7	.0061	.2162	7	.0069	.1957
8	.0012	.3480	8	.0014	.3212
9	.0002	.4973	9	.0002	.4673
10	.0000	.6437	10	.0000	.6145
11	.0000	.7691	11	.0000	.7444
12	.0000	.8636	12	.0000	.8452
13	.0000	.9267	13	.0000	.9145
14	.0000	.9642	14	.0000	.9570
15	.0000	.9841	15	.0000	.9803
16	.0000	.9936	16	.0000	.9918
17	.0000	.9976	17	.0000	.9969
18	.0000	.9992	18	.0000	.9989
19	.0000	.9998	19	.0000	.9997
20	.0000	.9999	20	.0000	.9999
21	.0000	1.0000	21	.0000	1.0000

LESSON 17. THE BINOMIAL DISTRIBUTION

You are quite aware of the work needed to compute probabilities for the hypergeometric distribution. We will see how to use the following mathematical theorem to get some simplification.

Theorem: *Let* $\lim\limits_{N \to \infty} D/N = \theta$ *for* $0 < \theta < 1$. *For each positive integer n and each* $x = 0(1)n$,

$$\lim_{N \to \infty} \begin{bmatrix} D \\ x \end{bmatrix} \begin{bmatrix} N - D \\ n - x \end{bmatrix} \div \begin{bmatrix} N \\ n \end{bmatrix} = \begin{bmatrix} n \\ x \end{bmatrix} \theta^x (1 - \theta)^{n - x}. \quad (*)$$

Proof: Expanding each of the combinatorials, we find

$$\begin{bmatrix} D \\ x \end{bmatrix} = \frac{D(D-1)(D-2) \quad \cdots \quad (D-x+1)}{x!} ;$$

$$\begin{bmatrix} N - D \\ n - x \end{bmatrix} = \frac{(N-D)(N-D-1)(N-D-2) \cdots (N-D-x+1)}{(n - x)!} ;$$

$$\begin{bmatrix} N \\ n \end{bmatrix} = \frac{N(N-1)\cdots(N-n+1)}{n!}$$

Then the ratio in the left–hand side of (*) can be written as

$$\frac{D}{N} \frac{D-1}{N-1} \frac{D-2}{N-2} \cdots \frac{D-x+1}{N-x+1} \text{ times}$$

$$\frac{N-D}{N-x} \frac{N-D-1}{N-x-1} \cdots \frac{N-D-n+x+1}{N-n+1} \text{ times } \begin{bmatrix} n \\ x \end{bmatrix}.$$

Each factor in the product beginning with D/N, converges to θ and there are a finite number, namely x, of them. Each factor in the product beginning with $(N - D)/(N - x)$ converges to $1 - \theta$ and there are $n - x$ of those; the combinatorial $\begin{bmatrix} n \\ x \end{bmatrix}$ is fixed. The conclusion follows by application of the law for the limit of a product.

The ratio in the left–hand side of (*) is also the PDF of the hypergeometric distribution so that for $m = n \wedge D$,

$$\sum_{x=0}^{m} \left[\begin{array}{c} D \\ x \end{array} \right] \left[\begin{array}{c} N-D \\ n-x \end{array} \right] \div \left[\begin{array}{c} N \\ n \end{array} \right] = 1 .$$

Taking the limit of this equality with $D/N \to \theta$, yields

$$\sum_{x=0}^{n} \left[\begin{array}{c} n \\ x \end{array} \right] \theta^{x} (1 - \theta)^{n-x} = 1 .$$

Actually, this is just a special case of the binomial theorem: for a positive integer n, (r and s fixed),

$$(r + s)^{n} = \sum_{x=0}^{n} \left[\begin{array}{c} n \\ x \end{array} \right] r^{x} \cdot s^{n-x} .$$

Exercise 1: Prove the binomial theorem by induction.

Definition: *Let a RV X have the density (PDF)*

$$P(X = x) = b_{X}(x \; ; \; n, \theta) = \left[\begin{array}{c} n \\ x \end{array} \right] \theta^{x} (1 - \theta)^{n-x}, \; x = 0(1)n.$$

Then X is called a binomial random variable.

Exercise 2: Show that a binomial RV has mean $n\theta$ and variance $n\theta(1 - \theta)$. Hint: recall the corresponding derivations for the hypergeometric in lesson 13.

The limit in the theorem is expressible as a ratio:

$$\frac{\left[\begin{array}{c} D \\ x \end{array} \right] \left[\begin{array}{c} N-D \\ n-x \end{array} \right] \div \left[\begin{array}{c} N \\ n \end{array} \right]}{\left[\begin{array}{c} n \\ x \end{array} \right] (D/N)^{x} (1 - D/N)^{n-x}}$$

converges to 1 as $N \to \infty$ and D/N converges to θ . Formally, we say that h(x ; N, D, n) and b(x ; n, θ) are *asymptotic* as $N \to \infty$ and $D/N \to \theta$ since the limit of their ratio is 1 . This is the form which yields an approximation

of the hypergeometric density $\left[\begin{array}{c} D \\ x \end{array} \right] \left[\begin{array}{c} N-D \\ n-x \end{array} \right] \div \left[\begin{array}{c} N \\ n \end{array} \right]$

by the binomial density $\left[\begin{array}{c} n \\ x \end{array} \right] (D/N)^{x} (1 - D/N)^{n-x}$.

Example: For N = 2500, D = 125, n = 12,

$$\begin{bmatrix} 125 \\ 0 \end{bmatrix} \begin{bmatrix} 2375 \\ 12 \end{bmatrix} \div \begin{bmatrix} 2500 \\ 12 \end{bmatrix} \approx \begin{bmatrix} 12 \\ 0 \end{bmatrix} (125/2500)^0 (2375/2500)^{12}$$

$$\approx .54036$$

Exercise 3: Approximate

a) $\begin{bmatrix} 100 \\ 10 \end{bmatrix} \begin{bmatrix} 100 \\ 10 \end{bmatrix} \div \begin{bmatrix} 200 \\ 20 \end{bmatrix}$ b) $\begin{bmatrix} 100 \\ 0 \end{bmatrix} \begin{bmatrix} 100 \\ 20 \end{bmatrix} \div \begin{bmatrix} 200 \\ 20 \end{bmatrix}$.

To simplify, we carry only four decimal places in the tables at the end of this lesson. The heading is n BINMOMIAL $\theta_o \theta_a$; the first column has values of $c-1$; the second column is $P(X \le c - 1 \mid \theta_o)$; the third column is $P(X \le c - 1 \mid \theta_a)$.

Example: Recall the testing problems of lesson 15 like

H_o : X is hypergeometric with N = 500 and D_o = 25 vs

H_a : X is hypergeometric with N = 500 and D_a = 100 .

Finding a test or critical region, {X $\ge$ c}, with given α and β requires solving (finding n and c such that)

$$P(X \ge c \mid H_o) \le \alpha \text{ and } P(X < c \mid H_a) \le \beta.$$

It is obviously easier to deal with the approximations:

$$\sum_{x=c}^{n} \begin{bmatrix} n \\ x \end{bmatrix} (.05)^x (.95)^{n-x} \le \alpha$$

and

$$\sum_{x=0}^{c-1} \begin{bmatrix} n \\ x \end{bmatrix} (.20)^x (.80)^{n-x} \le \beta .$$

Suppose we take α and β both .10 . Then we look for

$$P(X \le c - 1 \mid .05) \ge 1 - \alpha = .90$$

and

$$P(X \le c - 1 \mid .2) \le \beta = .10$$

Looking at the tables, we see that these inequalities are not satisfied for n = 10 or 20 or 30 but are satisfied when n = 40 and c–1 = 4. We then try:

n = 35   to find   c–1 = 3 ;
n = 32   yields   c–1 = 3 .

The inequalities are not satisfied for $n = 31$. Hence our sampling plan for testing is:

> take a sample of size 32 ; if the number of defective items is at most $c-1 = 3$, accept the lot; otherwise, reject the lot.

About 92% of the "good" lots will be accepted and about 91% of the "bad" lots will be rejected.

Exercise 4: a) Find n and c such that

$$\sum_{x=c}^{n} \binom{n}{x}(.1)^x(.9)^{n-x} \le .10 , \sum_{x=0}^{c-1} \binom{n}{x}(.2)^x(.8)^{n-x} \le .10 .$$

Hint: since .1 is closer to .05 than .2 is, start with larger sample sizes than in the example.

b)   Find a sampling plan to test the hypothesis $H_o : \theta = .2$ against the hypothesis $H_a : \theta = .4$ when the distribution is binomial and $\alpha = \beta = .05$.

The hypergeometric distribution was taken as a model for sampling from a finite population without replacements and without regard to order within the sample. Now consider sampling with replacement from a population consisting of D defective items and $N - D$ non–defective items. Let W be the number of defective items in the sample of size n also without regard to order. Such sampling with replacement "restores" the distribution "in the urn" so we are dealing with repeated trials.

Example: We continue this discussion with $n = 5$ trials.
$W = 0$ iff all trials result in non–defective items; by independence, $P(5 \text{ non–defectives}) = P(1 \text{ defective})^5$
$$= (1 - D/N)^5.$$

$W = 1$ iff exactly one of the trials turns up a defective item and the other four are non–defective; each of these $\binom{5}{1}$ cases has probability $(D/N)(1 - D/N)^4$ again by independence.

Similarly, $W = 2$ has probability $\binom{5}{2}(D/N)^2(1 - D/N)^3$.

$$P(W = 3) = \binom{5}{3}(D/N)^3(1 - D/N)^2 .$$

$$P(W = 4) = \begin{bmatrix} 5 \\ 4 \end{bmatrix} (D/N)^4 (1 - D/N) .$$

$$P(W = 5) = (D/N)^5 .$$

Exercise 5: Find the values of $P(W = w)$ when $N = 200$, $D = 10$, and $n = 6$ when sampling is with replacement without regard to order.

The conclusion of the limit form given earlier is expressed by saying that when N and D are "large", there is little difference between the probability distributions for the number of defectives when sampling with replacement and sampling without replacement. The following describes one kind of repeated trials more precisely.

Definition: *The experiment has sample space $\Omega = \{S, F\}$ with subsets $\phi, \{S\}, \{F\}, \Omega$ and probabilities $P(S) = \theta = 1 - P(F)$. On the first trial, the outcome is $X_1 = x_1$ (an S or an F); on the second trial the outcome is $X_2 = x_2$ (S or F); etc. The trials are independent so that $P(X_1 = x_1, X_2 = x_2, X_3 = x_3, \cdots)$*

$$= P(X_1 = x_1)P(X_2 = x_2)P(X_3 = x_3) \cdots$$

These are called Bernoulli (dichotomous) trials.

It will be convenient to represent the outcomes of such Bernoulli trials by 1 and 0 so that $X_i(S) = 1$ and $X_i(F) = 0$ defines a random variable with only two values. "Repeated trials" then specifies that these random variables, $X_1, X_2, \cdots, X_n$ be *independent and identically distributed* (IID).

Definition: *When the random variables $X_1, X_2, \cdots, X_n$ are IID, they are said to be a simple random sample of size n.*

Continuing with the Bernoulli zero–one trials, let

$$T = X_1 + X_2 + \cdots + X_n .$$

The observed value of T is t iff

$$\#\{ X_i = 1 \} = t \text{ and } \#\{ X_i = 0 \} = n - t .$$

This is exactly "t successes in n trials" so that the RV T has a binomial distribution with parameters

$$n \text{ and } \theta = P(X_i = 1) = P(S)$$

as in the earlier definition. This is the model often assumed in questions about single proportions.

Example: In a certain group, the proportion who will have an adverse reaction to vaccine BB is believed to be $\theta = .01$. Suppose that the group is large enough ($n/N \leq .10$) to use the binomial approximation for a sample of size $n = 20$.

    a)   What is the expected number of adverse reactions in the sample? $E[X] = n\theta = 20(.01) = 2$

    b)   What is the probability that there will be at least two adverse reactions in the sample?

$$P(X \geq 2) = 1 - P(X \leq 1)$$

$$= 1 - \left[\begin{array}{c} 20 \\ 0 \end{array} \right](.01)^0(.99)^{20} - \left[\begin{array}{c} 20 \\ 1 \end{array} \right](.01)^1(.99)^{19}$$

$$\approx .016859 .$$

Example: The proportion of students at a certain high school who have "tried drugs" is believed to be about $\theta = .55$. Consider a random sample of size $n = 100$.

    a)   What is the probability that at least half of the sample will have tried drugs?

$$P(X \geq 51) = \sum_{x=51}^{100} \left[\begin{array}{c} 100 \\ x \end{array} \right](.55)^x(.45)^{100-x}$$

$$\approx .7596 .$$

    b)   If it turned out that 90 of the 100 had tried drugs, would that be a "rare event" if θ is really .60?

$$P(X \geq 90) = \sum_{x=90}^{100} \left[\begin{array}{c} 100 \\ x \end{array} \right](.6)^x(.4)^{100-x}$$

$$\approx .0$$

so the answer is yes!

Exercise 6: If the probability of a defective is $\theta = .1$, what is the probability that in a random sample of size $n = 100$, there will be no defectives? at least one defective?

Exercise 7: Suppose that for a certain couple, the probability that any child will have blue eyes is 1/4. What is the probability that their next five children will all have blue eyes? Given that at least one of their five children has blue eyes, what is the probability that all five have blue eyes?

Example: Toss a penny.
If it turns up heads, answer yes or no to question 1:
    Have you had "bad" thoughts?
If your toss turns up tails, answer yes or no to question 2:
    Were you born in January, February, or March?
Presumably, you could and would answer truthfully since the recorder would not know which question you are answering. (It's your penny!) A sample space comes from

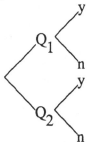

with the probability of "yes" given by

$$P(y \cap Q_1) + P(y \cap Q_2) = P(y \mid Q_1)P(Q_1) + P(y \mid Q_2)P(Q_2)$$

$$= \theta(1/2) + (1/4)(1/2)$$

where $\theta = P(y \mid Q_1)$ is the unknown of interest. If the relative frequency interpretation of probability is applicable in this kind of a problem, by counting the number of yes's in a large group,

we would have $\#y/n \approx P(y) = \theta(1/2) + 1/8$

and hence $\theta \approx (\#y/n - 1/8)2$.

Question 1 can contain any sensitive issue and question 2 can be any neutral question for which probabilities are known; one can use other kinds of coins. For further information on "randomized response", see Tamhane, 1981 .

INE–Exercises:
1.   Rewrite the six steps for testing in lesson 16 in terms of a binomial distribution.
2.   The ten components in a certain system are to be treated as Bernoulli trials with $P(\text{Fail}) = \theta = 1 - P(\text{Not fail})$. Given that at least one component has failed, what is the probability that at least two components have failed? Try $\theta = .1$ and $\theta = .01$.
3.   Let the RV X have PDF $p(x) = P(X = x)$. A *mode* of the distribution of X ia a value of X, say x_o, such that

$p(x_o) \geq p(x)$ for all x.

a)   Find a mode for the binomial distribution. Hint: for any RV with positive integral values, consider $p(x+1)/p(x)$.
b)   Note that there may not be a single mode even in this case. What does this tell you about using a mode to characterize a distribution?
4.   Let the RV X have the CDF F. A *median* of the distribution of X is any real number, say m_d, such that

$F(m_d) = P(X \leq m_d) \geq .5$ and $P(X \geq m_d) \geq .5$.

a)   Show that $P(X < m_d) \leq .5 \leq P(X \leq m_d)$

b)   Find a median of the distributions of exercises 4 and 5 in lesson 9.
c)   Find a median for each of the binomial distributions with $n = 20$ and $\theta = .3$ or $.5$ or $.8$.

5.   For n a positive integer, $\sum_{k=0}^{n} \binom{n}{k} x^k = (1 + x)^n$. Many interesting relations among binomial coefficients can be obtained by differentiating both sides of this equality with respect to x and then evaluating at $x = -1$ or 0 or 1 . Try zeroth, first and second derivatives.

Table   n BINOMIAL θ_o , θ_a

This table (generated using STSC–APL) contains values of biniomial distribution functions with parameters sample size n, proportions θ_o and θ_a . For example,

$P(X \leq 4 \mid n = 20, \theta_o = .05) = .9974$
$P(X \leq 4 \mid n = 20, \theta_a = .20) = .6296$.

10 BINOMIAL	.05	,	.20	20 BINOMIAL	.05	,	.20
x				x			
0	.5987		.1074	0	.3585		.0115
1	.9139		.3758	1	.7358		.0692
2	.9885		.6778	2	.9245		.2061
3	.9990		.8791	3	.9841		.4114
4	.9999		.9672	4	.9974		.6296
5	1.0000		.9936	5	.9997		.8042
6	1.0000		.9991	6	1.0000		.9133
7	1.0000		.9999	7	1.0000		.9679
8	1.0000		1.0000	8	1.0000		.9900
9	1.0000		1.0000	9	1.0000		.9974
10	1.0000		1.0000	10	1.0000		.9994
11	1.0000		1.0000	11	1.0000		.9999
12	1.0000		1.0000	12	1.0000		1.0000

30 BINOMIAL	.05	,	.20	40 BINOMIAL	.05	,	.20
x				x			
0	.2146		.0012	0	.1285		.0001
1	.5535		.0105	1	.3991		.0015
2	.8122		.0442	2	.6767		.0079
3	.9392		.1227	3	.8619		.0285
4	.9844		.2552	4	.9520		.0759
5	.9967		.4275	5	.9861		.1613
6	.9994		.6070	6	.9966		.2859
7	.9999		.7608	7	.9993		.4371
8	1.0000		.8713	8	.9999		.5931
9	1.0000		.9389	9	1.0000		.7318
10	1.0000		.9744	10	1.0000		.8392
11	1.0000		.9905	11	1.0000		.9125
12	1.0000		.9969	12	1.0000		.9568
13	1.0000		.9991	13	1.0000		.9806
14	1.0000		.9998	14	1.0000		.9921
15	1.0000		.9999	15	1.0000		.9971
16	1.0000		1.0000	16	1.0000		.9990
17	1.0000		1.0000	17	1.0000		.9997
18	1.0000		1.0000	18	1.0000		.9999
19	1.0000		1.0000	19	1.0000		1.0000

35 BINOMIAL	.05	,	.20	33 BINOMIAL	.05	,	.20
x				x			
0	.1661		.0004	0	.1840		.0006
1	.4720		.0040	1	.5036		.0059
2	.7458		.0190	2	.7728		.0268
3	.9042		.0605	3	.9192		.0808

4	.9710	.1435		4	.9770	.1821
5	.9927	.2721		5	.9946	.3290
6	.9985	.4328		6	.9989	.5004
7	.9997	.5993		7	.9998	.6657
8	1.0000	.7450		8	1.0000	.8000
9	1.0000	.8543		9	1.0000	.8932
10	1.0000	.9253		10	1.0000	.9492
11	1.0000	.9656		11	1.0000	.9784
12	1.0000	.9858		12	1.0000	.9918
13	1.0000	.9947		13	1.0000	.9972
14	1.0000	.9982		14	1.0000	.9992
15	1.0000	.9995		15	1.0000	.9998
16	1.0000	.9999		16	1.0000	.9999
17	1.0000	1.0000		17	1.0000	1.0000

32 BINOMIAL .05 , .20			31 BINOMIAL .05 , .20		
x			x		
0	.1937	.0008	0	.2039	.0010
1	.5200	.0071	1	.5366	.0087
2	.7861	.0317	2	.7992	.0374
3	.9262	.0931	3	.9329	.1070
4	.9796	.2044	4	.9821	.2287
5	.9954	.3602	5	.9961	.3931
6	.9991	.5355	6	.9993	.5711
7	.9999	.6982	7	.9999	.7300
8	1.0000	.8254	8	1.0000	.8492
9	1.0000	.9102	9	1.0000	.9254
10	1.0000	.9589	10	1.0000	.9673
11	1.0000	.9833	11	1.0000	.9873
12	1.0000	.9939	12	1.0000	.9956
13	1.0000	.9980	13	1.0000	.9987
14	1.0000	.9994	14	1.0000	.9996
15	1.0000	.9999	15	1.0000	.9999
16	1.0000	1.0000	16	1.0000	1.0000

40 BINOMIAL .10 , .20			50 BINOMIAL .10 , .20		
x			x		
0	.0148	.0001	0	.0052	.0000
1	.0805	.0015	1	.0338	.0002
2	.2228	.0079	2	.1117	.0013
3	.4231	.0285	3	.2503	.0057
4	.6290	.0759	4	.4312	.0185
5	.7937	.1613	5	.6161	.0480
6	.9005	.2859	6	.7702	.1034
7	.9581	.4371	7	.8779	.1904

8	.9845	.5931	8	.9421	.3073
9	.9949	.7318	9	.9755	.4437
10	.9985	.8392	10	.9906	.5836
11	.9996	.9125	11	.9968	.7107
12	.9999	.9568	12	.9990	.8139
13	1.0000	.9806	13	.9997	.8894
14	1.0000	.9921	14	.9999	.9393
15	1.0000	.9971	15	1.0000	.9692
16	1.0000	.9990	16	1.0000	.9856
17	1.0000	.9997	17	1.0000	.9937
18	1.0000	.9999	18	1.0000	.9975
19	1.0000	1.0000	19	1.0000	.9991
20	1.0000	1.0000	20	1.0000	.9997

60 BINOMIAL	.10 ,	.20	70 BINOMIAL	.10 ,	.20
x			x		
0	.0018	.0000	0	.0006	.0000
1	.0138	.0000	1	.0055	.0000
2	.0530	.0002	2	.0242	.0000
3	.1374	.0010	3	.0712	.0002
4	.2710	.0039	4	.1588	.0008
5	.4372	.0121	5	.2872	.0027
6	.6065	.0308	6	.4418	.0080
7	.7516	.0670	7	.5989	.0200
8	.8584	.1268	8	.7363	.0437
9	.9269	.2132	9	.8414	.0845
10	.9658	.3234	10	.9127	.1468
11	.9854	.4486	11	.9559	.2317
12	.9943	.5764	12	.9795	.3360
13	.9980	.6944	13	.9912	.4524
14	.9993	.7935	14	.9965	.5709
15	.9998	.8694	15	.9987	.6814
16	.9999	.9228	16	.9996	.7765
17	1.0000	.9573	17	.9999	.8519
18	1.0000	.9779	18	1.0000	.9075
19	1.0000	.9893	19	1.0000	.9455
20	1.0000	.9952	20	1.0000	.9697

80 BINOMIAL	.10 ,	.20	85 BINOMIAL	.10 ,	.20
x			x		
0	.0002	.0000	0	.0001	.0000
1	.0022	.0000	1	.0013	.0000
2	.0107	.0000	2	.0070	.0000
3	.0353	.0000	3	.0245	.0000
4	.0880	.0001	4	.0643	.0001

5	.1769	.0006	5	.1360	.0002
6	.3005	.0018	6	.2422	.0009
7	.4456	.0053	7	.3753	.0026
8	.5927	.0131	8	.5195	.0069
9	.7234	.0287	9	.6566	.0159
10	.8266	.0565	10	.7724	.0332
11	.8996	.1006	11	.8601	.0627
12	.9462	.1640	12	.9202	.1081
13	.9733	.2470	13	.9577	.1718
14	.9877	.3463	14	.9791	.2538
15	.9947	.4555	15	.9904	.3507
16	.9979	.5664	16	.9959	.4568
17	.9992	.6708	17	.9983	.5644
18	.9997	.7621	18	.9994	.6661
19	.9999	.8366	19	.9998	.7557
20	1.0000	.8934	20	.9999	.8296

90 BINOMIAL	.10 ,	.20	95 BINOMIAL	.10 ,	.20
x			x		
0	.0001	.0000	0	.0000	.0000
1	.0008	.0000	1	.0005	.0000
2	.0046	.0000	2	.0030	.0000
3	.0169	.0000	3	.0115	.0000
4	.0465	.0000	4	.0334	.0000
5	.1032	.0001	5	.0775	.0000
6	.1925	.0004	6	.1511	.0002
7	.3115	.0013	7	.2550	.0006
8	.4487	.0035	8	.3820	.0017
9	.5875	.0086	9	.5184	.0045
10	.7125	.0190	10	.6488	.0105
11	.8135	.0378	11	.7607	.0221
12	.8874	.0688	12	.8478	.0424
13	.9366	.1152	13	.9095	.0747
14	.9667	.1791	14	.9497	.1220
15	.9837	.2600	15	.9738	.1859
16	.9925	.3548	16	.9872	.2658
17	.9968	.4580	17	.9941	.3586
18	.9987	.5626	18	.9975	.4591
19	.9995	.6617	19	.9990	.5610
20	.9998	.7497	20	.9996	.6577

92 BINOMIAL	.10 ,	.20	91 BINOMIAL	.10 ,	.20
x			x		
0	.0001	.0000	0	.0001	.0000
1	.0007	.0000	1	.0008	.0000
2	.0039	.0000	2	.0042	.0000
3	.0145	.0000	3	.0157	.0000
4	.0408	.0000	4	.0436	.0000
5	.0922	.0001	5	.0976	.0001
6	.1750	.0003	6	.1836	.0003
7	.2880	.0009	7	.2996	.0011
8	.4214	.0027	8	.4349	.0031
9	.5598	.0067	9	.5736	.0076
10	.6874	.0150	10	.7000	.0169
11	.7931	.0306	11	.8034	.0340
12	.8723	.0569	12	.8800	.0626
13	.9265	.0972	13	.9317	.1059
14	.9605	.1542	14	.9637	.1663
15	.9802	.2283	15	.9820	.2438
16	.9907	.3175	16	.9916	.3359
17	.9959	.4171	17	.9963	.4374
18	.9983	.5208	18	.9985	.5417
19	.9993	.6219	19	.9994	.6419
20	.9998	.7141	20	.9998	.7321

93 BINOMIAL	.10 ,	.20	94 BINOMIAL	.10 ,	.20
x			x		
0	.0001	.0000	0	.0000	.0000
1	.0006	.0000	1	.0006	.0000
2	.0036	.0000	2	.0033	.0000
3	.0134	.0000	3	.0125	.0000
4	.0382	.0000	4	.0357	.0000
5	.0870	.0001	5	.0821	.0001
6	.1667	.0002	6	.1587	.0002
7	.2767	.0008	7	.2657	.0007
8	.4081	.0023	8	.3949	.0020
9	.5459	.0059	9	.5321	.0052
10	.6746	.0134	10	.6617	.0119
11	.7825	.0275	11	.7717	.0247
12	.8644	.0516	12	.8562	.0468
13	.9211	.0892	13	.9154	.0817
14	.9571	.1428	14	.9535	.1321
15	.9782	.2135	15	.9761	.1994
16	.9896	.2996	16	.9885	.2824
17	.9953	.3971	17	.9948	.3776

18	.9980	.5001	18	.9978	.4795
19	.9992	.6017	19	.9991	.5814

10 BINOMIAL	.20	, .40	20 BINOMIAL	.20	, .40
x			x		
0	.1074	.0060	0	.0115	.0000
1	.3758	.0464	1	.0692	.0005
2	.6778	.1673	2	.2061	.0036
3	.8791	.3823	3	.4114	.0160
4	.9672	.6331	4	.6296	.0510
5	.9936	.8338	5	.8042	.1256
6	.9991	.9452	6	.9133	.2500
7	.9999	.9877	7	.9679	.4159
8	1.0000	.9983	8	.9900	.5956
9	1.0000	.9999	9	.9974	.7553
10	1.0000	1.0000	10	.9994	.8725
11	1.0000	1.0000	11	.9999	.9435
12	1.0000	1.0000	12	1.0000	.9790
13	1.0000	1.0000	13	1.0000	.9935
14	1.0000	1.0000	14	1.0000	.9984
15	1.0000	1.0000	15	1.0000	.9997
16	1.0000	1.0000	16	1.0000	1.0000

30 BINOMIAL	.20	, .40	40 BINOMIAL	.20	, .40
x			x		
0	.0012	.0000	0	.0001	.0000
1	.0105	.0000	1	.0015	.0000
2	.0442	.0000	2	.0079	.0000
3	.1227	.0003	3	.0285	.0000
4	.2552	.0015	4	.0759	.0000
5	.4275	.0057	5	.1613	.0001
6	.6070	.0172	6	.2859	.0006
7	.7608	.0435	7	.4371	.0021
8	.8713	.0940	8	.5931	.0061
9	.9389	.1763	9	.7318	.0156
10	.9744	.2915	10	.8392	.0352
11	.9905	.4311	11	.9125	.0709
12	.9969	.5785	12	.9568	.1285
13	.9991	.7145	13	.9806	.2112
14	.9998	.8246	14	.9921	.3174
15	.9999	.9029	15	.9971	.4402
16	1.0000	.9519	16	.9990	.5681
17	1.0000	.9788	17	.9997	.6885

18	1.0000	.9917	18	.9999	.7911
19	1.0000	.9971	19	1.0000	.8702
20	1.0000	.9991	20	1.0000	.9256

39 BINOMIAL	.20 ,	.40	38 BINOMIAL	.20 ,	.40
x			x		
0	.0002	.0000	0	.0002	.0000
1	.0018	.0000	1	.0022	.0000
2	.0095	.0000	2	.0113	.0000
3	.0332	.0000	3	.0387	.0000
4	.0866	.0000	4	.0986	.0001
5	.1800	.0002	5	.2004	.0003
6	.3124	.0009	6	.3404	.0012
7	.4683	.0029	7	.5003	.0040
8	.6243	.0082	8	.6553	.0110
9	.7586	.0205	9	.7845	.0268
10	.8594	.0450	10	.8781	.0572
11	.9258	.0882	11	.9377	.1089
12	.9645	.1554	12	.9712	.1864
13	.9846	.2484	13	.9880	.2897
14	.9940	.3635	14	.9954	.4127
15	.9978	.4914	15	.9984	.5439
16	.9993	.6193	16	.9995	.6696
17	.9998	.7347	17	.9999	.7781
18	.9999	.8287	18	1.0000	.8624
19	1.0000	.8979	19	1.0000	.9216
20	1.0000	.9441	20	1.0000	.9591

35 BINOMIAL	.20 ,	.40	37 BINOMIAL	.20 ,	.40
x			x		
0	.0004	.0000	0	.0003	.0000
1	.0040	.0000	1	.0027	.0000
2	.0190	.0000	2	.0135	.0000
3	.0605	.0000	3	.0450	.0000
4	.1435	.0002	4	.1120	.0001
5	.2721	.0010	5	.2225	.0005
6	.4328	.0034	6	.3698	.0017
7	.5993	.0102	7	.5330	.0054
8	.7450	.0260	8	.6859	.0148
9	.8543	.0575	9	.8091	.0348
10	.9253	.1123	10	.8954	.0722
11	.9656	.1952	11	.9483	.1333
12	.9858	.3057	12	.9769	.2217
13	.9947	.4361	13	.9907	.3350
14	.9982	.5728	14	.9966	.4645

15	.9995	.7003	15	.9989	.5968
16	.9999	.8065	16	.9997	.7181
17	1.0000	.8857	17	.9999	.8180
18	1.0000	.9385	18	1.0000	.8920
19	1.0000	.9700	19	1.0000	.9414

20 BINOMIAL .01 , .10			20 BINOMIAL .20 , .50		
x			x		
0	.8179	.1216	0	.0115	.0000
1	.9831	.3917	1	.0692	.0000
2	.9990	.6769	2	.2061	.0002
3	1.0000	.8670	3	.4114	.0013
4	1.0000	.9568	4	.6296	.0059
5	1.0000	.9887	5	.8042	.0207
6	1.0000	.9976	6	.9133	.0577
7	1.0000	.9996	7	.9679	.1316
8	1.0000	.9999	8	.9900	.2517
9	1.0000	1.0000	9	.9974	.4119
10	1.0000	1.0000	10	.9994	.5881
11	1.0000	1.0000	11	.9999	.7483
12	1.0000	1.0000	12	1.0000	.8684
13	1.0000	1.0000	13	1.0000	.9423
14	1.0000	1.0000	14	1.0000	.9793
15	1.0000	1.0000	15	1.0000	.9941
16	1.0000	1.0000	16	1.0000	.9987
17	1.0000	1.0000	17	1.0000	.9998
18	1.0000	1.0000	18	1.0000	1.0000

20 BINOMIAL .30 , .60			20 BINOMIAL .40 , .70		
x			x		
0	.0008	.0000	0	.0000	.0000
1	.0076	.0000	1	.0005	.0000
2	.0355	.0000	2	.0036	.0000
3	.1071	.0000	3	.0160	.0000
4	.2375	.0003	4	.0510	.0000
5	.4164	.0016	5	.1256	.0000
6	.6080	.0065	6	.2500	.0003
7	.7723	.0210	7	.4159	.0013
8	.8867	.0565	8	.5956	.0051
9	.9520	.1275	9	.7553	.0171
10	.9829	.2447	10	.8725	.0480
11	.9949	.4044	11	.9435	.1133
12	.9987	.5841	12	.9790	.2277
13	.9997	.7500	13	.9935	.3920
14	1.0000	.8744	14	.9984	.5836

15	1.0000	.9490	15	.9997	.7625
16	1.0000	.9840	16	1.0000	.8929
17	1.0000	.9964	17	1.0000	.9645
18	1.0000	.9995	18	1.0000	.9924
19	1.0000	1.0000	19	1.0000	.9992
20	1.0000	1.0000	20	1.0000	1.0000

20 BINOMIAL .80 , .90

x					
0	.0000	.0000	10	.0026	.0000
1	.0000	.0000	11	.0100	.0001
2	.0000	.0000	12	.0321	.0004
3	.0000	.0000	13	.0867	.0024
4	.0000	.0000	14	.1958	.0113
5	.0000	.0000	15	.3701	.0432
6	.0000	.0000	16	.5886	.1330
7	.0000	.0000	17	.7939	.3231
8	.0001	.0000	18	.9308	.6083
9	.0006	.0000	19	.9885	.8784
			20	1.0000	1.0000

LESSON *18. MATCHING AND CATCHING

The following are traditional examples illustrating the meaning of the participles in the title.

Example: a) A secretary types 10 letters and addresses 10 envelopes and (who knows why?) puts one letter in each envelope "at random". What is the probability that no addressee gets the right letter?

b)   A young person wants to collect the 21 baseball pictures packaged individually and, presumably at random, in boxes of cereal. How many boxes must be purchased to have probability at least .90 of finding all 21 pictures?

In order to get a neat handle on these problems, it is convenient to use *indicator* functions with the following basic property. The proof should remind you of counting and the construction of sample spaces.

Lemma: *For any subset C of a set S,* $I_C(s) = \begin{cases} 1 & \text{if } s \in C \\ 0 & \text{if } s \notin C \end{cases}$. *Let B and C be subsets of S. Then, a)* $I_{B \cap C} = I_B \cdot I_C$;

$$b) \; I_{B \cup C} = I_B + I_C - I_{B \cap C}.$$

Proof: Consider the partition :

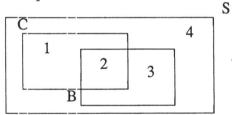

For point s in the region in the first column, the values of the indicator functions are given in the remaining columns:

	I_B	I_C	$I_B \cdot I_C$	$I_{B \cap C}$	$I_B + I_C - I_B \cdot I_C$	$I_{B \cup C}$
1	0	1	0	0	$0 + 1 - 0 = 1$	1
2	1	1	1	1	$1 + 1 - 1 = 1$	1
3	1	0	0	0	$1 + 0 - 0 = 1$	1
4	0	0	0	0	$0 + 0 - 0 = 0$	0

The equality of columns 4 and 5 validates a) ; that of columns 6,7 validates b) .

Exercise 1: a) Use indicator functions tos how that

$$(A \cup B)^C = A^C \cap B^C.$$

b)   Argue verbally that for subsets $A_1, A_2, \cdots, A_n$,

$$I_{\cap_{i=1}^n A_i} = \Pi_{i=1}^n I_{A_i}.$$

The main result is the following "union–intersection" principle.

Theorem: *For $A_1, A_2, \cdots$ subsets of S and each positive integer*

$$n, \ I_{\cup_{i=1}^n A_i} = 1 - \Pi_{i=1}^n (1 - I_{A_i})$$

$$= 1 - \Pi_{i=1}^n I_{A_i}{}^c = 1 - I_{\cap_{i=1}^n A_i{}^c}.$$

Proof: For $n = 1$, $I_{\cup_{i=1}^n A_i} = I_{A_1}$ and

$$1 - \Pi_{i=1}^1 (1 - I_{A_i}) = 1 - (1 - I_{A_1}) = I_{A_1}$$

so the conclusion holds for $n = 1$. Assume

$$I_{\cup_{i=1}^k A_i} = 1 - \Pi_{i=1}^k (1 - I_{A_i}).$$

Letting $B = \cup_{i=1}^k A_i$ and $C = A_{k+1}$ in the previous lemma yields: $I_{\cup_{i=1}^{k+1} A_i} = I_{(\cup_{i=1}^{k} A_i) \cup A_{k+1}}$

$$= I_{\cup_{i=1}^{k} A_i} + I_{A_{k+1}} - I_{\cup_{i=1}^{k} A_i} \cdot I_{A_{k+1}} .$$

Substituting the result of the induction hypothesis in the right side of this last equality, we get

$$(1 - \Pi_{i=1}^{k}(1 - I_{A_i})) + I_{A_{k+1}} - (1 - \Pi_{i=1}^{k}(1 - I_{A_i})) \cdot I_{A_{k+1}}$$

$$= 1 - \Pi_{i=1}^{k}(1 - I_{A_i}) + I_{A_{k+1}}$$

$$- I_{A_{k+1}} + \Pi_{i=1}^{k}(1 - I_{A_i}) \cdot I_{A_{k+1}}$$

$$= 1 - (\Pi_{i=1}^{k}(1 - I_{A_I}))(1 - I_{A_{k+1}}) = 1 - \Pi_{i=1}^{k+1}(1 - I_{A_i}) .$$

This completes the induction. The remaining equalities are notational.

Example : Since the indicators are real–valued functions, we may expand the right–hand side of the equality in the theorem by ordinary algebra. For $n = 4$, we get

$$1 - (1 - I_{A_1})(1 - I_{A_2})(1 - I_{A_3})(1 - I_{A_4})$$

$$= I_{A_1} + I_{A_2} + I_{A_3} + I_{A_4}$$

$$- I_{A_1} \cdot I_{A_2} - I_{A_1} \cdot I_{A_3} - I_{A_1} \cdot I_{A_4}$$

$$- I_{A_2} \cdot I_{A_3} - I_{A_2} \cdot I_{A_4} - I_{A_3} \cdot I_{A_4}$$

$$+ I_{A_1} \cdot I_{A_2} \cdot I_{A_3} + I_{A_1} \cdot I_{A_2} \cdot I_{A_4}$$

$$+ I_{A_1} \cdot I_{A_3} \cdot I_{A_4} + I_{A_2} \cdot I_{A_3} \cdot I_{A_4}$$

$$- I_{A_1} \cdot I_{A_2} \cdot I_{A_3} \cdot I_{A_4}$$

which can also be written as

$$I_{A_1} + I_{A_2} + I_{A_3} + I_{A_4}$$

$$- I_{A_1 \cap A_2} - I_{A_1 \cap A_3} - I_{A_1 \cap A_4}$$

$$- I_{A_2 \cap A_3} - I_{A_2 \cap A_4} - I_{A_3 \cap A_4}$$

$$+ I_{A_1 \cap A_2 \cap A_3} + I_{A_1 \cap A_2 \cap A_4}$$

$$+ I_{A_1 \cap A_3 \cap A_4} + I_{A_2 \cap A_3 \cap A_4}$$

$$- I_{A_1 \cap A_2 \cap A_3 \cap A_4}$$

or in an abbreviated form

$$\sum_{i=1}^{4} I_{A_i} - \sum_{i<j} I_{A_i \cap A_j}$$

$$+ \sum_{i<j<k} I_{A_i \cap A_j \cap A_k} - I_{A_1 \cap A_2 \cap A_3 \cap A_4}.$$

Then in general, (by informal induction),

$$I_{\cup_{i=1}^{n} A_i} = \sum_{i=1}^{n} I_{A_i} - \sum_{i<j} I_{A_i \cap A_j}$$

$$+ \sum_{i<j<k} I_{A_i \cap A_j \cap A_k} - \cdots$$

$$+ (-1)^{n-1} I_{A_1 \cap \cdots \cap A_n}. \quad (*)$$

When A is an event in a probability space $[S, \mathcal{F}, P]$,

$$P(A) = P(\{s : s \in A\}) = P(\{s : I_A(s) = 1\}) = P(I_A(s) = 1);$$

$$P(A^C) = P(I_A(s) = 0).$$

Thus the indicator function is a Bernoulli random variable and

$$E[I_A] = 1 \cdot P(I_A = 1) + 0 \cdot P(I_A = 0)$$

$$= 1 \cdot P(A) + 0 \cdot P(A^C) = P(A).$$

Corollary : *For* $A_1, A_2, \cdots, A_n$ *in a probability space* $[S, \mathcal{F}, P]$,

$$P(\cup_{i=1}^{n} A_i) = \Sigma_{i=1}^{n} P(A_i) - \Sigma_{i<j} P(A_i \cap A_j)$$
$$+ \Sigma_{i<j<k} P(A_i \cap A_j \cap A_k) - \cdots$$
$$+ (-1)^{n-1} P(A_1 \cap A_2 \cap \cdots \cap A_n) .$$

Proof: Since $E[I_A] = P(A)$, $P(\cup_{i=1}^{n} A_i) = E[I_{\cup_{i=1}^{n} A_i}]$ and a similar manipulation can be applied to each term on the right hand side of (*). The proof is completed by using a previous theorem (Lesson 12) for the expected value of a linear combination of random variables.

Note that this is the formula for the probability that at least one of the given events $A_1, A_2, \cdots, A_n$ occurs.

Exercise 2: Suppose that the probability is uniform on a finite set of size M. Multiply both sides of the conclusion in this corollary by M to obtain a formula for the counting function (mentioned in lesson 4).

To return to matching, let A_i be the event that the ith letter gets put into its correct envelope, i = 1(1)n; (we may as well be general.) Then the probability of no match is $1 - P(\cup_{i=1}^{n} A_i)$. The phrase "presumably at random" is interpreted as imposing a uniform distribution on the n! permutations. Then, for each i, the other letters can be permuted in n − 1 envelopes, so that

$$P(A_i) = (n - 1)!/n! ;$$

for each fixed i < j, the other letters are permuted in n − 2 envelopes, so that

$$P(A_i \cap A_j) = (n - 2)!/n! ;$$

for each fixed i < j < k , $P(A_i \cap A_j \cap A_k) = (n - 3)!/n! ;$

$\cdots$ $P(A_1 \cap A_2 \cap \cdots \cap A_n) = (n - n)!/n! = 1/n! .$

There are n single choices for A_i so that

$$\Sigma_{i=1}^{n} P(A_i) = nP(A_i) = n(n-1)!/n! = 1/1! .$$

There are $\begin{bmatrix} n \\ 2 \end{bmatrix}$ pairs with $i < j$ so that

$$\Sigma_{i<j} P(A_i \cap A_j) = \begin{bmatrix} n \\ 2 \end{bmatrix} P(A_i \cap A_j) = 1/2! .$$

There are $\begin{bmatrix} n \\ 3 \end{bmatrix}$ triples with $i < j < k$ so that

$$\Sigma_{i<j<k} P(A_i \cap A_j \cap A_k) = \begin{bmatrix} n \\ 3 \end{bmatrix} P(A_i \cap A_j \cap A_k) = 1/3! .$$

. . . .

Hence, P(at least one match)

$$= P(\cup_{i=1}^{n} A_i) = 1/1! - 1/2! + 1/3! - \cdots \pm 1/n! .$$

You should recognize this as a Taylor polynomial for $1 - e^{-1}$. The probability of no match equals

$$1 - P(\text{at least one match})$$

and is approximately $e^{-1} \approx .367879$ for $n \geq 6$. The exact value for ten letters is

$$1/2! - 1/3! + 1/4! - 1/5! + 1/6! - 1/7! + 1/8! - 1/9! + 1/10!$$

which approximates to .367879464 .

Exercise 3: Write out twelve terms of the Taylor polynomial for $e^x = \exp(x)$ about the fixed point x_0 .

Exercise 4:   An ESP experiment consists of matching two sequences of 5 different cards, one laid out by a person in "room A" and another laid out by the candidate in "room B". With no ESP, matching is "random".
   a)   What is the exact probability that there is at least one match?
   b)   If this experiment is *repeated* ten times, what is the probability that there will be at least one match in all ten cases?

c)   If you were a witness to this occurrence of ten, would you be impressed?

Now we turn to the probability of "catching" all 21 baseball pictures among 50 boxes of cereal and the like. Here the trick for counting is to look at the problem backwards, imagining the boxes to "fall" onto the pictures. A general statement is:

find the probability that when tossing at random $r \geq n$ distinguishable balls (boxes) into n distinguishable cells (pictures), no cell ends up empty.

Equivalently, we can toss r n–sided dice once or one n–sided die r times and seek the probability that all n sides "appear".

Let A_i be the event that the ith cell is empty. Then:

for each i, all balls go to other cells and $P(A_i) = (n - 1)^r/n^r$;

for each fixed $i < j$, $\qquad\qquad P(A_i \cap A_j) = (n - 2)^r/n^r$;

for each fixed $i < j < k$, $\qquad P(A_i \cap A_j \cap A_k) = (n - 3)^r/n^r$;

of course, $\qquad\qquad P(A_1 \cap A_2 \cap \cdots \cap A_n) = (n - n)^r/n^r = 0$.

Finally, $P(\cup_{i=1}^{n} A_i)$

$$= \begin{bmatrix} n \\ 1 \end{bmatrix}(n - 1)^r/n^r - \begin{bmatrix} n \\ 2 \end{bmatrix}(n - 2)^r/n^r$$

$$+ \begin{bmatrix} n \\ 3 \end{bmatrix}(n - 3)^r/n^r - \cdots + (-1)^{n-2}\begin{bmatrix} n \\ n-1 \end{bmatrix}(n - (n-1))^r/n^r$$

$$= \sum_{k=1}^{n-1} (-1)^{k-1} \begin{bmatrix} n \\ k \end{bmatrix}(1 - k/n)^r .$$

This is the probability that there is at least one empty cell (at least one face of the die doesn't appear, at least one picture is missing). Hence, the probability that there is no empty cell or that the young person has all the pictures is

$$p(0; r, n) = 1 - \Sigma_{k=1}^{n-1} (-1)^{k-1} \begin{bmatrix} n \\ k \end{bmatrix}(1 - k/n)^r .$$

Exercise 5: Evaluate $p(0;r, n)$ for $n = 6, 7, 8$ and $r = n+1$, $n+5$, $n+10$, $n+20$.

For $n = 21$, we get

r	p
21	.00000008
42	.03000
63	.34105
84	.69525
105	.88048
106	.88591
107	.89111
108	.89608
109	.90083
110	.90538

The young person must eat (at least buy) 109 boxes of cereal to have 90 to 10 odds of finding all 21 pictures among them.

We should consider the probability for the occurrence of exactly m of the events or at least m of the events (matches, catches, $\cdots$). Formulae for these can be found in books like Feller, 1968 and Neuts, 1973. In the following, we present another approach (Johnson, 1980) to verify a formula once it has been discovered.

At the end of lesson 3, we saw the partition of a sample space into 8 [16] subsets from 3 [4] given subsets. For $A_1, A_2, \cdots, A_n$, we have 2^n subsets of the form

$$A_1^{\delta_1} \cap A_2^{\delta_2} \cap \cdots \cap A_n^{\delta_n}$$

with

$$\delta_k \in \{-1, 1\}, \quad A^{-1} = A^c, \quad A^1 = A .$$

Since intersection operations are commutative (and associative), these disjoint events, say E_i , $i=1(1)2^n$, can be written as

$$E_i = A_{j_i(1)} \cap A_{j_i(2)} \cap \cdots \cap A_{j_i(o_i)} \cap A_{j_i(o_{i+1})} \cap \cdots \cap A_{j_i(n)}$$

where $j_i(1), \cdots, j_i(n)$ is a permutation of $1, 2, \cdots, n$ and o_i is the number of A's "occurring". If B_m is the event that exactly m of the events $A_1, \cdots, A_n$ occur, then B_m is the union of all the

E_i events for which $o_i = m$; let $\Omega = \{i : o_i = m\}$. Then,

$$P(B_m) = \sum_{i \in \Omega} P(E_i).$$

For $j_1, \cdots, j_k$ fixed, $P(A_{j_1} \cap \cdots \cap A_{j_k})$ is the sum of $P(E_i)$ for which $\{j_1, ..., j_k\} \subset \{j_{i(1)}, \cdots, j_{i(o_i)}\}$. Therefore, $\begin{bmatrix} o_i \\ k \end{bmatrix}$ of the terms $P(A_{j_1} \cap \cdots \cap A_{j_k})$ each contain a $P(E_i)$. The other way around,

$$S_k = \sum_{1 \le j_1 < \cdots < j_k \le n} P(A_{j_1} \cap \cdots \cap A_{j_k}) = \sum_{i=1}^{2^n} \begin{bmatrix} o_i \\ k \end{bmatrix} P(E_i) .$$

(Since we know the answer,) we now consider

$$\sum_{k=m}^{n} (-1)^{k-m} \begin{bmatrix} k \\ m \end{bmatrix} S_k = \sum_{k=m}^{n} (-1)^{k-m} \begin{bmatrix} k \\ m \end{bmatrix} \cdot \sum_{i=1}^{2^n} \begin{bmatrix} o_i \\ k \end{bmatrix} P(E_i)$$

$$= \sum_{i=1}^{2^n} \left[\sum_{k=m}^{n} (-1)^{k-m} \begin{bmatrix} k \\ m \end{bmatrix} \begin{bmatrix} o_i \\ k \end{bmatrix} \right] P(E_i).$$

Exercise 6: Show that the sum in square brackets, say c_i, is:

$$0 \text{ for } o_i < m \;;\; 1 \text{ for } o_i = m \;;\; 0 \text{ for } o_i > m .$$

Hint on the last case: rewrite the sum in terms of $s = k-m$.

The conclusion of all this discussion is

$$P(B_m) = \sum_{i \in \Omega} P(E_i) = \sum_{i=1}^{2^n} c_i P(E_i) = \sum_{k=m}^{n} (-1)^{k-m} \begin{bmatrix} k \\ m \end{bmatrix} S_k$$

for the probability that exactly m of the events $A_1, \cdots, A_n$ occur.

Exercise 7: Let C_m be the event that at least m of the events $A_1, \cdots, A_n$ occur.

a)   Prove by induction that

$$P(C_m) = \sum_{k=m}^{n} (-1)^{k-m} \binom{k-1}{m-1} S_k.$$

b)   In particular, show that $P(C_1) = S_1 - S_2 + \cdots \pm S_n$
      is the same result as obtained in the corollary above.

Additional results may be found also in Johnson and Kotz, 1973, and Leader, 1985.

LESSON 19. CONFIDENCE INTERVALS FOR A BERNOULLI θ

Let X be the total number of "successes" in a random sample of n Bernoulli trials with $P(S) = \theta = 1 - P(F)$; the distribution of X is binomial. (Usually "p" is used for this parameter; the Greek analogue π has another meaning so we use θ.) We could rewrite lessons 15 and 16 to develop a test for
$$H_o: \theta = \theta_o \text{ versus } H_a: \theta = \theta_a > \theta_o.$$
Instead, we consider a "two–sided hypothesis"
$$H_o : \theta = \theta_o , H_a : \theta \neq \theta_o ,$$
and develop a corresponding "two–tailed" test; this will lead to the notion of a confidence interval.

The following "filler" recently appeared in a newspaper. "Thirty percent of the deaths of children under the age of five are consequences of automobile accidents." To get some idea whether or not this figure was reasonable, we obtained a simple random sample of the death records of children (under 5) in New Mexico, 1984–1985. With n = 20, the observed number of auto–related deaths was 9. Now what?

This is one mode in which "binomial problems" appear in statistical methods books. Like our earlier discussion, the formalities of a solution focus on the mean. When H_o is true, $E[X] = n\theta_o = 20(.30) = 6$; when H_a is true, E[X] can be either smaller or larger than 6. We will agree that it is reasonable to reject H_o when the observed X is "too big or too small"; a critical (rejection) region takes the form
$$C = \{X \leq c_1\} \cup \{X \geq c_2\}$$

with the significance level or size of type I error
$$\alpha = P(C \mid H_o) = \sum_{x=0}^{c_1} \binom{20}{x}(.3)^x(.7)^{20-x}$$
$$+ \sum_{x=c_2}^{20} \binom{20}{x}(.3)^x(.7)^{20-x}.$$

Then the size of type II error $\beta = P(C^c \mid \theta \, \varepsilon \, H_a)$ is the complement of $\displaystyle\sum_{x=0}^{c_1-1} \left[\begin{array}{c} 20 \\ x \end{array}\right] (\theta)^x (1-\theta)^{20-x}$

$$+ \sum_{x=c_2}^{20} \left[\begin{array}{c} 20 \\ x \end{array}\right] (\theta)^x (1-\theta)^{20-x} .$$

Given c_1 and c_2, α is determined but the value of β depends on what the alternative θ is. A general testing procedure is: first determine c_1 and c_2; if for the observations, $c_1 < x < c_2$, do not reject H_o .

Suppose that we decide to ignore β momentarily and to limit α to .05. Then it is customary (but not necessary) to take

$$P(X \le c_1 \mid H_o) \le .025 \text{ and } P(X \ge c_2 \mid H_o) \le .025$$

where the inequalities result from the discreteness of X, its distribution function increasing only in "jumps". Using the CDF in our tables, we must look for c_1 and c_2 such that

$$P(X \le c_1 \mid n = 20, \theta = .3) \le .025$$

$$\text{and } P(X \le c_2 - 1 \mid n = 20, \theta = .3) \ge .975 .$$

Tables of the binomial CDF for selected sample sizes n and proportions θ (the heading is BINOMIAL n) are given on the pages at the end of the lesson. The first row lists

$$\theta = .05 \ .1 \ .2 \ .3 \ .4 \ .5 \ .6 \ .7 \ .8 \ .9 \ .95$$

The first column contains the values $c = 0(1)n$; the succeeding columns contain the values of $P(X \le c \mid n, \theta = .05)$, $P(X \le c \mid n, \theta = .1)$, and so on.

Exercise 1: Verify some of the columns for n = 5; that is, calculate things like $P(X \le 0 \mid 5, .05)$, $P(X \le 1 \mid 5, .05)$, $P(X \le 2 \mid 5, .1)$, etc.

Continuing with our example on auto–deaths, look at BINOMIAL 20 with $\theta_o = .3$.

For $P(X \le c_1 \mid 20, .3) \le .025$, we find $c_1 = 1$.

For $P(X \le c_2 - 1 \mid 20, .3) \ge .975$, we find $c_2 - 1 = 10$.

Thus the critical region C = { X ≤ 1} ∪ { X ≥ 11} . Since the observed value 9 is not in C, we do not reject H_o. On the basis of this sample, we cannot say that the proportion of such deaths (in New Mexico) is not (around) .3 .

Example: To resurrect an old friend, when is a coin fair? Obviously, when you toss the coin a reasonable number of times and get "about" equal numbers of heads and tails. Formally, on the basis of n = 40 tosses of a two–sided coin, decide which of $H_o : \theta = .5$ or $H_a : \theta \neq .5$ is reasonable. Let us take $\alpha = .01$ so that we look for

$$P(X \leq c_1 \mid 40, .5) \leq .005 \text{ and } P(X \leq c_2-1 \mid 40, .5) \geq .995.$$

We find $c_1 = 11$ and $c_2-1 = 28$. If in 40 tosses, we get no more than 11 heads or at least 29 heads, then we reject H_o and suggest that the coin may not be fair.

Exercise 2: For each sample from a Bernoulli population, test the hypothesis $H_o: \theta = .4$ against $H_a : \theta \neq .4$. Use the significance level $\alpha = .05$ in two equal tails.
- a)   n = 20 x = 12
- b)   n = 25 x = 15
- c)   n = 15 x = 12
- d)   n = 15 x = 6.

Going back to our example with children under 5, suppose that we had rejected H_o . Then another question would arise; since θ is probably not .3, what is it? We can get a reasonable answer by "reversing" the test.

First we will describe the procedure and notation; then we will give a theorem for justification. The results in our examples will be rather rough approximations since our θ values are rather rough; we should have finer (and more) values of θ like .001, .002, $\cdots$. In real life, there are nice graphs and tables which yield values quickly; we may even do a little interpolation.

Let α_1 and α_2 and n be given. For each θ, let $x_1 = x_1(\theta)$ be such that

$$\sum_{k=0}^{x_1} \begin{bmatrix} n \\ k \end{bmatrix} \theta^k (1-\theta)^{n-k} \le \alpha_1 < \sum_{k=0}^{x_1+1} \begin{bmatrix} n \\ k \end{bmatrix} \theta^k (1-\theta)^{n-k}.$$

For each θ, let $x_2 = x_2(\theta)$ be such that

$$\sum_{k=x_2}^{n} \begin{bmatrix} n \\ k \end{bmatrix} \theta^k (1-\theta)^{n-k} \le \alpha_2 < \sum_{k=x_2-1}^{n} \begin{bmatrix} n \\ k \end{bmatrix} \theta^k (1-\theta)^{n-k}.$$

In words, x_1 is the largest integer and x_2 is the smallest integer for which

$$P(X \le x_1 \mid n, \theta) \le \alpha_1 \text{ and } P(X \ge x_2 \mid n, \theta) \le \alpha_2 . \quad (*)$$

If x is the observed value of the binomial RV X, let $(\theta_L(x), \theta_U(x))$ be the interval containing all θ for which

$$x_1(\theta) + 1 \le x \le x_2(\theta) - 1 . \quad (**)$$

Example: Using the table of the CDF, for each θ, we find

$$x_1 = x_1(\theta) \text{ with } P(X \le x_1 \mid n, \theta) \le \alpha_1 = .025$$

and $x_2 = x_2(\theta)$ with $P(X \le x_2 - 1 \mid n, \theta) \ge 1 - \alpha_2 = .975$.

Then when x is observed, $\theta_L(x)$ and $\theta_U(x)$ are defined by examining $x_1 + 1 \le x \le x_2 - 1$.

a) The case $n = 10$. Look at the table to see that

$P(X \le x_1 \mid 10, \theta) \le .025$ has no solution for $\theta \le .3$.

Then $P(X \le x_1 \mid 10, .4) \le .025$ implies $x_1(.4) = 0$;

$P(X \le x_1 \mid 10, .5) \le .025$ implies $x_1(.5) = 1$; etc.

$P(X \le x_2 - 1 \mid 10, .05) \ge .975$ implies $x_2(.05) - 1 = 2$;

$P(X \le x_2 - 1 \mid 10, .1) \ge .975$ implies $x_2(.1) - 1 = 3$; etc.

Exercise 3: Check the table for $n = 10$ to verify the following summary.

θ	.05	.1	.2	.3	.4	.5	.6	.7	.8	.9	.95
x_1	–	–	–	–	0	1	2	3	4	6	7
x_2-1	2	3	5	6	7	8	9	10	10	10	10

Example: a) Continuing with n = 10, suppose that the observed value is x = 4. Then $x_1 + 1 \leq x = 4$ implies $x_1 \leq 3$; the largest such integer is therefore 3 and (**) implies $\theta_U(4) = .7$. Similarly, $x_2 - 1 \geq x = 4$ implies that the smallest such integer $x_2 = 5$; (**) implies $\theta_L(4) \approx .15$.

b) Consider n = 15; for each θ, choose x_1 so that

$P(X \leq x_1 \mid 15, \theta) \leq .025$ and choose x_2 such that

$P(X \leq x_2 - 1 \mid 15, \theta) \geq .975$.

θ	.05	.1	.2	.3	.4	.5	.6	.7	.8	.9	.95
x_1	–	–	–	0	2	3	5	6	8	10	11
x_2-1	3	4	6	8	10	11	13	14	15	15	15

If x = 4, $x_1 \leq 3$ and $\theta_U(4) = .5$; for, $x_2-1 \geq 4$, $\theta_L(4) = .1$.

If x = 10, $x_1 \leq 9$, $\theta_U(10) = .85$, $x_2-1 \geq 10$, $\theta_L(10) = .4$.

Exercise 4: a) For n = 20 and x = 15, find θ_L and θ_U with $\alpha_1 = \alpha_2 = .02$.

b) For n = 20 and x = 13, find θ_L and θ_U with

$\alpha_1 = \alpha_2 = .03$.

c) For n = 30 and x = 20, find θ_U and θ_L with $\alpha_1 = .02$ and $\alpha_2 = .03$.

The theorem is as follows but the proof will be delayed and somewhat drawnout.

Theorem: *For each value of the binomial random variable X, let the interval $(\theta_L(X), \theta_U(X))$ contain all θ such that*

$x_1(\theta) + 1 \leq X \leq x_2(\theta) - 1$ *(as described in (*) and (**)).*

Then this random interval has probability $\geq 1 - \alpha_2 - \alpha_1$.

The interpretation is this. Each time we observe the same binomial experiment (n and θ fixed), we obtain an interval $(\theta_L(x), \theta_U(x))$; overall, $100(1 - \alpha_2 - \alpha_1)\%$ of these intervals contain the "true" θ even when it is unknown. Of course, we never know which of the observed intervals contain an unknown θ and which do not, but, if $100(1 - \alpha_2 - \alpha_1)$ is "large", we will be *confident* that most of them do.

Definition: *For each of the repetitions of a binomial experiment, the interval $(\theta_L(x), \theta_U(x))$ is a $100(1-\alpha_2-\alpha_1)\%$ confidence interval for θ.*

Exercise 6: a) Find a 90% confidence interval for θ when n = 30, x = 9; take $\alpha_1 = \alpha_2 = .05$.
b)   Find a 99% confidence interval for θ when n = 30 and x = 10 ; take $\alpha_1 = \alpha_2 = .01$.

Exercise 7: a) For fixed integers $0 < x < n$ and proportion $0 < \theta < 1$, show that

$$R(\theta;x) = \sum_{k=x}^{n} \binom{n}{k}\theta^k(1-\theta)^{n-k} = n\binom{n-1}{x-1}\int_0^\theta b^{x-1}(1-b)^{n-x}\, db .$$

by integrating by parts.

b)   Show that for x and n fixed, $R(\theta;x)$ is a continuous and (strictly) increasing function of θ. That is, $\theta_1 < \theta_2$ implies $R(\theta_1;x) < R(\theta_2;x)$.

c)   Show that for x and n fixed, $F(x;\theta) = P(X \leq x \mid n, \theta)$ is a decreasing function of θ.

On the other hand, for n and θ fixed in the above sum, $R(\theta;x)$ is a decreasing function of x; that is, $R(\theta;x+1) < R(\theta;x)$.

Exercise 8: Show that for $0 < x < n$,

$$R(1;x) = F(x;0) = 1 \quad \text{while} \quad R(0;x) = F(x;1) = 0.$$

Proof of the theorem: Since $0 \leq R(\theta;x) \leq 1$, continuity of R in θ

yields (via the intermediate value theorem) a solution $\theta_L = \theta_L(x)$ of $R(\theta;x) = \alpha_2$ for each $0 < \alpha_2 < 1$. Monotonicity of R in x guarantees $x_2 > x$ with $R(\theta;x_2) \le \alpha_2 < R(\theta;x-1)$. Together these imply

$$R(\theta;x_2) \le R(\theta_L;x).$$

Similarly, for $0 < \alpha_1 < 1$, $F(x;\theta) = \alpha_1$ has a solution $\theta_U = \theta_U(x)$; with $F(x_1;\theta) \le \alpha_1 < F(x_1+1;\theta)$, this implies

$$F(x_1;\theta) \le F(x;\theta_U). \qquad (2)$$

Lemma: $\theta_L < \theta < \theta_U$ *iff* $x_1 < x < x_2$ *for each x.*

Proof: Otherwise, for some x, $x_1 \ge x$ and, since F is increasing in x, $F(x_1;\theta) \ge F(x;\theta)$. Then, since F is decreasing in θ,

$$F(x;\theta) > F(x;\theta_U).$$

Together, these imply $F(x_1;\theta) > F(x;\theta_U)$ which contradicts (2). Similarly, $R(\theta_L;x) < R(\theta;x_2)$ contradicts (1).

Continuing with the proof of the theorem: From the lemma, it follows that

$$P(\theta_L(X) < \theta < \theta_U(X)) = P(x_1(\theta) < X < x_2(\theta)).$$

This latter value is (at last!)

$$P(X \le x_2(\theta) - 1) - P(X \le x_1(\theta))$$

$$= 1 - R(\theta;x_2) - F(x_1;\theta) \ge 1 - \alpha_2 - \alpha_1.$$

The article by Blyth, 1986, contains a discussion of the accuracy of different approximations for F and R.

Exercise 9: a) In the integral obtained for the binomial probability, $R(\theta;x) = P(X \ge x)$, substitute $b = 1/[1 + (n-x+1)F/x]$. (It turns out that the non–negative functions of "b" and "F" are densities for the b(eta) and F(isher) distributions.) After the substituion, you should have:

$$\frac{n!}{(x-1)!(n-x)!}[(n-x+1)/x]^{n-x+1}\int_c^\infty F^{n-x}/[1+(n-x+1)F/x]^{n+1}dF$$

where $c = (1 - \theta)x/\theta(n - x + 1)$.

b) Evaluate the integral above when $\theta = 1$.

Table BINOMIAL n

This table (generated by STSC–APL) contains the cumulative distribution function of binomial random variables with parameters n and θ = .05 .20 .30 .30 .40 .50 .60 .70 .80 .90 .95 ; the latter are listed across the top row in each group. For example, $P(B \le 3 \mid n = 5, \theta = .3) = .9692$.

```
BINOMIAL    5
x   .0500   .1000   .2000   .3000   .4000   .5000
0   .7738   .5905   .3277   .1681   .0778   .0313
1   .9774   .9185   .7373   .5282   .3370   .1875
2   .9988   .9914   .9421   .8369   .6826   .5000
3 1.0000   .9995   .9933   .9692   .9130   .8125
4 1.0000 1.0000   .9997   .9976   .9898   .9688
5 1.0000 1.0000 1.0000 1.0000 1.0000 1.0000
BINOMIAL    5
x   .5000   .6000   .7000   .8000   .9000   .9500
0   .0313   .0102   .0024   .0003   .0000   .0000
1   .1875   .0870   .0308   .0067   .0005   .0000
2   .5000   .3174   .1631   .0579   .0086   .0012
3   .8125   .6630   .4718   .2627   .0815   .0226
4   .9688   .9222   .8319   .6723   .4095   .2262
5 1.0000 1.0000 1.0000 1.0000 1.0000 1.0000

BINOMIAL    10
x   .0500   .1000   .2000   .3000   .4000   .5000
0   .5987   .3487   .1074   .0282   .0060   .0010
1   .9139   .7361   .3758   .1493   .0464   .0107
2   .9885   .9298   .6778   .3828   .1673   .0547
3   .9990   .9872   .8791   .6496   .3823   .1719
4   .9999   .9984   .9672   .8497   .6331   .3770
5 1.0000   .9999   .9936   .9527   .8338   .6230
6 1.0000 1.0000   .9991   .9894   .9452   .8281
7 1.0000 1.0000   .9999   .9984   .9877   .9453
8 1.0000 1.0000 1.0000   .9999   .9983   .9893
9 1.0000 1.0000 1.0000 1.0000   .9999   .9990
10 1.0000 1.0000 1.0000 1.0000 1.0000 1.0000
```

BINOMIAL   10

x	.5000	.6000	.7000	.8000	.9000	.9500
0	.0010	.0001	.0000	.0000	.0000	.0000
1	.0107	.0017	.0001	.0000	.0000	.0000
2	.0547	.0123	.0016	.0001	.0000	.0000
3	.1719	.0548	.0106	.0009	.0000	.0000
4	.3770	.1662	.0473	.0064	.0001	.0000
5	.6230	.3669	.1503	.0328	.0016	.0001
6	.8281	.6177	.3504	.1209	.0128	.0010
7	.9453	.8327	.6172	.3222	.0702	.0115
8	.9893	.9536	.8507	.6242	.2639	.0861
9	.9990	.9940	.9718	.8926	.6513	.4013
10	1.0000	1.0000	1.0000	1.0000	1.0000	1.0000

BINOMIAL   12

x	.0500	.1000	.2000	.3000	.4000	.5000
0	.5404	.2824	.0687	.0138	.0022	.0002
1	.8816	.6590	.2749	.0850	.0196	.0032
2	.9804	.8891	.5583	.2528	.0834	.0193
3	.9978	.9744	.7946	.4925	.2253	.0730
4	.9998	.9957	.9274	.7237	.4382	.1938
5	1.0000	.9995	.9806	.8822	.6652	.3872
6	1.0000	.9999	.9961	.9614	.8418	.6128
7	1.0000	1.0000	.9994	.9905	.9427	.8062
8	1.0000	1.0000	.9999	.9983	.9847	.9270
9	1.0000	1.0000	1.0000	.9998	.9972	.9807
10	1.0000	1.0000	1.0000	1.0000	.9997	.9968
11	1.0000	1.0000	1.0000	1.0000	1.0000	.9998
12	1.0000	1.0000	1.0000	1.0000	1.0000	1.0000

BINOMIAL   12

x	.5000	.6000	.7000	.8000	.9000	.9500
0	.0002	.0000	.0000	.0000	.0000	.0000
1	.0032	.0003	.0000	.0000	.0000	.0000
2	.0193	.0028	.0002	.0000	.0000	.0000
3	.0730	.0153	.0017	.0001	.0000	.0000
4	.1938	.0573	.0095	.0006	.0000	.0000
5	.3872	.1582	.0386	.0039	.0001	.0000
6	.6128	.3348	.1178	.0194	.0005	.0000
7	.8062	.5618	.2763	.0726	.0043	.0002
8	.9270	.7747	.5075	.2054	.0256	.0022
9	.9807	.9166	.7472	.4417	.1109	.0196
10	.9968	.9804	.9150	.7251	.3410	.1184
11	.9998	.9978	.9862	.9313	.7176	.4596
12	1.0000	1.0000	1.0000	1.0000	1.0000	1.0000

BINOMIAL   15

x	.0500	.1000	.2000	.3000	.4000	.5000
0	.4633	.2059	.0352	.0047	.0005	.0000
1	.8290	.5490	.1671	.0353	.0052	.0005
2	.9638	.8159	.3980	.1268	.0271	.0037
3	.9945	.9444	.6482	.2969	.0905	.0176
4	.9994	.9873	.8358	.5155	.2173	.0592
5	.9999	.9978	.9389	.7216	.4032	.1509
6	1.0000	.9997	.9819	.8689	.6098	.3036
7	1.0000	1.0000	.9958	.9500	.7869	.5000
8	1.0000	1.0000	.9992	.9848	.9050	.6964
9	1.0000	1.0000	.9999	.9963	.9662	.8491
10	1.0000	1.0000	1.0000	.9993	.9907	.9408
11	1.0000	1.0000	1.0000	.9999	.9981	.9824
12	1.0000	1.0000	1.0000	1.0000	.9997	.9963
13	1.0000	1.0000	1.0000	1.0000	1.0000	.9995
14	1.0000	1.0000	1.0000	1.0000	1.0000	1.0000
15	1.0000	1.0000	1.0000	1.0000	1.0000	1.0000

BINOMIAL   15

x	.5000	.6000	.7000	.8000	.9000	.9500
0	.0000	.0000	.0000	.0000	.0000	.0000
1	.0005	.0000	.0000	.0000	.0000	.0000
2	.0037	.0003	.0000	.0000	.0000	.0000
3	.0176	.0019	.0001	.0000	.0000	.0000
4	.0592	.0093	.0007	.0000	.0000	.0000
5	.1509	.0338	.0037	.0001	.0000	.0000
6	.3036	.0950	.0152	.0008	.0000	.0000
7	.5000	.2131	.0500	.0042	.0000	.0000
8	.6964	.3902	.1311	.0181	.0003	.0000
9	.8491	.5968	.2784	.0611	.0022	.0001
10	.9408	.7827	.4845	.1642	.0127	.0006
11	.9824	.9095	.7031	.3518	.0556	.0055
12	.9963	.9729	.8732	.6020	.1841	.0362
13	.9995	.9948	.9647	.8329	.4510	.1710
14	1.0000	.9995	.9953	.9648	.7941	.5367
15	1.0000	1.0000	1.0000	1.0000	1.0000	1.0000

BINOMIAL   20

x	.0500	.1000	.2000	.3000	.4000	.5000
0	.3585	.1216	.0115	.0008	.0000	.0000
1	.7358	.3917	.0692	.0076	.0005	.0000
2	.9245	.6769	.2061	.0355	.0036	.0002
3	.9841	.8670	.4114	.1071	.0160	.0013
4	.9974	.9568	.6296	.2375	.0510	.0059
5	.9997	.9887	.8042	.4164	.1256	.0207

6	1.0000	.9976	.9133	.6080	.2500	.0577
7	1.0000	.9996	.9679	.7723	.4159	.1316
8	1.0000	.9999	.9900	.8867	.5956	.2517
9	1.0000	1.0000	.9974	.9520	.7553	.4119
10	1.0000	1.0000	.9994	.9829	.8725	.5881
11	1.0000	1.0000	.9999	.9949	.9435	.7483
12	1.0000	1.0000	1.0000	.9987	.9790	.8684
13	1.0000	1.0000	1.0000	.9997	.9935	.9423
14	1.0000	1.0000	1.0000	1.0000	.9984	.9793
15	1.0000	1.0000	1.0000	1.0000	.9997	.9941
16	1.0000	1.0000	1.0000	1.0000	1.0000	.9987
17	1.0000	1.0000	1.0000	1.0000	1.0000	.9998
18	1.0000	1.0000	1.0000	1.0000	1.0000	1.0000
19	1.0000	1.0000	1.0000	1.0000	1.0000	1.0000

BINOMIAL   20

x	.5000	.6000	.7000	.8000	.9000	.9500
0	.0000	.0000	.0000	.0000	.0000	.0000
1	.0000	.0000	.0000	.0000	.0000	.0000
2	.0002	.0000	.0000	.0000	.0000	.0000
3	.0013	.0000	.0000	.0000	.0000	.0000
4	.0059	.0003	.0000	.0000	.0000	.0000
5	.0207	.0016	.0000	.0000	.0000	.0000
6	.0577	.0065	.0003	.0000	.0000	.0000
7	.1316	.0210	.0013	.0000	.0000	.0000
8	.2517	.0565	.0051	.0001	.0000	.0000
9	.4119	.1275	.0171	.0006	.0000	.0000
10	.5881	.2447	.0480	.0026	.0000	.0000
11	.7483	.4044	.1133	.0100	.0001	.0000
12	.8684	.5841	.2277	.0321	.0004	.0000
13	.9423	.7500	.3920	.0867	.0024	.0000
14	.9793	.8744	.5836	.1958	.0113	.0003
15	.9941	.9490	.7625	.3704	.0432	.0026
16	.9987	.9840	.8929	.5886	.1330	.0159
17	.9998	.9964	.9645	.7939	.3231	.0755
18	1.0000	.9995	.9924	.9308	.6083	.2642
19	1.0000	1.0000	.9992	.9885	.8784	.6415
20	1.0000	1.0000	1.0000	1.0000	1.0000	1.0000

BINOMIAL   25

x	.0500	.1000	.2000	.3000	.4000	.5000
0	.2774	.0718	.0038	.0001	.0000	.0000
1	.6424	.2712	.0274	.0016	.0001	.0000
2	.8729	.5371	.0982	.0090	.0004	.0000
3	.9659	.7636	.2340	.0332	.0024	.0001
4	.9928	.9020	.4207	.0905	.0095	.0005

5	.9988	.9666	.6167	.1935	.0294	.0020
6	.9998	.9905	.7800	.3407	.0736	.0073
7	1.0000	.9977	.8909	.5118	.1536	.0216
8	1.0000	.9995	.9532	.6769	.2735	.0539
9	1.0000	.9999	.9827	.8106	.4246	.1148
10	1.0000	1.0000	.9944	.9022	.5858	.2122
11	1.0000	1.0000	.9985	.9558	.7323	.3450
12	1.0000	1.0000	.9996	.9825	.8462	.5000
13	1.0000	1.0000	.9999	.9940	.9222	.6550
14	1.0000	1.0000	1.0000	.9982	.9656	.7878
15	1.0000	1.0000	1.0000	.9995	.9868	.8852
16	1.0000	1.0000	1.0000	.9999	.9957	.9461
17	1.0000	1.0000	1.0000	1.0000	.9988	.9784
18	1.0000	1.0000	1.0000	1.0000	.9997	.9927
19	1.0000	1.0000	1.0000	1.0000	.9999	.9980
20	1.0000	1.0000	1.0000	1.0000	1.0000	.9995
21	1.0000	1.0000	1.0000	1.0000	1.0000	.9999
22	1.0000	1.0000	1.0000	1.0000	1.0000	1.0000
23	1.0000	1.0000	1.0000	1.0000	1.0000	1.0000
24	1.0000	1.0000	1.0000	1.0000	1.0000	1.0000
25	1.0000	1.0000	1.0000	1.0000	1.0000	1.0000

BINOMIAL   25

x	.5000	.6000	.7000	.8000	.9000	.9500
0	.0000	.0000	.0000	.0000	.0000	.0000
1	.0000	.0000	.0000	.0000	.0000	.0000
2	.0000	.0000	.0000	.0000	.0000	.0000
3	.0001	.0000	.0000	.0000	.0000	.0000
4	.0005	.0000	.0000	.0000	.0000	.0000
5	.0020	.0001	.0000	.0000	.0000	.0000
6	.0073	.0003	.0000	.0000	.0000	.0000
7	.0216	.0012	.0000	.0000	.0000	.0000
8	.0539	.0043	.0001	.0000	.0000	.0000
9	.1148	.0132	.0005	.0000	.0000	.0000
10	.2122	.0344	.0018	.0000	.0000	.0000
11	.3450	.0778	.0060	.0001	.0000	.0000
12	.5000	.1538	.0175	.0004	.0000	.0000
13	.6550	.2677	.0442	.0015	.0000	.0000
14	.7878	.4142	.0978	.0056	.0000	.0000
15	.8852	.5754	.1894	.0173	.0001	.0000
16	.9461	.7265	.3231	.0468	.0005	.0000
17	.9784	.8464	.4882	.1091	.0023	.0000
18	.9927	.9264	.6593	.2200	.0095	.0002
19	.9980	.9706	.8065	.3833	.0334	.0012
20	.9995	.9905	.9095	.5793	.0980	.0072
21	.9999	.9976	.9668	.7660	.2364	.0341

```
22 1.0000  .9996   .9910   .9018   .4629   .1271
23 1.0000  .9999   .9984   .9726   .7288   .3576
24 1.0000 1.0000   .9999   .9962   .9282   .7226
25 1.0000 1.0000  1.0000  1.0000  1.0000  1.0000
```

BINOMIAL   30

x	.0500	.1000	.2000	.3000	.4000	.5000
0	.2146	.0424	.0012	.0000	.0000	.0000
1	.5535	.1837	.0105	.0003	.0000	.0000
2	.8122	.4114	.0442	.0021	.0000	.0000
3	.9392	.6474	.1227	.0093	.0003	.0000
4	.9844	.8245	.2552	.0302	.0015	.0000
5	.9967	.9268	.4275	.0766	.0057	.0002
6	.9994	.9742	.6070	.1595	.0172	.0007
7	.9999	.9922	.7608	.2814	.0435	.0026
8	1.0000	.9980	.8713	.4315	.0940	.0081
9	1.0000	.9995	.9389	.5888	.1763	.0214
10	1.0000	.9999	.9744	.7304	.2915	.0494
11	1.0000	1.0000	.9905	.8407	.4311	.1002
12	1.0000	1.0000	.9969	.9155	.5785	.1808
13	1.0000	1.0000	.9991	.9599	.7145	.2923
14	1.0000	1.0000	.9998	.9831	.8246	.4278
15	1.0000	1.0000	.9999	.9936	.9029	.5722
16	1.0000	1.0000	1.0000	.9979	.9519	.7077
17	1.0000	1.0000	1.0000	.9994	.9788	.8192
18	1.0000	1.0000	1.0000	.9998	.9917	.8998
19	1.0000	1.0000	1.0000	1.0000	.9971	.9506
20	1.0000	1.0000	1.0000	1.0000	.9991	.9786
21	1.0000	1.0000	1.0000	1.0000	.9998	.9919
22	1.0000	1.0000	1.0000	1.0000	1.0000	.9974
23	1.0000	1.0000	1.0000	1.0000	1.0000	.9993
24	1.0000	1.0000	1.0000	1.0000	1.0000	.9998
25	1.0000	1.0000	1.0000	1.0000	1.0000	1.0000
26	1.0000	1.0000	1.0000	1.0000	1.0000	1.0000
27	1.0000	1.0000	1.0000	1.0000	1.0000	1.0000
28	1.0000	1.0000	1.0000	1.0000	1.0000	1.0000
29	1.0000	1.0000	1.0000	1.0000	1.0000	1.0000
30	1.0000	1.0000	1.0000	1.0000	1.0000	1.0000

BINOMIAL   30

x	.5000	.6000	.7000	.8000	.9000	.9500
0	.0000	.0000	.0000	.0000	.0000	.0000
1	.0000	.0000	.0000	.0000	.0000	.0000
2	.0000	.0000	.0000	.0000	.0000	.0000
3	.0000	.0000	.0000	.0000	.0000	.0000
4	.0000	.0000	.0000	.0000	.0000	.0000

5	.0002	.0000	.0000	.0000	.0000	.0000
6	.0007	.0000	.0000	.0000	.0000	.0000
7	.0026	.0000	.0000	.0000	.0000	.0000
8	.0081	.0002	.0000	.0000	.0000	.0000
9	.0214	.0009	.0000	.0000	.0000	.0000
10	.0494	.0029	.0000	.0000	.0000	.0000
11	.1002	.0083	.0002	.0000	.0000	.0000
12	.1808	.0212	.0006	.0000	.0000	.0000
13	.2923	.0481	.0021	.0000	.0000	.0000
14	.4278	.0971	.0064	.0001	.0000	.0000
15	.5722	.1754	.0169	.0002	.0000	.0000
16	.7077	.2855	.0401	.0009	.0000	.0000
17	.8192	.4215	.0845	.0031	.0000	.0000
18	.8998	.5689	.1593	.0095	.0000	.0000
19	.9506	.7085	.2696	.0256	.0001	.0000
20	.9786	.8237	.4112	.0611	.0005	.0000
21	.9919	.9060	.5685	.1287	.0020	.0000
22	.9974	.9565	.7186	.2392	.0078	.0001
23	.9993	.9828	.8405	.3930	.0258	.0006
24	.9998	.9943	.9234	.5725	.0732	.0033
25	1.0000	.9985	.9698	.7448	.1755	.0156
26	1.0000	.9997	.9907	.8773	.3526	.0608
27	1.0000	1.0000	.9979	.9558	.5886	.1878
28	1.0000	1.0000	.9997	.9895	.8163	.4465
29	1.0000	1.0000	1.0000	.9988	.9576	.7854
30	1.0000	1.0000	1.0000	1.0000	1.0000	1.0000

BINOMIAL 40

x	.0500	.1000	.2000	.3000	.4000	.5000
0	.1285	.0148	.0001	.0000	.0000	.0000
1	.3991	.0805	.0015	.0000	.0000	.0000
2	.6767	.2228	.0079	.0001	.0000	.0000
3	.8619	.4231	.0285	.0006	.0000	.0000
4	.9520	.6290	.0759	.0026	.0000	.0000
5	.9861	.7937	.1613	.0086	.0001	.0000
6	.9966	.9005	.2859	.0238	.0006	.0000
7	.9993	.9581	.4371	.0553	.0021	.0000
8	.9999	.9845	.5931	.1110	.0061	.0001
9	1.0000	.9949	.7318	.1959	.0156	.0003
10	1.0000	.9985	.8392	.3087	.0352	.0011
11	1.0000	.9996	.9125	.4406	.0709	.0032
12	1.0000	.9999	.9568	.5772	.1285	.0083
13	1.0000	1.0000	.9806	.7032	.2112	.0192
14	1.0000	1.0000	.9921	.8074	.3174	.0403
15	1.0000	1.0000	.9971	.8849	.4402	.0769

16	1.0000	1.0000	.9990	.9367	.5681	.1341
17	1.0000	1.0000	.9997	.9680	.6885	.2148
18	1.0000	1.0000	.9999	.9852	.7911	.3179
19	1.0000	1.0000	1.0000	.9937	.8702	.4373
20	1.0000	1.0000	1.0000	.9976	.9256	.5627
21	1.0000	1.0000	1.0000	.9991	.9608	.6821
22	1.0000	1.0000	1.0000	.9997	.9811	.7852
23	1.0000	1.0000	1.0000	.9999	.9917	.8659
24	1.0000	1.0000	1.0000	1.0000	.9966	.9231
25	1.0000	1.0000	1.0000	1.0000	.9988	.9597
26	1.0000	1.0000	1.0000	1.0000	.9996	.9808
27	1.0000	1.0000	1.0000	1.0000	.9999	.9917
28	1.0000	1.0000	1.0000	1.0000	1.0000	.9968
29	1.0000	1.0000	1.0000	1.0000	1.0000	.9989
30	1.0000	1.0000	1.0000	1.0000	1.0000	.9997
31	1.0000	1.0000	1.0000	1.0000	1.0000	.9999
32	1.0000	1.0000	1.0000	1.0000	1.0000	1.0000
33	1.0000	1.0000	1.0000	1.0000	1.0000	1.0000
34	1.0000	1.0000	1.0000	1.0000	1.0000	1.0000
35	1.0000	1.0000	1.0000	1.0000	1.0000	1.0000
36	1.0000	1.0000	1.0000	1.0000	1.0000	1.0000
37	1.0000	1.0000	1.0000	1.0000	1.0000	1.0000
38	1.0000	1.0000	1.0000	1.0000	1.0000	1.0000
39	1.0000	1.0000	1.0000	1.0000	1.0000	1.0000

BINOMIAL   40

x	.5000	.6000	.7000	.8000	.9000	.9500
0	.0000	.0000	.0000	.0000	.0000	.0000
1	.0000	.0000	.0000	.0000	.0000	.0000
2	.0000	.0000	.0000	.0000	.0000	.0000
3	.0000	.0000	.0000	.0000	.0000	.0000
4	.0000	.0000	.0000	.0000	.0000	.0000
5	.0000	.0000	.0000	.0000	.0000	.0000
6	.0000	.0000	.0000	.0000	.0000	.0000
7	.0000	.0000	.0000	.0000	.0000	.0000
8	.0001	.0000	.0000	.0000	.0000	.0000
9	.0003	.0000	.0000	.0000	.0000	.0000
10	.0011	.0000	.0000	.0000	.0000	.0000
11	.0032	.0000	.0000	.0000	.0000	.0000
12	.0083	.0001	.0000	.0000	.0000	.0000
13	.0192	.0004	.0000	.0000	.0000	.0000
14	.0403	.0012	.0000	.0000	.0000	.0000
15	.0769	.0034	.0000	.0000	.0000	.0000
16	.1341	.0083	.0001	.0000	.0000	.0000
17	.2148	.0189	.0003	.0000	.0000	.0000
18	.3179	.0392	.0009	.0000	.0000	.0000

19	.4373	.0744	.0024	.0000	.0000	.0000
20	.5627	.1298	.0063	.0000	.0000	.0000
21	.6821	.2089	.0148	.0001	.0000	.0000
22	.7852	.3115	.0320	.0003	.0000	.0000
23	.8659	.4319	.0633	.0010	.0000	.0000
24	.9231	.5598	.1151	.0029	.0000	.0000
25	.9597	.6826	.1926	.0079	.0000	.0000
26	.9808	.7888	.2968	.0194	.0000	.0000
27	.9917	.8715	.4228	.0432	.0001	.0000
28	.9968	.9291	.5594	.0875	.0004	.0000
29	.9989	.9648	.6913	.1608	.0015	.0000
30	.9997	.9844	.8041	.2682	.0051	.0000
31	.9999	.9939	.8890	.4069	.0155	.0001
32	1.0000	.9979	.9447	.5629	.0419	.0007
33	1.0000	.9994	.9762	.7141	.0995	.0034
34	1.0000	.9999	.9914	.8387	.2063	.0139
35	1.0000	1.0000	.9974	.9241	.3710	.0480
36	1.0000	1.0000	.9994	.9715	.5769	.1381
37	1.0000	1.0000	.9999	.9921	.7772	.3233
38	1.0000	1.0000	1.0000	.9985	.9195	.6009
39	1.0000	1.0000	1.0000	.9999	.9852	.8715
40	1.0000	1.0000	1.0000	1.0000	1.0000	1.0000

LESSON 20. THE POISSON DISTRIBUTION

In lesson 17, we saw the use of the binomial as an approximation to the hypergeometric for large populations. When n is large in the binomial, we have similar problems of computation:

what is the value of $\begin{bmatrix} 1000 \\ 200 \end{bmatrix}(.01)^{200}(.99)^{1800}$?

The following mathematical theorem yields one approximation. Another will be seen in part III.

Theorem: *For each n = 1(1)∞, let p_n be a real number. Suppose that np_n converges to a finite λ > 0 as n tends to ∞. Then for each fixed non-negative integer x,*

$$\begin{bmatrix} n \\ x \end{bmatrix} p_n^{\ x} (1 - p_n)^{n - x} \text{ converges to } e^{-\lambda} \lambda^x/x! \ .$$

Proof: Rewrite $\begin{bmatrix} n \\ x \end{bmatrix} p_n^{\ x} (1 - p_n)^{n - x}$ as the product

$$\frac{1}{x!} \left[\frac{n}{n} \ \frac{(n-1)}{n} \ \frac{(n-2)}{n} \cdots \frac{(n-x+1)}{n} \right] (np_n)^x (1-p_n)^n (1-p_n)^{-x} \ .$$

Each of the factors in the large brackets converges to 1 and there are a finite number, namely x , of them. Since np_n converges to the finite number λ , p_n itself must converge to 0 ; hence, $(1 - p_n)^{-x}$ converges to 1. The finite product $(np_n)^x$ converges to λ^x. In an exercise, you to show that $(1 - p_n)^n$ converges to $e^{-\lambda}$. Putting all these together yields the conclusion.

Formally, the approximation is the limit of a ratio as in the earlier work. Informally, when θ is "small" and n is "large",

$$\begin{bmatrix} n \\ x \end{bmatrix} \theta^x (1 - \theta)^{n-x} \approx e^{-n\theta}(n\theta)^x/x! \ .$$

The approximation is "good" when θ < .1 and n ≥ 50; of course, since X and n − X have complementary distributions, we also have, for $\theta > .9$,

$$\begin{bmatrix} n \\ n-x \end{bmatrix} (1 - \theta)^{n-x}\theta^x \approx e^{-n(1-\theta)}(n(1-\theta))^{n-x}/(n-x)!.$$

At the end of this lesson, we have included two tables for comparisons using

$$\theta = .001 \ .002 \ .005 \ .01 \ .02 \ .03 \ .04 \ .05 \ .1 \ .2 \ .3 \ .$$

Exercise 1: Prove that $\lim_{n\to\infty} np_n = \lambda > 0$ implies

$$\lim_{n\to\infty} (1 - p_n)^n = e^{-\lambda} \ .$$

Hint: there is a Maclaurin polynomial for $\log(1 - x)$ lying around in this.

This numerical trick introduces another complication. If the approximation were really accurate, we would have

$$\sum_{x=0}^{n} e^{-n\theta} (n\theta)^x/x! \approx 1 \ .$$

But $\displaystyle\sum_{x=0}^{n} (n\theta)^x/x!$ is only part of the Maclaurin series for $e^{n\theta}$.

More precisely, $e^{-\lambda}\lambda^x/x!$ is positive and $\displaystyle\sum_{x=0}^{\infty} e^{-\lambda}\lambda^x/x! = 1$.

This looks like a probability density function except for the fact that our rules apply to finite sets only. The solution is simple —we change the rules—without losing anything that we already have.

Definition: *Let S be a countable set with elements $s_1, s_2, s_3, \cdots$. Let $\mathscr{F}$ be the collection of all subsets of S (the power set including ϕ and S). Let P be a set function on $\mathscr{F}$ such that:*

1) for all $A \in \mathscr{F}, P(\phi) = 0 \le P(A) \le 1 = P(S)$;

2) for $A_1, A_2, A_3, \cdots \in \mathscr{F},$ with $A_i \cap A_j = \phi$ when $i \ne j$,

$$P(\cup_{i=1}^{\infty} A_i) = \Sigma_{i=1}^{\infty} P(A_i) \ .$$

Then [S, $\mathscr{F}$,P] is a probability space.

This does include the earlier results when S is finite. For then, $\mathcal{F}$ also has a finite number of elements so that $\cup_{i=1}^{\infty} A_i =$ $\lim_{n \to \infty} \cup_{i=1}^{n} A_i$ must reduce to $\cup_{i=1}^{M} A_i$ for some (large?) M and the duplicates, $A_i = A_k$ for $i \neq k$, are discarded.. It follows that $\Sigma_{i=1}^{\infty} P(A_i) = \Sigma_{i=1}^{M} P(A_i)$.

Exercise 2: For the probability space $[S, \mathcal{F}, P]$ in the definition above, show that $P(s_i) \geq 0$ and $\Sigma_{i=1}^{\infty} P(s_i) = 1$.

When the elements of S are real numbers, we consider the identity function $X(s) = s$ as a discrete random variable. This is what our limit theorem really suggests.

Definition: *Let the random variable X have range $X(S) = \{0,1,2,3, \cdots \}$. Let $\mathcal{F}$ be the collection of all subsets of this range. Let $P(X = x \mid \lambda) = e^{-\lambda} \lambda^x / x!$ for some $\lambda > 0$. Then X is called a Poisson random variable.*

There is a short table of the Poisson CDF $P(X \leq x \mid \lambda)$ at the end of this lesson.

Exercise 3: Let X be a Poisson random variable. Complete the evaluations as in:
$$P(X \leq 2 \mid \lambda = .1) = .9998$$
$$P(X > 5 \mid \lambda = 1) = 1 - P(X \leq 5 \mid \lambda = 1) = 1 - .9994.$$
a) $P(X \leq 1 \mid \lambda = .05)$
b) $P(X \leq 2 \mid \lambda = .005)$
c) $P(X \geq 10 \mid \lambda = 1)$
d) $P(X = 20 \mid \lambda = 5)$
e) $P(3 \leq X \leq 10 \mid \lambda = 20)$
f) $P(3 < X < 20 \mid \lambda = 10)$.

As you may have anticipated, we also need to extend the definition of expectation.

Definition: *For the discrete type RV X with countable range*

$\mathscr{X} = \{x_1, x_2, x_3, \cdots\}$, *let $\mathscr{F}$ be the power set and let P be the probability function:* $P(X = x_i) = p_X(x_i)$. *Let h be a real valued function on $\mathscr{X}$; then the expectation or expected value of h(X) is*

$$E[h(X)] = \sum_{i=1}^{\infty} h(x_i)p_X(x_i) \quad \text{whenever this series is absolutely}$$

convergent.

Example: For the Poisson RV, we can let $x_1 = 0$, $x_2 = 1$, $x_3 = 2$,

etc. Then, $E[X] = \sum_{i=1}^{\infty} x_i p_X(x_i) = \sum_{k=0}^{\infty} ke^{-\lambda}\lambda^k/k!$. This

reduces to $\sum_{k=1}^{\infty} ke^{-\lambda}\lambda^k/k!$ and then to

$$\sum_{k=1}^{\infty} e^{-\lambda}\lambda^k/(k-1)! \quad \text{as } k/k! = (k-1)! \text{ for } k \geq 1 .$$

Let $k - 1 = w$. Then, $\sum_{k=1}^{\infty} e^{-\lambda}\lambda^k/(k-1)!$

$$= \sum_{w=0}^{\infty} e^{-\lambda}\lambda^{w+1}/w! = \lambda \sum_{w=0}^{\infty} e^{-\lambda}\lambda^w/w! .$$

But this last sum is 1 so that $E[X] = \lambda$. You may also have anticipated this for we started out with $np_n \approx \lambda \approx n\theta$, the mean of the binomial RV.

In the next exercise, you will use the following theorems on series. If $\sum_{i=1}^{\infty} a_i$ converges to the finite number A

and $\sum_{i=1}^{\infty} b_i$ converges to the finite number B,

then $\sum_{i=1}^{\infty} (a_i + b_i) = A + B.$

If $\sum a_i^2$ and $\sum b_i^2$ are finite, then

$$\left[\sum a_i b_i \right]^2 \leq \sum a_i^2 \cdot \sum b_i^2 .$$

Exercise 4:   Let X be a discrete type RV as in the above definition.

a)   Show that $E[X^2] < \infty$ implies $E[\,|X|\,]$ is finite whence also $E[X] < \infty$. Hint: first show that

$$\left[\sum |x_i| p_X(x_i) \right]^2 \leq \sum x_i^2 p_X(x_i) \cdot \sum p_X(x_i) .$$

b)   Then show that $E[X^2] = E[X(X-1)] + E[X]$.

c)   Find the variance of the Poisson RV (distribution).

The next example shows the necessity of the condition of absolute convergence.

Exercise 5:   Read "An elementary proof of $\sum_{n=1}^{\infty} 1/n^2 = \pi^2/6$" by Choe, 1987.

Example: Now we can discuss a random variable X with range the positive integers and probability $P(X = k) = 6/\pi^2 k^2$. Let $h(x) = (-1)^x x$. Then,

$$\sum_{k=1}^{\infty} (-1)^k k(6/\pi^2 k^2) = (6/\pi^2) \sum_{k=1}^{\infty} (-1)^k /k . \qquad (*)$$

Now $S = \sum_{k=1}^{\infty} (-1)^k /k$ is the convergent harmonic series so that the sum in (*) is finite. But $\sum_{k=1}^{\infty} 1/k$ is the divergent harmonic series so that the sum in (*) is not absolutely convergent. This leads to inconsistencies as we now show. Write out a few terms of S:

$$S = -1 + 1/2 - 1/3 + 1/4 - 1/5 + 1/6 - 1/7 + 1/8 - \cdots \qquad (**)$$

Multiplying a convergent series by a constant does not affect the condition of convergence:

$$(1/2)S = -1/2 + 1/4 - 1/6 + 1/8 - 1/10 + 1/12 - 1/14 + \cdots .$$

Inserting or deleting zeroes in a convergent series does not affect the condition of convergence:

$$(1/2)S = 0 - 1/2 + 0 + 1/4 + 0 - 1/6 + 0 + 1/8 + 0 - 1/10$$
$$+ 0 + 1/12 + 0 - 1/14 + 0 \cdots$$

(***)

As noted above, convergent series may be added term by term so that when we "add" (**) and (***), we get

$$(3/2)S = -1 + 0 - 1/3 + 1/2 - 1/5 + 0 - 1/7 + \cdots \quad (****)$$

But when we now delete the zeroes on the right hand side of (****) we see a rearrangement of our original S; we seem to have (3/2)S = S. The lesson is: non–absolutely convergent series can not be rearranged without destroying the value of the sum. (Rearrangements of absolutely convergent series preserve the value of the sum.)

Just for fun, we list a few partial sums:

number of terms	100	500	1000	5000
(**)	–.68817	–.69215	–.69265	–.69305
(***)	–.34162	–.34558	–.34607	–.34647
(****)	-1.03971	-1.03807	-1.04051	-1.04003

In later discussion of the theory, absolute convergence will follow from other considerations. But if we allowed non–absolutely convergent series in expectation, the mean of a random variable (the mean of its distribution) would depend on the order in which the values were labeled or arranged; the definition eliminates this confusing peculiarity.

The following is one of the more popular applications of the Poisson distribution. "Raisins" in cookies, can be replaced by "chocolate chips"; one can also consider telephone calls in a fixed period of time, people arriving for service at a bank, etc.

Example: An advertisement claims that "Now there are more raisins per cookie". (Usually, they don't say "More than last week", or "More than last year" or "More than URS".) When pressed, the manufacturer might say something like "On the average, 10 raisins per cookie." We translate this to: assuming that the number of raisins per cookie is a Poisson random variable with mean λ, test $H_o : \lambda = 10$ against $H_a : \lambda < 10$. Since our hypothesis is one–sided, our critical region will be one–tailed: reject H_o when the observed value of the Poisson RV is "too small".

Now suppose that counting raisins leads to the observed value 8. Then $P(X \le 8 \mid \lambda = 10) = .3328$ so that this is not a "rare" event. We accept H_o. From the distribution function, we see that when $\lambda = 10$, the "rare events" are five or fewer raisins.

Exercise 5:   Continue the discussion of the Poisson RV.
   a)   Devise a test of $H_o : \lambda = 10$ vs $H_a : \lambda > 10$ based on
        one observation; take $\alpha = .05$.

   b)   Formalize a test of $H_o : \lambda = 10$ vs $H_a : \lambda = 20$.
        Include $P(\text{Reject } H_o \mid H_o \text{ true}) = \alpha \le .02$ and
        $P(\text{Do not reject } H_o \mid H_o \text{ false}) = \beta \le .03$.

INE–Exercises:
1.   Suppose that X is a non–negative discrete type RV with finite mean; then $X(s) \ge 0$ for all s in S and $E[X] < \infty$. Suppose that Y is another discrete type RV with $0 \le Y(s) \le X(s)$ for all s in S. Show that $E[Y] < \infty$.

2.   Let X be a discrete type RV such that $E[|X|^k] < \infty$ for some positive constant k. Show that for all $0 \le j \le k$, $E[|X|^j] < \infty$. Hint: first argue that $|x|^j \le |x|^k + 1$ by considering the cases $|x| \le 1$ and $|x| > 1$.

3.   This game is played with a fair coin. Able tosses the coin first; if it lands heads–up, Able wins and the game is over. Otherwise, Baker tosses the coin; if it lands heads–up, he wins and the game is over. Otherwise Charlie tosses, winning on heads–up and stopping the game. Otherwise, A tosses the coin, etc. What is the probability that each wins the game?

4.   Suppose that the game in 3. is played with an unfair coin. What are the probabilities that each of A, B, C wins when $P(H) = \theta \ne .5$? Who has the best position when $\theta < .5$?

5. Show that the gamma function $\Gamma(r) = \int_0^\infty g^{r-1} e^{-g} dg$ is finite for all real numbers $r > 0$.

6. a) Show that for each non–negative integer x,

$$\sum_{k=0}^{x} e^{\lambda} \lambda^k / k! = 1 - \int_0^{\lambda} g^x e^{-g} dg / \Gamma(x+1)$$

$$= 1 - \int_0^{2\lambda} c^{(x+1)-1} e^{-c/2} dc / 2^{x+1} \Gamma(x+1).$$

   b) So what?—Relate this to the Poisson distribution. (It turns out that the non–negative functions of g and c are densities for the g(amma) and c(hisquare) distributions.)

7. A book with N pages contains "on the average " λ misprints per page. Find the probability that at least M pages contain more than K misprints?

8. Find a median and a mode for the Poisson distribution.

Table POISSON λ

This table (generated using STSC–APL) contains (part of) the cumulative distribution functions for Poisson random variables with parameter λ = mean . For values of x greater than those listed, the value of the CDF is 1 . For example,
$$P(X \leq 3 \mid \lambda = .5000) = .9982 .$$

x	.0010	.0050	.0100	.0500	.1000	.5000
0	.9990	.9950	.9900	.9512	.9048	.6065
1	1.0000	1.0000	1.0000	.9988	.9953	.9098
2	1.0000	1.0000	1.0000	1.0000	.9998	.9856
3	1.0000	1.0000	1.0000	1.0000	1.0000	.9982
4	1.0000	1.0000	1.0000	1.0000	1.0000	.9998
5	1.0000	1.0000	1.0000	1.0000	1.0000	1.0000

x	λ					
	1	2	3	5	10	20
0	.3679	.1353	.0498	.0067	.0000	.0000
1	.7358	.4060	.1991	.0404	.0005	.0000
2	.9197	.6767	.4232	.1247	.0028	.0000
3	.9810	.8571	.6472	.2650	.0103	.0000
4	.9963	.9473	.8153	.4405	.0293	.0000
5	.9994	.9834	.9161	.6160	.0671	.0001
6	.9999	.9955	.9665	.7622	.1301	.0003
7	1.0000	.9989	.9881	.8666	.2202	.0008
8	1.0000	.9998	.9962	.9319	.3328	.0021
9	1.0000	1.0000	.9989	.9682	.4579	.0050
10	1.0000	1.0000	.9997	.9863	.5830	.0108
11	1.0000	1.0000	.9999	.9945	.6968	.0214
12	1.0000	1.0000	1.0000	.9980	.7916	.0390
13	1.0000	1.0000	1.0000	.9993	.8645	.0661
14	1.0000	1.0000	1.0000	.9998	.9165	.1049
15	1.0000	1.0000	1.0000	.9999	.9513	.1565
16	1.0000	1.0000	1.0000	1.0000	.9730	.2211
17	1.0000	1.0000	1.0000	1.0000	.9857	.2970
18	1.0000	1.0000	1.0000	1.0000	.9928	.3814
19	1.0000	1.0000	1.0000	1.0000	.9965	.4703
20	1.0000	1.0000	1.0000	1.0000	.9984	.5591
21	1.0000	1.0000	1.0000	1.0000	.9993	.6437
22	1.0000	1.0000	1.0000	1.0000	.9997	.7206
23	1.0000	1.0000	1.0000	1.0000	.9999	.7875
24	1.0000	1.0000	1.0000	1.0000	1.0000	.8432
25	1.0000	1.0000	1.0000	1.0000	1.0000	.8878
26	1.0000	1.0000	1.0000	1.0000	1.0000	.9221
27	1.0000	1.0000	1.0000	1.0000	1.0000	.9475
28	1.0000	1.0000	1.0000	1.0000	1.0000	.9657
29	1.0000	1.0000	1.0000	1.0000	1.0000	.9782
30	1.0000	1.0000	1.0000	1.0000	1.0000	.9865
31	1.0000	1.0000	1.0000	1.0000	1.0000	.9919
32	1.0000	1.0000	1.0000	1.0000	1.0000	.9953
33	1.0000	1.0000	1.0000	1.0000	1.0000	.9973
34	1.0000	1.0000	1.0000	1.0000	1.0000	.9985
35	1.0000	1.0000	1.0000	1.0000	1.0000	.9992
36	1.0000	1.0000	1.0000	1.0000	1.0000	.9996
37	1.0000	1.0000	1.0000	1.0000	1.0000	.9998
38	1.0000	1.0000	1.0000	1.0000	1.0000	.9999
39	1.0000	1.0000	1.0000	1.0000	1.0000	.9999
40	1.0000	1.0000	1.0000	1.0000	1.0000	1.0000

Table BINPOIS n

This table (generated using STSC–APL) contains some values of the BINOMIAL cumulative distribution function and its POISSON approximation, in alternating columns. The first row of each group lists the parameters θ , λ = n·θ .

BINPOIS n = 20

x			θ , n·θ			
	.0010,	*.0200*	*.0020,*	*.0400*	*.0050,*	*.1000*
0	.9802	.9802	.9608	.9608	.9046	.9048
1	.9998	.9998	.9993	.9992	.9955	.9953
2	1.0000	1.0000	1.0000	1.0000	.9999	.9998
3	1.0000	1.0000	1.0000	1.0000	1.0000	1.0000
4	1.0000	1.0000	1.0000	1.0000	1.0000	1.0000
5	1.0000	1.0000	1.0000	1.0000	1.0000	1.0000
x	*.0100,*	*.2000*	*.0200,*	*.4000*	*.0300,*	*.6000*
0	.8179	.8187	.6676	.6703	.5438	.5488
1	.9831	.9825	.9401	.9384	.8802	.8781
2	.9990	.9989	.9929	.9921	.9790	.9769
3	1.0000	.9999	.9994	.9992	.9973	.9966
4	1.0000	1.0000	1.0000	.9999	.9997	.9996
5	1.0000	1.0000	1.0000	1.0000	1.0000	1.0000
6	1.0000	1.0000	1.0000	1.0000	1.0000	1.0000
x	*.0400,*	*.8000*	*.0500,*	*1.0000*	*.1000,*	*2.0000*
0	.4420	.4493	.3585	.3679	.1216	.1353
1	.8103	.8088	.7358	.7358	.3917	.4060
2	.9561	.9526	.9245	.9197	.6769	.6767
3	.9926	.9909	.9841	.9810	.8670	.8571
4	.9990	.9986	.9974	.9963	.9568	.9473
5	.9999	.9998	.9997	.9994	.9887	.9834
6	1.0000	1.0000	1.0000	.9999	.9976	.9955
7	1.0000	1.0000	1.0000	1.0000	.9996	.9989
8	1.0000	1.0000	1.0000	1.0000	.9999	.9998
9	1.0000	1.0000	1.0000	1.0000	1.0000	1.0000
10	1.0000	1.0000	1.0000	1.0000	1.0000	1.0000

BINPOIS n = 50

x	θ , $n\cdot\theta$					
	.0010,	*.0500*	*.0020,*	*.1000*	*.0050,*	*.2500*
0	.9512	.9512	.9047	.9048	.7783	.7788
1	.9988	.9988	.9954	.9953	.9739	.9735
2	1.0000	1.0000	.9999	.9998	.9979	.9978
3	1.0000	1.0000	1.0000	1.0000	.9999	.9999
4	1.0000	1.0000	1.0000	1.0000	1.0000	1.0000
5	1.0000	1.0000	1.0000	1.0000	1.0000	1.0000

x	*.0100,*	*.5000*	*.0200,*	*1.0000*	*.0300,*	*1.5000*
0	.6050	.6065	.3642	.3679	.2181	.2231
1	.9106	.9098	.7358	.7358	.5553	.5578
2	.9862	.9856	.9216	.9197	.8108	.8088
3	.9984	.9982	.9822	.9810	.9372	.9344
4	.9999	.9998	.9968	.9963	.9832	.9814
5	1.0000	1.0000	.9995	.9994	.9963	.9955
6	1.0000	1.0000	.9999	.9999	.9993	.9991
7	1.0000	1.0000	1.0000	1.0000	.9999	.9998
8	1.0000	1.0000	1.0000	1.0000	1.0000	1.0000
9	1.0000	1.0000	1.0000	1.0000	1.0000	1.0000
10	1.0000	1.0000	1.0000	1.0000	1.0000	1.0000

x	*.0400,*	*2.0000*	*.0500,*	*2.5000*	*.1000,*	*5.0000*
0	.1299	.1353	.0769	.0821	.0052	.0067
1	.4005	.4060	.2794	.2873	.0338	.0404
2	.6767	.6767	.5405	.5438	.1117	.1247
3	.8609	.8571	.7604	.7576	.2503	.2650
4	.9510	.9473	.8964	.8912	.4312	.4405
5	.9856	.9834	.9622	.9580	.6161	.6160
6	.9964	.9955	.9882	.9858	.7702	.7622
7	.9992	.9989	.9968	.9958	.8779	.8666
8	.9999	.9998	.9992	.9989	.9421	.9319
9	1.0000	1.0000	.9998	.9997	.9755	.9682
10	1.0000	1.0000	1.0000	.9999	.9906	.9863

LESSON *21. THE NEGATIVE BINOMIAL DISTRIBUTION

We begin with a new problem: a blood bank continues to purchase blood from individuals until it collects 6 pints of type B negative needed for an upcoming operation. How many pints should they be prepared to buy?

In building a model for any problem, we have to make assumptions. Here we have:

1) the individuals (blood types) are independent responses,
2) with the same probability θ = P(B negative) = P(S);
3) money, storage facilities, etc. are such that the procedure can be carried out (more or less).

We can now talk about Bernoulli trials but note that the sample size is not fixed; it is, in fact, the random variable of interest: Y is the number of the trial on which the sixth success S occurrs. First, note that Y must be at least 6 and since we are assuming repeated trials,

$$P(Y = 6) = P(SSSSSS) = P(S)P(S)P(S)P(S)P(S)P(S) = \theta^6.$$

Now Y = 7 means that the seventh trial is S and there are exactly 5 successes among the first 6 trials;

$$P(Y = 7) = P(5 \text{ successes out of } 6 \text{ trials}) \cdot P(S)$$

$$= \left[\begin{bmatrix} 6 \\ 5 \end{bmatrix} (\theta)^5 (1 - \theta)^1 \right] (\theta).$$

$$P(Y = 8) = P(5 \text{ successes in } 7 \text{ trials}) \cdot P(S)$$

$$= \left[\begin{bmatrix} 7 \\ 5 \end{bmatrix} (\theta)^5 (1 - \theta)^2 \right] (\theta).$$

In the mathematical world, this could go on forever so that the range of Y is infinite.

$$P(Y = y) = \left[\begin{bmatrix} y-1 \\ 5 \end{bmatrix} \theta^5 (1 - \theta)^{y-6} \right] \theta, \ y = 6(1)\infty.$$

Exercise 1: a) Evaluate P(Y = y) for y = 6, 7, 8, 9, 10, θ = .4 .
b) Do the same when θ = .5 .

The general form is in the

Definition: *Let r be a fixed positive integer. For IID trials with*
P(S) = θ, let Y be the number of the trial on which the rth
success occurs. Then Y has the negative binomial distribution
with

$$P(Y = y \mid r, \theta) = \left[\begin{array}{c} y\text{-}1 \\ r\text{-}1 \end{array} \right] \theta^r (1 - \theta)^{y\text{-}r}, \, y = r(1)\infty \, .$$

Sometimes, the number W of failures among the Y trials is
called the negative binomial. For W = Y − r,

$$P(W = w \mid r, \theta) = P(Y - r = w \mid r, \theta) = P(Y = w + r \mid r, \theta)$$

$$= \left[\begin{array}{c} w+r-1 \\ r-1 \end{array} \right] \theta^r (1 - \theta)^w, \, w = 0(1)\infty \, .$$

This leads to the reason for the name. We now modify the
meaning of $\left[\begin{array}{c} a \\ k \end{array} \right]$ and in so doing we are limited no longer to
combinations.

Definition: *Let a be any real number; let k be a non-negative*
integer. Then, $\left[\begin{array}{c} a \\ 0 \end{array} \right] = 1$ *and for k ≥ 1,*

$$\left[\begin{array}{c} a \\ k \end{array} \right] = \frac{a(a\text{-}1)(a\text{-}2) \cdots (a\text{-}k+1)}{k!} \, .$$

Exercise 2: a)Evaluate

$$\left[\begin{array}{c} 1.5 \\ 3 \end{array} \right], \left[\begin{array}{c} -3 \\ 4 \end{array} \right], \left[\begin{array}{c} -1.7 \\ 3 \end{array} \right], \left[\begin{array}{c} 4 \\ 5 \end{array} \right], \left[\begin{array}{c} -1/2 \\ 7 \end{array} \right] \, .$$

b)   Show that for positive integers a, k,

$$\left[\begin{array}{c} -a \\ k \end{array} \right] = (-1)^k \left[\begin{array}{c} a + k - 1 \\ k \end{array} \right] \, .$$

(You may use the formula and omit the induction.)

For the density of W above,

$$\sum_{w=0}^{\infty} \left[\begin{array}{c} w + r - 1 \\ r - 1 \end{array} \right] \theta^r (1 - \theta)^w = 1 \text{ or}$$

$$\sum_{w=0}^{\infty} \left[\begin{array}{c} w + r - 1 \\ r - 1 \end{array} \right] (1 - \theta)^w = \theta^{-r} \, .$$

So by the last exercise, $\displaystyle\sum_{w=0}^{\infty} \left[\begin{array}{c} -r \\ w \end{array}\right] (-1)^W (1 - \theta)^W = \theta^{-r}$. The

left–hand side of this last equality is a special case of the binomial series:

$$(1 + x)^{-r} = \sum_{w=0}^{\infty} \left[\begin{array}{c} -r \\ w \end{array}\right] x^W \text{ converges for } |x| < 1 .$$

The original question (about the number of pints to buy) might be interpreted as finding $E[Y] = E[W] + r$. Now

$$(1 - \theta)^W = (\theta - 1)^W (-1)^W$$

so that $\displaystyle E[W] = \sum_{w=0}^{\infty} w \left[\begin{array}{c} -r \\ w \end{array}\right] (-1)^W \theta^r (1 - \theta)^W$

$$= \sum_{w=0}^{\infty} w \left[\begin{array}{c} -r \\ w \end{array}\right] (\theta - 1)^W \theta^r$$

$$= \theta^r (\theta - 1) \sum_{w=0}^{\infty} \left[\begin{array}{c} -r \\ w \end{array}\right] w (\theta - 1)^{W-1}$$

$$= \theta^r (\theta - 1) \sum_{w=0}^{\infty} \left[\begin{array}{c} -r \\ w \end{array}\right] \frac{d}{d\theta} (\theta - 1)^W$$

$$= \theta^r (\theta - 1) \frac{d}{d\theta} \sum_{w=0}^{\infty} \left[\begin{array}{c} -r \\ w \end{array}\right] (\theta - 1)^W$$

$$= \theta^r (\theta - 1) \frac{d}{d\theta} (\theta^{-r}) .$$

Here we have employed the following from analysis.

Theorem: *a) The power series* $\displaystyle\sum_{n=0}^{\infty} a_n x^n$ *converges absolutely uniformly for* $|x| < 1/\lim_{n\to\infty} |a_{n+1}/a_n| = \rho$. *(ρ is called the radius of the circle of convergence even when x is real.)*
b)   A power series may be differentiated or integrated term by term within its circle of convergence.

Exercise 3:   Show that $\sum_{w=0}^{\infty} \left[\begin{array}{c} -r \\ w \end{array} \right] (\theta - 1)^W$ has $\rho = 1$; that is,

this series converges for $|\theta - 1| < 1$.

Continuing with $\theta^r(\theta - 1)\dfrac{d}{d\theta}(\theta^{-r})$ we have

$$E[W] = \theta^r(\theta - 1)(-r)(\theta)^{-r-1} = r(1 - \theta)/\theta .$$

Hence, $E[Y] = E[W] + r = r/\theta$.

If in the original problem $\theta = P(B \text{ negative}) = .001$, then the bank can expect to buy $6/.001 = 6000$ pints; rare blood is expensive!

Exercise 4:   a) Find $E[W(W-1)]$ by manipulating the appropriate series to involve $d^2/d\theta^2$.
b)   Find the variance of the distributions of W and Y.

Another interpretation of "being prepared" is to find c such that $P(Y \le c \mid r, \theta)$ is large. Say that the bank wants to be "almost certain" of getting 6 "B negative" before buying c pints of blood. The solution of

$$\sum_{y=r}^{c} \left[\begin{array}{c} y-1 \\ r-1 \end{array} \right] \theta^r(1 - \theta)^{y-r} \ge 1 - \alpha \ \ (\text{say } .975)$$

is accomplished by trial but not here.
An interesting special case of this distribution occurs when $r = 1$.

Definition: *The random variable G is called geometric if*
$$P(G = g \mid \theta) = \theta(1 - \theta)^{g-1}, \ g = 1(1)\infty .$$

Exercise 5:   Find the mean and variance of the geometric distribution by using results for Y.

Here is a simple application. Inspection of items along an assembly line continues until a "defective" item is found; this suggests adjusting the process, shutting–down the line, or whatever. This can also be based on the negative binomial with r > 1; then, no action is taken until r defectives have been found. Rejecting H_o (say, $\theta \leq .01$) when H_o is true results in an unnecessary interruption but not rejecting H_o when H_a (say, $\theta \geq .05$) is true results in a failure to make needed adjustments. Note that corresponding means are $r/.01 = 100r$ and $r/.05 = 20r$; one rejects the smaller θ when Y is small. Given r, another procedure to determine a confidence interval for θ (these are Bernoulli trials) could be developed.

The geometric distribution has an interesting property loosely described as being memoryless. First we need to inspect the use of conditional probabilities for random variables.

For an integer–valued RV like Y or W or G, the probability is defined on the induced power set of its range; these are nevertheless sets so our previous rules for probability carry over. In other words,

$$P(\{G \leq 7\} \mid \{G \geq 3\}) = P(\{G \leq 7\} \cap \{G \geq 3\}) \div P(\{G \geq 3\})$$

$$= P(\{3 \leq G \leq 7\}) \div P(\{G \geq 3\})$$

is a perfectly respectable instance of

$$P(A \mid B) = P(A \cap B) \div P(B) .$$

Exercise 6: Evaluate the probabilities on G in the paragraph above when $\theta = .01$ and when $\theta = .05$. Hint: you may want to (re)develop the formula for the sum of a geometric series first.

Theorem: *Let X be a RV with range $\mathscr{X} = 1(1)\infty$ and*

$$0 < \theta = P(X = 1) < 1 .$$

If $P(X = 1 + r \mid X > 1) = P(X = r)$ for $r = 1(1)\infty$, then

$$P(X = x) = (1 - \theta)^{x-1}\theta \ for \ x = 1(1)\infty .$$

Proof: The condition $P(X = 1 + r \mid X > 1) = P(X = r)$ implies that

$$P(X = 1 + r) \div P(X > 1) = P(X = r) \ or$$

$$P(X = 1 + r) = P(X > 1)P(X = r) . \qquad (*)$$

This is to hold for all positive integers r, so for r = 1,

$$P(X = 2) = P(X > 1)P(X = 1) = (1 - \theta)\theta .$$

This is the conclusion for x = 2 and $P(X = 1) = \theta$ is the conclusion for x = 1. Form the induction hypothesis that

$$P(X = s) = (1 - \theta)^{s-1}\theta .$$

Then (*) yields $P(X = 1 + s)$

$$= P(X > 1)P(X = s) = (1 - \theta)\left[(1 - \theta)^{s-1}\theta)\right] = (1- \theta)^{s}\theta$$

and the proof is complete.

"Lack of memory" is seen by noting that the probability that there are r + 1 failures given that there is at least one failure is just the probability that there are r failures. The converse is also true so that this property *characterizes* the geometric distribution: it is the only distribution on the positive integers "without memory".

Exercise 7:   Let X be a geometric RV; show that for positive integers k and r, $P(X = k + r \mid X > k) = P(X = r)$.

In terms of expectations, we interpret the conditioning as follows. The mean number of trials until the first success is $1/\theta$. If it is known that there are at least k successes, then the mean number of trials until the next success is still $1/\theta$. The alert reader will note that we have fudged a bit here, never having discussed conditional expectation. It is now time to cease these small quasi–historical adjustments of rules and examine the contemporary foundations of probability theory.

Exercise 8:   Find a mode and a median for the negative binomial distribution.

INE–Exercise: Even if you haven't discussed the catching problem in lesson 18, consider sampling with replacement from an urn containing n distinct objects. The first selection yields the first object and it is certainly "distinct". Let X_1 be the number of succeeding trials needed to obtain a second distinct object; that is, X_1 = k entails k–1 of the first object and a new object on trial k. Then, $P(X_1 = k) = (1/n)^{k-1}((n-1)/n)$. Similarly, let X_{j-1} be the number of trials after "j–1" successes up to and including the next, jth, new object. The total sample size required to get r objects is $T_{r,n} = 1 + X_1 + X_2 + \cdots + X_{r-1}$.

a)   Convince yourself that
$$P(X_{j-1} = k) = ((j-1)/n)^{k-1}((n-j+1)/n).$$

b)   Find $E[X_{j-1}]$.

c)   Show that $E[T_{r,n}]$
$$= n(1/n + 1/(n-1) + \cdots + 1/(n-r+1)) .$$

d)   Find $E[T_{n,n}]$ for $n = 5,10,21$.

PART II: PROBABILITY AND EXPECTATION

Overview

Lesson 1 contains various properties of sequences of subsets of a set Ω building up the notion of a Borel σ–field, $\mathscr{B}$, a class of sets closed under countable unions, intersections, and complementations. These two form a measurable space $[\Omega, \mathscr{B}]$.

With this basis, general axioms and theorems of probability on $\mathscr{B}$ are laid out in lesson 2. A principal result is continuity:

$$\text{when } \lim_{n \to \infty} A_n \text{ exists, } P(\lim_{n \to \infty} A_n) = \lim_{n \to \infty} P(A_n) \, .$$

At the end of this lesson, Lebesgue measure on the interval $(0,1)$ is shown to be a special case.

The cumulative distribution function (CDF) relates a random variable (RV) and its probability thru $\mathscr{B}$. Lesson 3 contains the basic properties of a CDF $F(x) = P(X \leq x)$ for any real RV.

Since virtually all the RVs in Part I were discrete type, examples of (absolutely) continuous type RVs are given in lesson 4; their CDFs can be expressed in terms of Riemann integrals. The very much used (and abused) normal distribution is explained in lesson 5.

Lesson 6 shows some of the care that is needed in dealing with general random variables.

Lessons 7 and 8 on convergence contain material which will be useful in inferential statistics but, at this point, they are principally additional preparation for integration.

As will be seen several times, expectation is a necessary part of probability theory and is not restricted to plain ordinary means. While lesson 11 is almost nothing but a list of the properties, lessons 9 and 10 contain precise clear details on the basic definition of expected value, namely, the Lebesgue–Stieltjes integral.

For many calculations, it is convenient to have the Stieltjes integral as explained in lesson 12. This is only a bit more general than the Riemann integral of the usual calculus course:

$$\int_{-\infty}^{\infty} h(x) \, dx \text{ is a special case of } \int_{-\infty}^{\infty} h(x) \, dF(x) \, .$$

Lesson 13 is certainly one of the most important as it contains Fubini's theorem on iterated integrals without which the independence of RVs would not be much help in solving problems in probability and statistics.

LESSON 1. SOME SET THEORY

We continue in what is called the "naive viewpoint" wherein a set is "primitive" and well–understood. Thus, for a given "set" Ω, with "points" ω, there are such things as:

subsets $A \subset B$, read "A is a subset of B",
if for each ω in A, ω is also in B or $\omega \in A$ implies $\omega \in B$ where $\in$ is read "belongs to";

sequences of subsets where for each $i = 1(1)\infty$, A_i is a subset of Ω; we write $\{A_i : i=1(1)\infty\}$ with $:$ read "such that";

arbitrary *families* of subsets where for each i in an(other) index set I, A_i is a subset of Ω.

Note that $A \subset B$ contains the possible equality $A = B$ when $B \subset A$. Of course, when Ω is finite, all families can have only a finite number of "distinct" subsets. Unless noted otherwise, "countable" will include the possibility of finite.

Example: Let Ω be the real line.
a)   The collection of all *half–open* intervals
$$(-\infty, \omega_0] = \{\omega : -\infty < \omega \le \omega_0\}$$
is a family indexed by the real numbers ω_0.

Similarly, we can have:
closed intervals $[a, b] = \{ \omega : a \le \omega \le b\}$;
half–open intervals $[a, b) = \{ \omega : a \le \omega < b\}$.

b)   The collection of all intervals with integral endpoints is a countable family of subsets :
$$(n,n+1] = \{x : n < x \le n+1\}, n = 0, \pm 1, \pm 2, \cdots .$$

c)   The collection of intervals $r \le x < r+1$ where r is a rational number is also countable; this follows by extending the following scheme attributed to Cantor.

The rational numbers can be imagined in an array for which we will consider only only positive values for simplicity:

1/1	1/2	1/3	1/4	1/5	1/6	1/7	1/8	1/9	$\cdots$
2/1	2/2	2/3	2/4	2/5	2/6	2/7	2/8	2/9	$\cdots$
3/1	3/2	3/3	3/4	3/5	3/6	3/7	3/8	3/9	$\cdots$
4/1	4/2	4/3	4/4	4/5	4/6	4/7	4/8	4/9	$\cdots$
5/1	5/2	5/3	5/4	5/5	5/6	5/7	5/8	5/9	$\cdots$ etc.

One countable arrangement is :

$$1/1 \quad 2/1 \quad 1/2 \quad 3/1 \quad 2/2 \quad 1/3 \quad 4/1 \quad 3/2 \quad 2/3 \quad 1/4$$
$$5/1 \quad 4/2 \quad 3/3 \quad 2/4 \quad \cdots.$$

In this ordering, we legitimately ignore equivalent values to

obtain $1/1 \quad 2/1 \quad 1/2 \quad 3/1 \quad 1/3 \quad 4/1 \quad 3/2 \quad 2/3 \quad 1/4 \quad 5/1 \quad 1/5 \cdots$.

In a more general context, a countable union (like rows $1,2, \cdots$) of countable sets (like columns $1,2, \cdots$) is countable. Now if the collection of real numbers were countable, we would have a similar array. For example, let x_{ij} be digits $0, 1, \cdots, 9$; then in $(0, 1)$, we would have a list

$$.x_{11}x_{12}x_{13}x_{14}\cdots$$

$$.x_{21}x_{22}x_{23}x_{24}\cdots$$

$$.x_{31}x_{32}x_{33}x_{34}\cdots \quad \text{etc.}$$

Construct a number with digits $x_{ii} + 1$ except that $9 + 1$ is

replaced by 0: $.(x_{11}+1)(x_{22}+1)(x_{33}+1)(x_{44}+1)\cdots$.

This number differs from each number in the list by at least one digit (that in position ii); therefore, this number is not in the list. Therefore, the set of real numbers cannot be so listed and is said to be *uncountable*.

The basic operations of union, intersection and complementation are given in the

Definition: *Let $\{A_i : i \in I\}$ be a non-empty set of subsets of Ω. $\cup_I A_i$ is the set containing all points in Ω which belong to any (some, at least one) A_i:*

$\omega \in \cup_I A_i$ *iff $\omega \in A_i$ for some i in I;*

$\cap_I A_i$ *is the set containing all points in Ω which belong to all (every, each) A_i:*

$\omega \in \cup_I A_i$ *iff $\omega \in A_i$ for all i in I;*

for each i in I, $A_i^{\,c}$ is the set of points in Ω which are not in A_i: $\omega \in A_i^{\,c}$ *iff $\omega \notin A_i$ (does not belong to).*

We have already seen and used some consequences of these definitions for families with finite index.

Theorem: *For A_1, A_2, A_3 subsets of Ω,*

a) $A_1 \cup A_2 = A_2 \cup A_1$ *and* $A_1 \cap A_2 = A_2 \cap A_1$

b) $A_1 \cap (A_2 \cup A_3) = (A_1 \cap A_2) \cup (A_1 \cap A_3)$

c) $A_1 \cup (A_2 \cap A_3) = (A_1 \cup A_2) \cap (A_1 \cup A_3)$

d) $(A_1 \cap A_2)^C = A_1^{\ C} \cup A_2^{\ C}$ *and*

   $(A_1 \cup A_2)^C = A_1^{\ C} \cap A_2^{\ C}$

e) $A_1 \cup (A_2 \cup A_3) = (A_1 \cup A_2) \cup A_3$ *and*

   $A_1 \cap (A_2 \cap A_3) = (A_1 \cap A_2) \cap A_3$.

Exercise 1:   Prove that for any family $\{A_i\}$ of subsets, DeMorgan's law holds: $(\cap_I A_i)^C = \cup_I A_i^{\ C}$;

equivalently, $(\cup_I A_i)^C = \cap_I A_i^{\ C}$.

In the special case of a sequence of sets, we see other notation.

Definition: *For* $I^+ = \{1, 2, 3, \cdots \}$ = *the set of positive integers, and the collection of subsets* $\{A_i : i = 1(1)\infty\}$,

$$\cup_{I^+} A_i = \cup_{i=1}^{\infty} A_i = \lim_{n \to \infty} \cup_{i=1}^{n} A_i \text{ and}$$

$$\cap_{I^+} A_i = \cap_{i=1}^{\infty} A_i = \lim_{n \to \infty} \cap_{i=1}^{n} A_i$$

Examples: Let Ω be the real line.

a)   Let $A_i = (1-1/i, 1+1/i)$, $i \in I^+$; e.g., $A_1 = (0, 2)$
   $A_2 = (1/2, 3/2)$ $A_3 = (2/3, 4/3)$ $A_4 = (3/4, 5/4) \cdots$.

Then, $\cap_{i=1}^{\infty} A_i = \lim_{n \to \infty} \cap_{i=1}^{n} A_i$

$$= \lim_{n \to \infty} (1-1/n, 1+1/n) = \{1\} .$$

Or, $\omega \in \cap_{i=1}^{\infty} A_i$ iff $\omega \in A_i$ for all i (in I^+) iff

$$1-1/i < \omega < 1+1/i \quad \text{for all i} .$$

The real number sequence $1-1/i$ converges up to 1 and the real number sequence $1+1/i$ converges down to 1. It follows that

$$\cup_{i=1}^{\infty} A_i = (0, 2) \quad \text{since } A_1 \supset A_2 \supset A_3 \supset A_4 \supset \cdots .$$

b) Let $A_i = (0, 1/i)$ for i even in I^+

$$= (-1/i, 0) \text{ for i odd in } I^+ .$$

Now $(0,1/2n) \cap (-1/2n+1, 0) = \phi$ so that

$$\cap_{i=1}^{\infty} A_i = \phi \quad \text{while} \quad \cup_{i=1}^{\infty} A_i = (-1, 0) \cup (0, 1/2) .$$

Exercise 2: Let Ω be the real line.

a) Let $A_i = (-1/i, i)$ for $i \in I^+$. Find $\cup_{i=1}^{\infty} A_i$, $\cap_{i=1}^{\infty} A_i$,

$\cup_{i=1}^{\infty} A_i^c$, $\cap_{i=1}^{\infty} A_i^c$.

b) Let $A_i = (-\infty, 1+1/i]$, $B_i = (-\infty, 1-1/i]$ for $i \in I^+$.

Find $\cap_{i=1}^{\infty} A_i$, $\cup_{i=1}^{\infty} A_i$, $\cap_{i=1}^{\infty} B_i$, $\cup_{i=1}^{\infty} B_i$,

$\cap_{i=1}^{\infty} (A_i \cup B_i)$, $\cup_{i=1}^{\infty} (A_i \cap B_i)$.

The following formalizes the concept of the limit of sequences of sets in the same way as that for sequences of real numbers. More importantly, these "limit points" appear again in studying convergence in probability spaces. Incidentally, hardly anybody actually says *limit supremum* or *limit infimum*.

Definition: *Let* $\{A_i\}$ *be a sequence of subsets of* Ω. *Let* I_n^+ *be the set of integers* $\geq$ *the positive integer* n. *Let*

$$B_n = \cup_{I_n^+} A_i = \cup_{i=n}^{\infty} A_i = \lim_{N \to \infty} \cup_{i=n}^{N} A_i \text{ and}$$

$$C_n = \cap_{I_n^+} A_i = \cap_{i=n}^{\infty} A_i = \lim_{N \to \infty} \cap_{i=n}^{N} A_i.$$

Then $\limsup \{A_i\} = \cap_{n=1}^{\infty} B_n$ *and* $\liminf \{A_i\} = \cup_{n=1}^{\infty} C_n$.

Also, if $\limsup A_i = \liminf A_i = A$, *then* $\lim A_i = A$.

Theorem: *In the definition above,*

    a) $\{B_n\}$ *is a non-increasing sequence:*

$$B_n \supset B_{n+1} \supset B_{n+2} \supset \cdots ;$$

    b) $\{C_n\}$ *is a non-decreasing sequence:*

$$C_n \subset C_{n+1} \subset C_{n+2} \subset \cdots ;$$

    c) $\omega \in \limsup A_i$ *iff* ω *belongs to infinitely many* A_i;

    d) $\omega \in \liminf A_i$ *iff* ω *belongs to all but a finite number of* A_i.

Partial proof: a) $\omega \in B_{n+1}$ iff $\omega \in A_i$ for some $i \geq n+1$; hence $\omega \in A_i$ for some $i \geq n$. Therefore, $B_{n+1} \subset B_n$.

c) ω belongs to infinitely many A_i iff ω does not belong to only finitely many A_i iff ω belongs to $\cup_{i=n}^{\infty} A_i$ for all n

($\in I^+$) iff ω belongs to $\cap_{n=1}^{\infty} \cup_{i=n}^{\infty} A_i = \limsup A_i$.

Exercise 3: Complete the proof of the theorem. Hint: for d), use DeMorgan's rule.

In probabilistic language, the event $\limsup A_i$ occurs iff infinitely many of the events A_i occur: $\limsup A_i = \{A_i \text{ i.o.}\}$.

Theorem: *If the sequence $\{A_i\}$ of subsets of Ω is nondecreasing,*

then $\lim A_i$ exists and equals $\cup_{i=1}^{\infty} A_i$.

Proof: We have $A_1 \subset A_2 \subset A_3 \subset \cdots \subset A_n \subset \cdots \subset A_N \subset \cdots$ so

that $\cap_{i=n}^{N} A_i = A_n$ and $\lim\limits_{N \to \infty} \cap_{i=1}^{N} A_i = A_n$. Hence,

$$\liminf A_i = \cup_{n=1}^{\infty} \cap_{i=n}^{\infty} A_i = \cup_{n=1}^{\infty} A_n$$

and

$$\limsup A_i = \cap_{n=1}^{\infty} \cup_{i=n}^{\infty} A_i \subset \cup_{i=1}^{\infty} A_i .$$

It is obvious that $\liminf A_i \subset \limsup A_i$ so that the equality
does hold.

Exercise 4: Prove the corresponding:

Theorem: *If $\{A_i\}$ is a nonincreasing sequence of subsets of Ω,*

then $\lim A_i$ exists and equals $\cap_{i=1}^{\infty} A_i$.

In general, such limit theorems are used more for
theoretical results than for applications. Here are some other

Examples: a) Let $A_n = \{\omega : 2n \leq \omega \leq 2n+1\}$. Then $\limsup A_n$ is
the intersection of

$$B_1 = [2, 3] \cup [4, 5] \cup [6, 7] \cup [8, 9] \cup \cdots$$
$$B_2 = \qquad\quad [4, 5] \cup [6, 7] \cup [8, 9] \cup \cdots$$
$$B_3 = \qquad\qquad\qquad\quad [6, 7] \cup [8, 9] \cup \cdots$$
$$\cdots$$

Obviously, $\limsup A_n = \phi$ so $\liminf A_n = \phi$ and $\lim A_n = \phi$.

b) Let the sets A_n be the singletons (a_n) where $\{a_n\}$ is an
ordering of the rational numbers in the real line R. A basic
property of R is that each real number r is a limit of some
subsequence $\{b_n\}$ of $\{a_n\}$ which may be taken as

$$b_n \leq b_{n+1} \leq \cdots \text{ with } \lim_n b_n = r.$$

Then limsup $\{(b_n)\} = \cap_{n=1}^{\infty} \cup_{i=n}^{\infty} \{b_i\}$

$$= \cap_{n=1}^{\infty} \{b_n, b_{n+1}, b_{n+2}, \cdots\} = r.$$

This implies that limsup $A_n = R$. But,

$$\text{liminf } A_n = \cup_{n=1}^{\infty} \cap_{i=n}^{\infty} A_i = \cup_{n=1}^{\infty} \phi = \phi.$$

Exercise 5: Find limsup and liminf.

    a) $A_n = [0, 1]$ for n odd and $A_n = [1, 2]$ for n even.

    b) $A_n = [-n, 0]$ for n odd and $A_n = [1/n, n]$ for n even.

    c) $A_n = (0, 1-1/n]$ for n odd and $A_n = [1/n, 1)$ for n even.

The reader will recall the inclusion of the power set (the collection of all subsets) in the definiton of probability on finite or countable sets. There are technical reasons which prevent the use of the power set in general; it turns out that we are limited to a σ–field "$\mathscr{B}$". (Other writers may use slightly different words and arrange the discussion in a different order.) A number of properties of $\mathscr{B}$ follow quite simply from its definition:

Definition: *Let $\mathscr{B}$ be a non-empty collection of subsets of a set Ω. Then $\mathscr{B}$ is a σ-field if $\mathscr{B}$ is closed under complements and countable unions;*   *i) $A \in \mathscr{B}$ implies $A^C \in \mathscr{B}$*

$$ii) \ \{A_i\} \in \mathscr{B} \text{ implies } \cup_{i=1}^{\infty} A_i \in \mathscr{B}.$$

The pair $[\Omega, \mathscr{B}]$ is a measurable space.

Corollary: *Let $\mathscr{B}$ be a σ-field.*

    a)   *The empty set ϕ and the "universe" set Ω belong to $\mathscr{B}$.*

Proof: Since $\mathscr{B}$ is non–empty, there is some subset A in $\mathscr{B}$. Then by i), A^C is in $\mathscr{B}$. Let $A_1 = A$ and $A_i = A^C$ for $i \geq 2$;

then by ii), $\cup_{i=1}^{\infty} A_i = A \cup A^c \in \mathscr{B}$ or $\Omega \in \mathscr{B}$. Whence by i)

again, $\Omega^c = \phi \in \mathscr{B}$

    b)   *If for i = 1(1)∞, $A_i \in \mathscr{B}$, then $\cap_{i=1}^{\infty} A_i \in \mathscr{B}$.*

Proof: For $A_i \in \mathscr{B}$, $A_i^c \in \mathscr{B}$ and so $\cup_{i=1}^{\infty} A_i^c \in \mathscr{B}$. Hence,

$(\cup_{i=1}^{\infty} A_i^c)^c \in \mathscr{B}$ or $\cap_{i=1}^{\infty} A_i \in \mathscr{B}$.

As in the previous extension from finite to countable probability spaces, we do not lose any fundamental properties. When Ω is finite, the power set contains only finitely many elements and is closed under complements and finite unions. Collections of subsets with these closure properties are called fields; a σ–field is a field.

Exercise 6: Show that when Ω is countable and $\mathscr{B}$ is the power set, $\mathscr{B}$ is a σ–field.

INE–Exercises:
1.   Let $\mathscr{B}_i$, $i \in I$, be a family of σ–fields of Ω.
    a)   Show that $\cap_{i \in I} \mathscr{B}_i$ is also a σ–field.
    b)   Show that $\mathscr{B}_1 \cup \mathscr{B}_2$ is not a σ–field by giving a counter–example.
2.   Let $\mathscr{C}$ be a collection of subsets of Ω.
    a)   Show that there is at least one σ–field containing $\mathscr{C}$.
    b)   Show that there is a smallest σ–field containing $\mathscr{C}$.
3.   Let $[\phi, \mathscr{A}]$ and $[\Omega, \mathscr{B}]$ be measurable spaces; let $f : \phi \to \Omega$.
    a)   Is $f(\mathscr{A}) = \{f(A) : A \in \mathscr{A}\}$ a σ–field for Ω?
    b)   Show that $f^{-1}(\mathscr{B}) = \{ f^{-1}(B) : B \in \mathscr{B}\}$ is a σ–field for ϕ.

LESSON 2. BASIC PROBABILITY THEORY

Except for the fact that the domain is a σ–field, the following is a repeat of our earlier definition. (Again, variation will be found across texts.)

Definition: *Let Ω be a set and let $\mathscr{B}$ be a σ-field of subsets of Ω. A probability measure P is a set function on $\mathscr{B}$ such that*
  i)   *for each $A \in \mathscr{B}$,* $P(\phi) = 0 \le P(A) \le 1 = P(\Omega)$
  ii) *for $A_i \in \mathscr{B}$ and $A_i \cap A_j = \phi$ when $i \ne j$,*

$$P(\cup_{i=1}^{\infty} A_i) = \sum_{i=1}^{\infty} P(A_i) .$$

$[\Omega, \mathscr{B}, P]$ is a probability space; $[\Omega, \mathscr{B}]$ is a measurable space. Elements of $\mathscr{B}$ are the events.

We note that the properties of a Borel σ–field assure that ϕ, Ω, $\cup_{i=1}^{\infty} A_i$ are in $\mathscr{B}$ so that their probabilities are defined. The axiom ii), called countable additivity or σ–additivity, has "finite additivity" as a special case:

  for n fixed and $E_1, E_2, \cdots, E_n$ disjoint sets in $\mathscr{B}$,

  let $A_i = E_i$ when $i \le n$ and $A_i = \phi$ when $i > n$. Then,

$$\cup_{i=1}^{n} E_i = \cup_{i=1}^{\infty} A_i \quad \text{and} \quad P(\cup_{i=1}^{n} E_i) = P(\cup_{i=1}^{\infty} A_i) = \Sigma_{i=1}^{\infty} P(A_i)$$

$$= \Sigma_{i=1}^{n} P(A_i) = \Sigma_{i=1}^{n} P(E_i)$$

(just as if Ω were finite and $\mathscr{B}$ were the power set) since by i) $P(\phi) = 0$.

Exercise 1: Prove the following:

Corollary: *Let $A_1, A_2, A_3, \cdots \in \mathscr{B}$ of a probability space $[\Omega, \mathscr{B}, P]$.*

  a)   *Then for each i, $A_i^c \in \mathscr{B}$, $P(A_i^c) = 1 - P(A_i)$ and*

$$P(\cap_j A_j) \le P(A_i) \le P(\cup_j A_j);$$

  b)   *For $A_1 \subset A_2$, $P(A_2 - A_1) = P(A_2) - P(A_1)$ which*

$$implies\ P(A_1) \le P(A_2);$$

c)   (sub-additivity) $P(\cup_{i=1}^{n} A_i) \le \Sigma_{i=1}^{n} P(A_i)$ for each positive integer n.

d)   (sub-σ-additivity) $P(\cup_{i=1}^{\infty} A_i) \le \Sigma_{i=1}^{\infty} P(A_i).$

One of the most useful concepts in classical analysis is that of continuity. The lemma and two theorems following contain a corresponding concept for the probability set function.

Lemma: *Let* $\{A_i\}$ *be a sequence of subsets of some* Ω. *Let* $B_1 = A_1$ *and for* $i \ge 2$, *let*

$$B_i = A_i - (\cup_{j=i}^{i-1} A_j) = A_i \cap A_{i-1}^{c} \cap A_{i-2}^{c} \cap \cdots \cap A_1^{c}.$$

Then,

  a) $B_i \cap B_j = \phi$ *for* $i \ne j$;

  b) $\cup_{i=1}^{n} A_i = \cup_{i=1}^{n} B_i$ *for each positive integer n;*

  c) $\cup_{i=1}^{\infty} A_i = \cup_{i=1}^{\infty} B_i$.

Proof: $\cup_{i=1}^{n} B_i = B_1 \cup B_2 \cup \cdots \cup B_n$

$$= A_1 \cup (A_2 \cap A_1^{c}) \cup (A_3 \cap A_2^{c} \cap A_1^{c}) \cup \cdots$$

$$\cup (A_n \cap A_{n-1}^{c} \cap \cdots \cap A_1^{c})$$

$$= \cup_{i=1}^{n} A_i \text{ and the rest is "obvious".}$$

(Some writers use $B_1 + B_2 + \cdots + B_n$ for such a "disjoint" union.)

Exercise 2:   Draw Venn diagrams with shading to illustrate this lemma when

  a) $A_1 \supset A_2 \supset A_3 \supset A_4$;

b) $A_1 \subset A_2 \subset A_3 \subset A_4$;

c) A_1, A_2, A_3, A_4 are "arbitrary".

Theorem: *Let $\{A_i\}$ be a nonincreasing sequence of elements of $\mathcal{B}$*

in $[\Omega, \mathcal{B}, P]$ so that $\lim\limits_{n \to \infty} A_i = \cap_{i=1}^{\infty} A_i \in \mathcal{B}$. Then,

$$\lim\limits_{i \to \infty} P(A_i) = P(\lim\limits_{i \to \infty} A_i).$$

Proof: From $A_1 \supset A_2 \supset A_3 \supset \cdots$,we get

$$A_1^{\,c} \subset A_2^c \subset A_3^{\,c} \subset \cdots$$

and $\lim\limits_{i \to \infty} A_i^{\,c} = \cup_{i=1}^{\infty} A_i^{\,c}$. Now, $(\cap_{i=1}^{\infty} A_i)^c = \cup_{i=1}^{\infty} A_i^{\,c} = \cup_{i=1}^{\infty} B_i$

where $B_1 = A_1^{\,c}$, and $B_i = A_i^{\,c} - A_{i-1}^{\,c}$ for $i \geq 2$ (as in, but

simpler than, the lemma). Since $B_i \cap B_j = \phi$ for $i \neq j$,

$$1 - P(\cap_{i=1}^{\infty} A_i)) = P((\cap_{i=1}^{\infty} A_i)^c) = P(\cup_{i=1}^{\infty} A_i^c) \qquad (*)$$

$$= P(\cup_{i=1}^{\infty} B_i) = \Sigma_{i=1}^{\infty} P(B_i)$$

$$= \lim\limits_{n \to \infty} \Sigma_{i=1}^{n} P(B_i) = \lim\limits_{n \to \infty} P(\cup_{i=1}^{n} B_i)$$

by finite additivity. By substitution, this last limit is

$$\lim\limits_{n \to \infty} P(A_1^{\,c} \cup (A_2^{\,c} - A_1^{\,c}) \cup \cdots \cup (A_n^{\,c} - A_{n-1}^{\,c}))$$

$$= \lim\limits_{n \to \infty} P(A_n^{\,c}) = \lim\limits_{n \to \infty} (1 - P(A_n)). \qquad (**)$$

The long string from (*) to (**) reduces to

$$1 - P(\cap_{i=1}^{\infty} A_i) = \lim\limits_{n \to \infty} (1 - P(A_n)).$$

Since P is finite, we can subtract to obtain:

$$P(\cap_{i=1}^{\infty} A_i) = \lim_{i \to \infty} P(A_i) \text{ or } P(\lim_{i \to \infty} A_i) = \lim_{i \to \infty} P(A_i).$$

Exercise 3: Prove the following:

Theorem: *If* $\{A_i\}$ *is a nondecreasing sequence of events for* $[\Omega, \mathcal{B}, P]$, *then* $\lim_{i \to \infty} P(A_i) = P(\lim_{i \to \infty} A_i)$.

In the following definition, conditional probability is only a bit more general than in Lesson 9, Part I.

Definition: *Let* A *and* B *be events in a probability space* $[\Omega, \mathcal{B}, P]$. *Then* $P(A|B)$ *is any finite number such that*

$$P(A \cap B) = P(A|B) \cdot P(B).$$

When $P(B) > 0$, there is only one $P(A|B)$, namely, $P(A \cap B)/P(B)$. But when $P(B) = 0$, $P(A \cap B) = 0$ since $A \cap B \subset B$; then $P(A|B)$ can be any finite number. The definition of mutually stochastically independent events (also "mutually independent" or often just "independent") is the same as in the earlier lesson.

Definition: *The events* A_1, A_2, $\cdots$ *of a probability space are mutually stochastically independent (msi) iff for any selection of* $n \geq 2$ *(distinct) events,*

$$P(A_{i_1} \cap A_{i_2} \cap \cdots \cap A_{i_n}) = P(A_{i_1})P(A_{i_2}) \cdots P(A_{i_n}).$$

INE–Exercise 4: Prove the following theorem (as in Lesson 9, Part I).

Theorem: *Events* A *and* B *in a probability space* $[\Omega, \mathcal{B}, P]$ *are independent iff* A^C *and* B *are independent iff* A *and* B^C *are independent iff* A^C *and* B^C *are independent. In fact,* $\{A_i\}$ *are mutually independent iff* $\{A_i^C\}$ *are mutually independent.*

The following is another "classical" result as evidenced by that fact that it has a name.

Theorem *(Borel-Cantelli Lemma): Let $\{A_i\}$ be a sequence of events in the probability space $[\Omega, \mathcal{B}, P]$ and let $A = \limsup A_i$ be the set of ω belonging to infinitely many of the A_i.*

i) *If $\sum_{i=1}^{\infty} P(A_i) < \infty$, then $P(A) = 0$.*

ii) *If $A_1, A_2, \cdots$ are also msi, then $\sum_{i=1}^{\infty} P(A_i) = \infty$, implies $P(A) = 1$.*

Proof: i) $A = \cap_{n=1}^{\infty} \cup_{i=n}^{\infty} A_i$; hence, $A \subset \cup_{i=n}^{\infty} A_i$ for each n and

$$P(A) \le P(\cup_{i=1}^{\infty} A_i) \le \sum_{i=n}^{\infty} P(A_i). \qquad (*)$$

Since the series of real numbers $\sum_{i=1}^{\infty} P(A_i)$ is convergent, given any $\varepsilon > 0$, the right–hand side of (*) can be made $< \varepsilon$ by taking n "large". On the other side, P(A) is free of epsilon so P(A) must be zero.

ii) $1 - P(A) = P(A^c) = P(\cup_{n=1}^{\infty} \cap_{i=n}^{\infty} A_i^c) \le \sum_{n=1}^{\infty} P(\cap_{i=1}^{\infty} A_i^c)$.

But, $\cap_{i=n}^{N} A_i^c$ is a "decreasing" sequence as $N \to \infty$ and, by independence, $P(\cap_{i=n}^{N} A_i^c) = \Pi_{i=1}^{N} P(A_i^c)$. It follows that

$$P(\cup_{i=n}^{\infty} A_i^c) = \lim_{N \to \infty} P(\cap_{i=n}^{N} A_i^c) = \Pi_{i=n}^{\infty} P(A_i^c) .$$

Therefore, $0 \le 1 - P(A) \le \sum_{n=1}^{\infty} \Pi_{i=n}^{\infty} (1 - P(A_i))$. Since, $\sum_{i=1}^{\infty} P(A_i) = \infty$, the infinite product diverges to 0 for each n: $\Pi_{i=n}^{\infty} (1 - P(A_i)) = 0$. (See the lemma below!) Thus, P(A) = 1.

Lemma: *If $0 \le a_n \le 1$, then $\sum_{n=1}^{\infty} a_n = \infty$ implies*

$$\Pi_{n=1}^{\infty}(1 - a_n) = 0 .$$

Partial proof: First recall a mean value theorem for the natural logarithm function: $\log(x) = -1/(1 - \xi)$ for $|\xi| < |x| < 1$. Suppose all $a_n < 1$. Now

$$\sum_{n=1}^{N} \log(1 - a_n) = \sum_{n=1}^{N'}(-a_n/(1 - \xi_n))$$

where the prime indicates that terms with $a_n = 0$ are omitted. Since $1/(1 - \xi_n) \ge 1$, this last sum diverges to $-\infty$ as $N' \to \infty$. Therefore, its antilog diverges to 0.

Exercise 5: Write out complete details for the proof of this last lemma; do not forget the case that some a_n may be 1.

Details for the following discussion can be found in texts on measure theory but will not be required herein.

Given a collection $\mathscr{G}$ of subsets of any set Ω, there is a minimal σ–field $\mathscr{B}(\mathscr{G})$ (the smallest collection of subsets closed under complements and countable unions) containing $\mathscr{G}$. If there is a set function Q on $\mathscr{G}$ having properties i and ii of the definition of probability, then this function can be extended (uniquely) to a probability measure P on $\mathscr{B}(\mathscr{G})$ such that $P(A) = Q(A)$ for all A in $\mathscr{G}$.

Example: Let $\Omega = [0, 1]$ and let $\mathscr{G}$ be the collection of intervals $(a, b] = \{\omega : a < \omega \le b\}$ for $0 \le a < b \le 1$. Let
$$Q(\ (a, b]\ ) = b - a .$$
Then the corresponding P is called the *Lebesgue measure* on $[0, 1]$. It turns out that the σ–field $\mathscr{B}(\mathscr{G})$ is the same whether $\mathscr{G}$ is taken as the collection of intervals (a, b), or $[a, b]$, or $[a, b)$; this is the Borel σ–field of Ω.

For the real line, say $\Omega = R$, the Borel σ–field is generated by the intervals $(a, b]$ for a and b finite real numbers. $Q(a,b] = b - a$ can be extended to a *measure function* Λ like P in all aspects except those depending on the finiteness of P since $\Lambda(R) = \infty$; this is the general *Lebesgue measure* on R. There are infinitely many Q which generate other measures.

Exercise 6: For $i = 1,2$, let $\mathcal{B}_i$ be a σ–field of Ω_i . Let $f : \Omega_1 \to \Omega_2$ be such that $A \in \mathcal{B}_2$ implies

$$f^{-1}(A) = \{\, b : f(b) \in A \,\} \in \mathcal{B}_1 .$$

Let P be a probability measure on $[\Omega_1, \mathcal{B}_1]$. Show that

$Q(A) = P(f^{-1}(A))$ defines a probability measure on $[\Omega_2, \mathcal{B}_2]$.

INE–Exercises:

1. Let $\{\, \mathcal{B}_i : i \in I \,\}$ be a collection of σ–fields of Ω. Show that $\cap_{i \in I} \mathcal{B}_i$ is also a σ–field. Is $\mathcal{B}_{i_2} \cup \mathcal{B}_{i_1}$ always a σ–field?

2. Let $\mathcal{C}$ be a collection of subsets of Ω.
   a) Show that there exists at least one σ–field containg $\mathcal{C}$.
   b) Show that there exists a smallest σ–field containing $\mathcal{C}$.

3. Let $f : U \to \Omega$. Let $\mathcal{A}$ be a σ–field of Ω. Let $\mathcal{C}$ be a σ–field of U .Show that $\{f^{-1}(A) : A \in \mathcal{A}\}$ is a σ–field of U.

4. Let $\mathcal{B}$ be a σ–field of Ω. Let A be a fixed set in $\mathcal{B}$ Show that $\{B \subset \Omega : B \cap A \in \mathcal{B}\}$ is a σ–field.

LESSON 3. THE CUMULATIVE DISTRIBUTION FUNCTION

All of the arithmetic associated with testing in the hyper–geometric distribution and with confidence intervals in the binomial distribution (respectively, Lessons 15, 19, Part I) was based on their CDFs. Here the extension of this latter concept will be made in two steps, one emphasizing the type of function which is "random", the other emphasizing its "distribution"; the first step (definition) is actually a technical matter needed in the second: only measurable functions have distributions.

Definition: *A real valued function X on Ω of a measurable space $[\Omega, \mathscr{B}]$ is a (real) random variable (RV) iff*

$$\{\omega : X(\omega) \leq x\} = X^{-1}(-\infty, x] \in \mathscr{B} \text{ for each real number } x.$$

Exercise 1: Check the "monotonicity" of X^{-1}:

for real numbers $x_1 < x_2$, $X^{-1}(-\infty, x_1] \subset X^{-1}(-\infty, x_2]$.

Definition: *Let X be a RV on Ω of a probability space $[\Omega, \mathscr{B}, P]$. Then the c(umulative) (probability) d(istribution) f(unction) of X is the function on the real line, R, given by*

$$F(x) = P(X \leq x) = P(X(\omega) \leq x) =$$

$$P(\{\omega : X(\omega) \leq x\}) = P(X^{-1}(-\infty, x]) .$$

The name is usually abbreviated to CDF .

Example: A simple case is a Bernoulli experiment. Let $X = 1$ if a coin toss turns up "heads" and let $X = 0$ otherwise. The set Ω is {heads, otherwise} and $\mathscr{B}$ is the power set. Suppose that $P(X = 1) = P(\text{heads}) = \theta$ and $P(X = 0) = 1 - \theta$. Then:

$F(x) = 0$ for x less than 0 since then $X^{-1}(-\infty, x]$ is ϕ;

     $= 1-\theta$ for $x \geq 0$ but < 1 since then $X^{-1}(-\infty, x]$ is "otherwise";

     $= 1$ for $x \geq 1$ since then $X^{-1}(-\infty, x] = \Omega$.

We write: $F(x) = 0$ for $x < 0$
                        $= 1 - \theta$ for $0 \leq x < 1$
                        $= 1$ for $1 \leq x$.

The following are the characteristic properties of a CDF.

Theorem: *For F as in the definiton above,*

    a) $F(-\infty) = \lim\limits_{x \to -\infty} F(x) = 0$

Partial proof: As the sequence $\{x_n\}$ "converges down to $-\infty$", the sequence $X^{-1}(-\infty, x_n]$ decreases to ϕ. By earlier continuity results for P, $F(x_n) = P(X^{-1}(-\infty, x_n])$ decreases to $P(\phi) = 0$. Of course, $X^{-1}(-\infty, x] \to \phi$ as $x \to -\infty$ in any mode.

    b) $F(\infty) = \lim\limits_{x \to \infty} F(x) = 1.$

    c)   *F is non-decreasing:* $x_1 < x_2$ *implies* $F(x_1) \le F(x_2).$

Proof: When $x_1 < x_2$, the interval $(-\infty, x_1] \subset (-\infty, x_2]$ so that

$$X^{-1}(-\infty, x_1] \subset X^{-1}(-\infty, x_2] \text{ and}$$
$$P(X^{-1}(-\infty, x_1]) \le P(X^{-1}(-\infty, x]),$$

    d)   *F is right-hand continuous:* $\lim\limits_{h \downarrow 0} F(x + h) = F(x).$

Proof: Let $h_1 > h_2 > h_3 > \cdots$ be a real sequence decreasing to 0. Then $A_i = X^{-1}(-\infty, x + h_i]$ is a "decreasing" sequence of sets with limit $\cap_{i=1}^{\infty} A_i = X^{-1}(-\infty, x]$. Hence, $F(x + h_i) = P(A_i)$ has limit $P(X^{-1}(-\infty, x]) = F(x)$.

Definition: *Jump points of a CDF F are values x such that*
$$P(X = x) = F(x) - F(x - 0) > 0$$
where
$$F(x - 0) = \lim\limits_{h \downarrow 0} F(x - h) .$$
Continuity points of F are values x such that $P(X = x) = 0$ *so that* $\quad F(x) = F(x - 0) = F(x + 0)$ *where*

$$F(x + 0) = \lim_{h \downarrow 0} F(x + h).$$

Exercise 2: a) Verify the four properties of the CDF for the Bernoulli RV of the example above. What are the jump points?
b)   Do the same for a binomial RV with n = 4.

Exercise 3: Write out the details to complete the proof of the theorem above including an argument justifying the results about "any mode".

If we had taken the CDF as $G(x) = P(X < x)$, the only difference in the above would be that the continuity would be left–handed; some writers do it this way.

Theorem: *Let C_F be the set of continuity points of the CDF F as above. Then $C_F^{\,c} = R - C_F$, the set of discontinuity points of F, is at most countable.*

Proof: For each positive integer n, let S_n be the set of points x at which F has a jump of at least 1/n: $F(x) - F(x - 0) \geq 1/n$. Since

$$S_n \subset S_{n+1} \, , \, C_F^{\,c} = \cup_{n=1}^{\infty} S_n = \lim_{n \to \infty} S_n.$$

For a fixed n, let $x_1 < x_2 < \cdots < x_v$ be some points in S_n. Then, $P(X = x_i) = F(x_i) - F(x_i - 0) \geq 1/n$ for i = i(1)v . Since

$\sum_{i=1}^{v} P(X = x_i) \leq 1$, it follows that $v/n \leq 1$. That is, the number of discontinuity points in S_n is at most n. Finally, recall (lesson 1, Part II) that the countable union (n = 1, 2, 3 $\cdots$) of countable, including finite, sets is countable.

Exercise 4: Sketch a graph of the CDF in each case; note the right–hand continuity at the points of discontinuity.
a)   A negative binomial RV X is the number of the independent Bernoulli trials on which the rth "S" appears: $P(X = k) = \begin{bmatrix} k - 1 \\ r - 1 \end{bmatrix} (1 - p)^{k-r} p^r.$

Take the case $r = 6$ so that $k = 6(1)\infty$; consider $p = .5, .2, .6$.

b)   A Poisson RV is the number of "arrivals" in a fixed unit of "time" with probability :

$$P(X = k) = e^{-\lambda} \lambda^k / k!, \; k = 0(1)\infty, \; \lambda > 0.$$

Take $\lambda = .1, 1, 4$.

Now we introduce a way of keeping account of the "jumping part" of a CDF.

Example: The simplest RV is *degenerate*.
For a fixed real number t, let

$$\delta_t(x) = 0 \text{ for } x < t$$
$$= 1 \text{ for } x \geq t.$$

Note that δ_t is the "indicator function" of the set of real numbers $\geq t$. Obviously δ_t has all the properties of a CDF. Let Ω be the real line and $\mathscr{B}$ be the aforementioned Borel σ–field generated by the intervals $(a,b]$. Let X be the identity function on Ω. Then, $P(X = t) = \delta_t(t) - \delta_t(t - 0) = 1 - 0 = 1$. Note that the derivative of δ_t is 0 for $x \neq t$. This function is equivalent to the famous Dirac delta function.

Exercise 5: a) Check that δ_t above is a CDF.

b)   Sketch graphs of the functions $3\delta_2$, $4\delta_5$, their sum, and their product on different real axes.

Suppose that $C_F{}^c = \{a_i\}$, the countable set of jump points of a CDF F, is not empty; say, $F(a_i) - F(a_i - 0) = b_i > 0$.

Let $F_d(x) = \Sigma_i b_i \delta_{a_i}(x) = \sum_{a_i \leq x} b_i$; that is, $F_d(x)$ is the sum of all the jumps at points in the half–line $(-\infty, x]$. It is easy to see that $F_d(-\infty) = 0$ and $F_d(\infty) = \Sigma_i b_i \leq 1$. This leads to:

Theorem: *Let* $F_c(x) = F(x) - F_d(x)$ *where F is a CDF and* F_d *is* *defined in terms of F as in the discussion above. Then F_c is non-negative, non-decreasing, and continuous.*

Proof: For $x < x_1$, $F_d(x_1) - F_d(x)$

$$= \sum_{a_j \le x_1} b_j - \sum_{a_j \le x} b_j = \sum_{x < a_j \le x_1} b_j$$

$$= \sum_{x < a_j \le x_1} (F(a_j) - F(a_j - 0))$$

$$\le F(x_1) - F(x) .$$

Rearranging the first and last differences of this string, we get

$$F(x) - F_d(x) \le F(x_1) - F_d(x_1) \tag{*}$$

or $$F_c(x) \le F_c(x_1)$$

which proves the monotonicity.

Letting $x \to -\infty$ in (*), we get $0 \le F(x_1) - F_d(x_1) = F_c(x_1)$.

Now F_d is right–hand continuous since each δ_{a_i} is right hand continuous and the series defining F_d converges uniformly in x. (See below). Moreover,

$$F(x) - F(x - 0) = F_d(x) - F_d(x - 0) = b_j \text{ when } x = a_j$$

and both differences equal 0 otherwise. Hence for each x,

$$F_c(x) - F_c(x - 0) = F(x) - F(x - 0) - (F_d(x) - F_d(x - 0)) = 0 .$$

This says that F_c is left–hand continuous; being right–hand continous as the difference of two such functions, F_c becomes continuous.

The following lemma is another result from real analysis.

Lemma: *Let $\{g_n\}$ be real valued right-hand continuous functions* on an interval $(a, b) \subset R$ and let $G(x) = \sum_{n=1}^{\infty} g_n(x)$ be uniformly convergent for x in (a, b). Then G is right-hand continuous on (a, b). [Uniformly convergent means that for each $\varepsilon > 0$, there is an N_ε such that $N > N_\varepsilon$ implies $|\Sigma_n^N g_n(x) - G(x)| < \varepsilon$ for all x in (a, b).]

Exercise 6: Show that the lemma applies to $F_d(x) = \sum_{j=1}^{\infty} b_j \delta_j(x)$

for x in any finite interval (a, b). Hint: first show that
$$\Sigma_j b_j \delta_j \leq \Sigma_j b_j .$$

Definition: *A CDF that can be represented in the form*
$$F(x) = \sum_j b_j \delta_j(x)$$
where $\{a_j\}$ is at most countable, $b_j > 0$ for each j and $\Sigma_j b_j = 1$ is called a discrete distribution function. The corresponding RV X with $P(X \leq x) = F(x)$ is also said to be discrete (type).
A CDF that is continuous everywhere is called a continuous distribution function and the corresponding RV is continuous (type).

Some examples of continuous type RV's will appear in the next lesson. For a continuous CDF, the corresponding F_d is identically 0; for a discrete CDF, the corresponding F_c is identically 0.
If for a given CDF F, the corresponding F_d and F_c are not both identically zero, then
$$0 < F_d(\infty) = b < 1 \text{ and } 0 < F_c(\infty) = 1 - b < 1 .$$
It is easy to see that $F_1 = F_d/b$ is a discrete CDF and that $F_2 = F_c/(1 - b)$ is a continuous CDF. Then $F = bF_1 + (1 - b)F_2$ is a *mixture* of two CDF's.

Exercise 7: a) Prove that F_1, F_2 described just above are CDFs.

b)   Prove that this decomposition of a CDF as a mixture is unique: if $F = bF_1 + (1 - b)F_2$ and $F = aF_3 + (1 - a)F_4$ where F_1 and F_3 are discrete CDF's and F_2 and F_4 are continuous CDF's with constants $0 < a < 1$, $0 < b < 1$, then $F_1 = F_3$ and $F_2 = F_4$.

Hint: $bF_1 - aF_3 = (1-a)F_4 - (1-b)F_2$.

The following theorem establishes a kind of one–to–one correspondence between measures and bounded distribution functions on R; since the proof is measure theoretic, we refer the reader to such books as Chung (1974). The situation on R^n is much more involved than would appear from this discussion. In particular, one needs to "complete" the measures so that every subset of a set of measure zero also has measure zero. Loosely speaking, the theorem says that the identity function $X(\omega) = \omega$ can be treated as a RV with $P(X \leq x) = F(x)$.

Theorem: *Let Ω be the real line with the Borel σ-field generated by the intervals $(a,b]$. For every function F on Ω, with properties a,b,c,d) of the first theorem, one can generate a complete measure μ such that $\mu(a,b] = F(b) - F(a)$ for all intervals $(a,b]$.*

LESSON 4. SOME CONTINUOUS CDFs

All the random variables in Part I were discrete; in the symbolism of the previous lesson, their CDFs were all "F_d"s. It turns out that there are two kinds of "F_c"s: those which are differentiable "almost everywhere" and those which are differentiable nowhere; the latter are of mathematical interest but are not discussed herein. In this lesson, we look at examples of the former which are given by Riemann integrals. We will return to "almost everywhere" in a later lesson.

General Example: Let the function f be non–negative and Riemann integrable on the real line R, say $\int_{-\infty}^{+\infty} f(x)\ dx = 1$.

Then the identity function X on R, with its Borel σ–field, can be viewed as a random variable with CDF

$$F(x) = P(X \le x) = \int_{-\infty}^{x} f(t)\ dt\ .$$

When F is differentiable at some x_o, $F'(x_o) = f(x_o)$ is the value of the *probability density function* (PDF) f at x_o. Most importantly, $P(a < X \le b) = F(b) - F(a) = \int_{a}^{b} f(x)\ dx$ which is, of course, the "area under f" from a to b. Note that by properties of this integral, $P(a < X \le b) = P(a < X < b)$
$$= P(a \le X < b) = P(a \le X \le b)$$
because $P(X = a) = 0 = P(X = b)$.

Definition: *The Uniform Distribution (family indexed by the real parameters $\alpha < \beta$) is given by the CDF*

$$
\begin{aligned}
F_U(x) &= 0 & &\text{for } x < \alpha \\
&= (x - \alpha)/(\beta - \alpha) & &\text{for } \alpha \le x < \beta \\
&= 1 & &\text{for } \beta \le x\ .
\end{aligned}
$$

A graph of F_U looks like

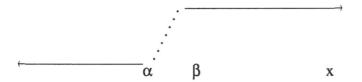

This distribution might also be given by the density since F_U is differentiable except at α and β :

$$F_U{}'(x) = f_U(x) = 0 \qquad \text{for } x < \alpha$$
$$= 1/(\beta - \alpha) \quad \text{for } \alpha < x < \beta$$
$$= 0 \qquad \text{for } \beta < x \ .$$

Its graph looks like

and the constant "equal" density suggests the name. (At times it may be necessary to keep track of the parameters with symbols like $F_U(x;\alpha,\beta)$, $f_U(x;\alpha,\beta)$.) Some like to view such distributions in terms of the *indicator function*: $I_A(x) = 1$ iff x is in the set A. Then, the PDF is

$$f_U(x) = I_{(\alpha,\beta)}(x)/(\beta - \alpha) \ \text{ and the CDF is}$$

$$F_U(x) = \int_{-\infty}^{x} I_{(\alpha,\beta)}(t)dt/(\beta - \alpha)$$

$$= (x - \alpha)I_{(\alpha,\beta)}(x)/(\beta - \alpha) + I_{(\beta,\infty)}(x) \ .$$

Example: Consider $\alpha = 0$ and $\beta = 1$ with $f_U(x) = I_{(0,1)}(x)$ and $F_U(x) = xI_{(0,1)}(x) + I_{[1,\infty)}(x)$.

    a)   Find $P(U \leq .6)$. Making use of the density function

we have $P(U \leq .6) = \int_{-\infty}^{.6} f_U(x) \, dx = \int_{0}^{.6} 1 \, dx = .6$

Making use of the distribution function we have

$$P(U \leq .6) = F_U(.6) = .6 \cdot I_{(0,1)}(.6) + I_{[1,\infty)}(.6) = .6 \ .$$

b)   For $P(U \geq .2)$ we have either

$$\int_{.2}^{\infty} f_U(x) \, dx = \int_{.2}^{1} 1 \, dx = 1 - .2 = .8 \text{ , or,}$$

$$P(U \geq .2) = 1 - P(U \leq .2) = 1 - F_U(.2) = 1 - .2 = .8.$$

Exercise 1:   Bees released at "A" fly off in a direction with respect to N(orth) measured by X (degrees). For this particular group of bees, "home" is to the southwest from "A". If some bee has been "disoriented" by an artificial sunlight–darkness scheme, its X should be uniformly distributed on the interval
$$(\alpha = 0, \ \beta = 360) \ .$$
Then, what is the probability that such a bee still flies "close to home", say between 120 and 150? Also find the values of $P(X > 270)$, $P(X \leq 100)$.

This may be the place to recall that we are dealing with mathematical models–always simplifications. Some argue that in "real life", nothing is continuous or, even if it were, we couldn't detect it. Computers are certainly limited in precision;even our best optical instruments can see only "so deep". Mathematically, the probability that the bee flies off at exactly 136 ($P(X = 136)$) is zero but we would get values like this. However, if we made our model to have a uniform distribution on the intervals which a compass could measure, say 0, 1, $\cdots$, 359, we would not have this anomolay. On the other hand, since $1/360 \approx .002777 \cdots$, up to two or even three decimal places, this discrete model may not be worth the extra effort. Such remarks apply to all continuous type RVs where a probability 0 does not mean impossible; but, $P(\text{impossible}) = P(\phi) = 0.$

Example: Here, we start with a density and derive the CDF.
    a)   The function $f(x) = cx$ for $0 < x < 1$
                    $= c(2 - x)$ for $1 < x < 2$
                    $= 0$ elsewhere

will be non–negative when $c \geq 0$. The total integral
$$\int_{-\infty}^{\infty} f(x) \, dx \quad \text{reduces to} \quad \int_{0}^{2} f(x) \, dx \quad \text{by the very}$$
definition of an integral. This must be evaluated in two parts because f "comes in two parts":

$$\int_0^1 cx\ dx\ +\ \int_1^2 c(2-x)\ dx$$

$$=\ cx^2/2\ \Big]_0^1\ +\ -c(2-x)^2/2\ \Big]_1^2$$

$$=\ c/2\ +\ c/2\ =\ c.$$

Therefore, $c = 1$.

b)   Now $F(x) = P(X \le x) = \int_{-\infty}^x f(t)\ dt$ is 0 for $x < 0$.

For $0 \le x < 1$, we want the area of the striped region

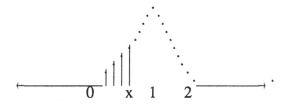

In particular, $P(X \le .4)$

$$=\ F(.4)\ =\ \int_0^{.4} f(t)\ dt$$

$$=\ \int_0^{.4} t\ dt = (.4)^2/2\ .$$

For $1 \le x < 2$, the area looks like:

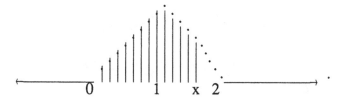

In particular,

$$F(1.4) = \int_0^{1.4} f(t)\ dt = \int_0^1 t\ dt + \int_1^{1.4} (2-t)\ dt$$

$$= t^2/2 \Big]_0^1 + -(2-t)^2/2 \Big]_1^{1.4}$$

$$= 1/2 - 0 - (.6)^2/2 + 1/2 = 1 - .18 = .82 .$$

c)   The use of indicators emphasizes the piecewise procedure:

$$f(x) = xI_{(0,1)}(x) + (2-x)I_{(1,2)}(x) .$$

For $0 \le a < 1$, $P(X \le a) = F(a) = \displaystyle\int_{-\infty}^{a} f(x)\,dx$

$$= \int_{-\infty}^{a} xI_{(0,1)}(x)\,dx + \int_{-\infty}^{a} (2-x)I_{(1,2)}(x)\,dx$$

$$= \int_0^a x\,dx + \int_0^a (2-x)(0)\,dx = a^2/2 .$$

For $1 \le a < 2$, $P(X \le a) = F(a) = \displaystyle\int_{-\infty}^{a} f(x)\,dx$

$$= \int_{-\infty}^{a} xI_{(0,1)}(x)\,dx + \int_{-\infty}^{a} (2-x)I_{(1,2)}(x)\,dx$$

$$= \int_0^1 x\,dx + \int_1^a (2-x)\,dx = 2a - a^2/2 - 1 .$$

Definition: *The Symmetric Triangular Distribution (family indexed by the real parameters* α, β *with* $\beta > 0$*) has a PDF*

$$f_T(x) = (x - \alpha + \beta)I_{[\alpha-\beta,\alpha)}(x)/\beta^2 +$$

$$(\alpha + \beta - x)I_{[\alpha,\alpha+\beta)}(x)/\beta^2$$

and a CDF

$$F_T(x) = \{(x - \alpha + \beta)^2/2\beta^2\}I_{[\alpha-\beta,\alpha)}(x)$$

$$+ \{1 - (\alpha + \beta - x)^2/2\beta^2\}I_{[\alpha,\alpha+\beta)}(x) + I_{[\alpha+\beta,\infty)}(x) .$$

Exercise 2: Following the example, verify that f_T is a density and its antiderivative is F_T. Sketch both graphs.

Of course, we can find these areas using triangles and trapezoids but the integration process illustrates a general procedure. If f were defined in terms of quadratics, cubics, $\cdots$, simple geometry would not suffice.

Exercise 3: "Crushing strength" of concrete is sometimes modeled by a triangular distribution with α the specified pressure. Say $\alpha = 5$, min $= \alpha - \beta = 2$, max $= \alpha + \beta = 8$ (all in Kpsi.) For this particular mix, call the random strength T; find the probability that
    a) $T < 3$ b) $T > 6$ c) $3 < T \le 6$
    d) $3 \le T < 7$ e) $T \ge 5.5$.

Exercise 4: Let c be a fixed positive constant. The function given by $f(x) = c(1-x^2)$ for $0 < |x| < 1$ and zero elsewhere is non–negative. Find the value of c so that f will be a probability density. Then find the CDF F .

Exercise 5: The function given by

$$g(x) = x^2 I_{(0,1)}(x) + (x - 2)^2 I_{(1,2)}(x)$$

is non–negative. Find the constant c such that $f(x) = cg(x)$ is (the value of) a PDF on the real line. Then find the corresponding CDF .

Definition: *The Gamma Distribution (family indexed by the positive parameters α and β) is given by the PDF*

$$f_G(x) = I_{[0,\infty)}(x)\beta^\alpha x^{\alpha-1} e^{-\beta x}/\Gamma(\alpha) .$$

We assume (via advanced calculus) that the gamma function exists (and is finite) for

$$\alpha > 0: \Gamma(\alpha) = \int_0^\infty x^{\alpha-1}e^{-x}dx .$$

Unfortunately, there is no closed form for the CDF

$$F_G(g) = \beta^\alpha \int_0^g x^{\alpha-1} e^{-\beta x} dx/\Gamma(\alpha) \text{ when } g > 0 .$$ The integral

$$\int_0^x t^{\alpha-1} e^{-t} dt$$ is an incomplete gammma function for which

there are some tables (but not here).

Exercise 6: Assume existence of the integrals.
   a)   Show, (integrate by parts) that $\Gamma(\alpha) = (\alpha-1)\Gamma(\alpha-1)$.
   b)   Show that $\Gamma(\alpha + 1) = \alpha!$ when α is a positive integer.

The gamma family has been used to model "lifetimes" in the treatment of diseases, evaluating electrical/mechanical components, forming learning curves, $\cdots$. A special case having $\alpha = 1$ is called the *exponential* distribution.

Exercise 7: Find the CDF F_E when the PDF is given by

$$f_E(x) = \beta\, e^{-\beta x} I_{(0,\infty)}(x) .$$

The exponential distribution has a characteristic "memoryless" property similar to that of the geometric distribution. The proof of the following theorem is relegated to an exercise after the succeeding lemma.

Theorem: *Suppose that* $P(X > a + b \mid X > a) = P(X > b)$ *, where* $P(X > x) = 1 - F(x)$ *is right-hand continuous, and* $F(0) = 0$ *. Then for some* $\beta > 0$,

$$F(x) = \{1 - e^{-\beta x}\} I_{[0,\infty)}(x).$$

Lemma: *a) If* $z(x + y) = z(x) + z(y)$ *where* z *is a right-hand continuous function on the real line, then* $z(x) = cx$ *for some constant* c .
b)   If w *is right-hand continuous and* $w(x + y) = w(x) \cdot w(y) > 0$, *then for some c,* $w(x) = exp(cx)$.

Partial proof: For $y = 0$, $z(x + y) = z(x) + z(y)$ becomes
$$z(x) = z(x) + z(0) \text{ and therefore, } z(0) = 0.$$
With $y = -x$, $z(0) = z(x) + z(-x)$ so that $z(-x) = -z(x)$.
For m a positive integer,

$$\begin{aligned} z(mx) &= z((m-1)x + x) = z((m-1)x) + z(x) \\ &= z((m-2)x) + z(x) + z(x) \\ &= \cdots = mz(x) \text{ (without formal induction). } (*) \end{aligned}$$

Let $x = n/m$ for positive integers n and m. Then,

$$z(n) = z(m(n/m)) = mz(n/m) \text{ by } (*) .$$

Therefore, $z(n/m) = z(n)/m = z(n \cdot 1)/m = nz(1)/m$. Now, let n/m $\downarrow$ x. Then by righthand continuity, $z(x) = xz(1)$ for $x > 0$. For x < 0, $z(x) = z(-(-x)) = -z(-x)$ with $-x > 0$. Hence,

$z(x) = -(-x)z(1) = xz(1)$. This shows that $c = z(1)$.

Exercise 8: a) Complete the proof of the lemma and the last theorem.
b)   Complete the characterization of the exponential: for f_E in

exercise 7, show that $P(X > a + b \mid X > a) = P(X > b)$.

INE–Exercises:
1.   Find the constant c which makes the following a PDF for α and $\beta > 0$. (W is for Weibull.)

$$f_W(x) = cx^{\beta - 1} e^{-\alpha x^{\beta}} I_{(0,\infty)}(x) .$$

2.   a) Show that for α and $\beta > 0$,

$$\int_0^\infty z^{\alpha+\beta-1} e^{-z} \, dz \int_0^1 w^{\alpha-1} (1 - w)^{\beta-1} \, dw$$

$$= \int_0^\infty x^{\alpha-1} e^{-x} \, dx \int_0^\infty y^{\beta-1} e^{-y} \, dy .$$

Hint : express the right–hand side as a double integral and integrate by substitution: $x + y = z$ and $x = w/z$.
b)   Write the corresponding density function
$f_B(w)$, $0 < w < 1$, in the most compact form.

(B is for beta.)

3.   Verify that each of the following is a PDF.

a)   (Cauchy) $f(x; \alpha) = \alpha/\pi(\alpha^2 + x^2)$ for $\alpha > 0$ and x real.

b)   (Laplace) $f(x) = e^{-|x|}/2$ for x real.

LESSON 5. THE NORMAL DISTRIBUTION

You should recall that one view of the binomial distribution is as an approximation to the hypergeomtric; also, for "large" n and "small" p, the binomial can be approximated by the Poisson. Here we introduce a third approximation for the binomial in terms of "large" n and "moderate" p using the so–called normal density

$$e^{-x^2/2}/\sqrt{2\pi} = \{\exp(-x^2/2)\}/(2\pi)^{1/2}.$$

This distribution is often named for Gauss, reflecting the importance of his careful studies, but it had been treated over 50 years earlier by DeMoivre. As you might guess, we need some more analysis.

Lemma: *a) (Stirling's approximation for factorials) For a positive integer m,*

$$m! = (2\pi m)^{1/2} m^m e^{-m + \theta(m)} \text{ where } 0 < \theta(m) < 1/12m.$$

b)   For $0 < |x| < 1/2$, $\log(1 + x) = x - x^2/2 + \rho x^3$ where

$|\rho| < 8/3$ *and "log" is natural logarithm.*

Exercise 1:   a) Check "An Elementary Proof of Stirling's Formula", by Diaconis and Freedman, 1986.
b)   Verify the Taylor formula for $\log(1 + x)$.

For the binomial, when J is the number of successes and K   is the number of failures in   n   trials,   J + K = n. One mathematical basis of the normal approximation is given below.

Exercise 2: Fill in all the missing steps/reasons in the proof of the following theorem.

Theorem: *For $0 < p < 1$ and $0 < q < 1$ fixed, let $j + k = n$,*
$\delta_n = j - np$, $b_n = n! p^j q^k / j! k!$, $\varphi_n = [\exp(-\delta_n^2/2npq)]/(2\pi npq)^{1/2}$.
If $\delta_n^3/n^2 \to 0$ as $n \to \infty$, then $b_n/\varphi_n \to 1$.

Proof: a) For $0 < \theta(n) < 1/12n$, $0 < \theta(j) < 1/12j$,

and $0 < \theta(k) < 1/12k$, $n! = \sqrt{2\pi n}\, n^n e^{-n+\theta(n)}$,

$$k! = \sqrt{2\pi\ k}\ k^k\ e^{-k+\theta(k)},$$

$$j! = \sqrt{2\pi\ j}\ j^j\ e^{-j+\theta(j)}.$$

Substituting these in b_n yields the value

$$\sqrt{(n/2\pi\ jk)}\ (np/j)^j\ (nq/k)^k\ e^{\theta(n)-\theta(j)-\theta(k)}.$$

Fortunately, $e^{\theta(n)-\theta(j)-\theta(k)} \to 1$ as $n \to \infty$.

b) $n/(jk) = n/(np + \delta_n)(nq - \delta_n)$

$$= 1/(npq + (q-p)\delta_n - \delta_n^2/n) \text{ so that}$$

$$\frac{n/jk}{1/npq} = 1/(1 + (q-p)\delta_n/npq - \delta_n^2/n^2pq).$$

c)   From $\delta_n^3/n^2 \to 0$, we get $\delta_n/n = (\delta_n^3/n^2)^{1/3}(n^{-1/3}) \to 0$

and $\delta_n^2/n^2 \to 0$. Combined with b), this yields

$$(n/2\pi\ jk)^{1/2}/(1/2\pi\ npq)^{1/2} \to 1.$$

d)   For $|\delta_n/np| < 1/2$,

$-j\log(j/np) = -(np + \delta_n)\log(1 + \delta_n/np)$

$$= -(np + \delta_n)(\delta_n/np - \delta_n^2/2n^2p^2 + \rho_n\delta_n^3/n^3p^3)$$

where $|\rho_n| < 8/3$. This last product simplifies to

$$-\delta_n - \delta_n^2/2np + \delta_n^3(1/2p^2 - \rho_n/p^2 - \rho_n\delta_n/np^3)/n^2.$$

Similarly for $|\delta_n/nq| < 1/2$, $-k\log(k/nq)$

$$= \delta_n - \delta_n^2/2nq - \delta_n^3(1/2q^2 - \sigma_n/q^2 + \sigma_n\delta_n/nq^3)/n^2$$

where $|\sigma_n| < 8/3$. Hence,

$$\log(np/j)^j(nq/k)^k = -\delta_n^2/2npq + B_n$$

$$\text{where } B_n = \delta_n^{\,3}\{(1/p^2 - 1/q^2)/2 - (\rho_n/p^2 - \sigma_n/q^2)$$
$$- \delta_n(\rho_n/p^3 + \sigma_n/q^3)/n\}/n^2 \to 0$$

(Note that this is where the hypothesis on δ_n originated!)

e)   From d), we get $(np/j)^j(nq/k)^k = \exp(-\delta_n^{\,2}/2npq + B_n)$.

The conclusion follows on combining all five parts.

The approximation is valid for "large" n and j; that is,

$$P(J = j) = b_n \approx e^{-(j-np)^2/2npq}/(2\pi\,npq)^{1/2} .$$

The following table (obtained using MIT–MACSYMA) contains both good and bad "results" illustrating the vagaries of numerical calculcations.

j	n	$\begin{bmatrix} n \\ j \end{bmatrix}(.4)^j(.6)^{n-j}$	$e^{-(j-.4n)^2/.48n}/\sqrt{.48\pi\,n}$
4	10	.2508	.2575
5	10	.2007	.2091
6	10	.1115	.1119
6	20	.1249	.1200
8	20	.1797	.1821
12	20	.0355	.0344
16	20	.0003	.0002
30	100	.01000	.01014
50	100	.01034	.01014
70	100	9.05056×10^{-10}	5.85845×10^{-10}
90	100	1.60405×10^{-25}	1.95569×10^{-24}
110	500	5.70182×10^{-18}	8.01458×10^{-17}
200	500	.036399	.036418
300	500	8.9529×10^{-20}	2.9222×10^{-20}
400	500	0	0

Later we shall see a normal approximation for the CDF $P(J \le j)$ which will be easier to use and more accurate.

Definition: *A random variable Z has the standard normal distribution if its PDF is*

$$\varphi_Z(z) = \{exp(-z^2/2)\}/\sqrt{2\pi} \ \ for \ \text{-}\infty < z < \infty \ .$$

Exercise 3: Show that $\int_{-\infty}^{\infty} \varphi_Z(z) \ dz = 1$ by first showing that the

product $\int_0^{\infty} e^{-x^2/2} \ dx \cdot \int_0^{\infty} e^{-y^2/2} \ dy = \pi/2$. Hint: write the

product as a double integral and integrate by substitution into polar coordinates.

Unfortunately, $P(Z \le z) = \phi(z) = \int_{-\infty}^{z} e^{-t^2/2} \ dt/\sqrt{2\pi}$

does not have an elementary form; as in the gamma case, the integral cannot be expressed as a finite combination of roots, polynomials, ratios, trigonometric and hyperbolic functions of z.

This is also true of $\int_0^{z} e^{-t^2/2} \ dt/\sqrt{2\pi}$ but we can approximate

this integral by some numerical technique. This has been done here using STSC_APL .

The standard normal density is symmetric about 0 so for $z > 0$, $P(Z \le \text{-}z) = 1 - P(Z \le z) = P(Z \ge z)$. Thus we have to calculate only "half" the integrals. The table at the end of this volume contains some values of $P(Z \le z)$ for $z > 0$.

Example: a) Reading the table directly, we find

   $P(Z \le 1.00) = .841345 = P(Z \ge \text{-}1.00)$;

   $P(Z \le 2.00) = .977250 = P(Z \ge \text{-}2.00)$;

   $P(Z > 1.65) = 1 - P(Z \le 1.65) = 1 - .950529 = .009071$;

   $P(1.30 < Z \le 2.80) = P(Z \le 2.80) - P(Z \le 1.30)$;

   $\qquad\qquad = .997445 - .903200 = .09425$;

b)   Reading the table in reverse, we find

   $P(Z \le ?) = .841345$ implies $? = 1.00$;

   $P(Z > ?) = .009071$ implies $? = 1.65$;

   $P(Z > ?) = .950929$ implies $? = \text{-}1.65$;

   $P(Z > ?) = .903200$ implies $z = \text{-}1.30$;

P(Z < ?) = .997445 implies z = 2.80;

c)   By symmetry, P(–a < Z < a) = .950004 implies

P(Z < –a) = (1 – .950004)/2 which then implies

P(Z < a) = P(–a < Z < a) + P(Z < –a)

$$= .950004 + .024998 = .975002$$

so that a = 1.96.

Exercise 4: a) Find P(Z ≤ 3.01), P(Z > –2.99),

P(1.36 < Z < 1.86), P(–1.36 < Z < 1.87),

P(–1.87 < Z < –1.36), P($|Z|$ < 2.23) .

b)   Find "c" when

P(Z ≤ c) = .868643 ; P(Z ≤ c) = .131357 ;

P(Z < c) = .868643 ; P(–c ≤ z ≤ c) = .892620 .

Although we do not know an underlying probability space [Ω, $\mathcal{B}$,P], we can still consider the meaning of the sets

$$\{\omega : Z(\omega) \le z\} ;$$

in fact for any RV Z on this space,

$$P(Z \le z) = P(\{\omega : Z(\omega) \le z\}) .$$

Now let μ,σ be fixed real numbers with σ > 0. Let X = μ + σZ, that is, X(ω) = μ + σZ(ω) for all ω in Ω . Then,

$$\{\omega : X(\omega) \le x\} = \{\omega : Z(\omega) \le (x - \mu)/\sigma\}$$

and

$$(X \le x) = P(\{ \omega : X(\omega) \le x\}) = P(\{\omega : Z(\omega) \le (x - \mu)/\sigma\})$$

$$= P(Z \le (x - \mu)/\sigma) .$$

Using more words, if Z is a real RV, the simple affine transformation X = μ + σZ defines a new RV X with CDF P(X ≤ x) = P(Z ≤ (x – μ)/σ) . In particular, when Z is normal, we get the:

Definition: *A Normal Distribution (family indexed by the parameters μ (real) and σ (positive)) is given by the CDF*

$$P(X \le x) = P(Z \le (x - \mu)/\sigma) = \int_{-\infty}^{(x-\mu)/\sigma} e^{-z^2/2}\, dz/\sqrt{2\pi} .$$

Exercise 5: a) Find the PDF of the normal distribution family by differentiating the CDF (above) with respect to x.

b)   Show that each normal density has maximum at $x = \mu$ and points of inflection at $x = \mu \pm \sigma$.

Exercise 6: Evaluate: a) $\int_{-\infty}^{\infty} x\{\exp(-(x - \mu)^2/2\sigma^2)\}dx$;

b) $\int_{-\infty}^{\infty} (x-\mu)^2\{\exp(-(x - \mu)^2/2\sigma^2\}dx$;

c) $\int_{-\infty}^{\infty} (x - \mu)^n e^{-(x - \mu)^2/2\sigma^2} dx$ for positive integers n. Hint:

for the last part, consider n odd and n even.

Exercise 7: Let $G(x) = \int_{-\infty}^{X} e^{-(t - \mu)^2/2\sigma^2} dt/(2\pi \sigma^2)^{1/2}$; find:

$$G(3.00) \quad \text{when } \mu = 2 \text{ and } \sigma^2 = 4;$$
$$G(3.00) \quad \text{when } \mu = 3 \text{ and } \sigma^2 = 4;$$
$$G(6.80) \quad \text{when } \mu = 2 \text{ and } \sigma^2 = 4;$$
$$G(11.40) \quad \text{when } \mu = 3 \text{ and } \sigma^2 = 9.$$

Hint: integrate by substitution and use the table for $P(Z \le (x - \mu)/\sigma)$.

INE–Exercises:
1.   Let U be uniformly distributed on [0, 1]:

$$P(U \le x) = 0 \text{ for } x < 0$$
$$= x \text{ for } 0 \le x < 1$$
$$= 1 \text{ for } 1 \le x .$$

Let $\alpha < \beta$ be constants. Show that $X = \alpha + (\beta - \alpha)U$ is a RV with a uniform distribution on $[\alpha, \beta]$.

2.   Let X be a normal RV with parameters μ and σ^2. Find the PDF of the distribution of $Y = e^X$.
Hint: when $y > 0$, $P(Y \le y) = P(X \le \log(y))$.

3.   Show that the following is a PDF (a log–normal distribution):

$$I_{(0,\infty)}(x) \, (1/x\sigma\sqrt{2\pi})\exp(-\log((x) - \mu)^2/2\sigma^2) .$$

Table Standard Normal

This table (generated by STSC–APL) contains values of the standard normal CDF . For example, $P(Z \le z = .49) = .687933$.

z		z		z	
.01	.503989	.41	.659097	.81	.791030
.02	.507978	.42	.662757	.82	.793892
.03	.511966	.43	.666402	.83	.796731
.04	.515953	.44	.670031	.84	.799546
.05	.519939	.45	.673645	.85	.802337
.06	.523922	.46	.677242	.86	.805105
.07	.527903	.47	.680822	.87	.807850
.08	.531881	.48	.684386	.88	.810570
.09	.535856	.49	.687933	.89	.813267
.10	.539828	.50	.691462	.90	.815940
.11	.543795	.51	.694974	.91	.818589
.12	.547758	.52	.698468	.92	.821214
.13	.551717	.53	.701944	.93	.823814
.14	.555670	.54	.705401	.94	.826391
.15	.559618	.55	.708840	.95	.828944
.16	.563559	.56	.712260	.96	.831472
.17	.567495	.57	.715661	.97	.833977
.18	.571424	.58	.719043	.98	.836457
.19	.575345	.59	.722405	.99	.838913
.20	.579260	.60	.725747	1.00	.841345
.21	.583166	.61	.729069	1.01	.843752
.22	.587064	.62	.732371	1.02	.846136
.23	.590954	.63	.735653	1.03	.848495
.24	.594835	.64	.738914	1.04	.850830
.25	.598706	.65	.742154	1.05	.853141
.26	.602568	.66	.745373	1.06	.855428
.27	.606420	.67	.748571	1.07	.857690
.28	.610261	.68	.751748	1.08	.859929
.29	.614092	.69	.754903	1.09	.862143
.30	.617911	.70	.758036	1.10	.864334
.31	.621720	.71	.761148	1.11	.866500
.32	.625516	.72	.764238	1.12	.868643
.33	.629300	.73	.767305	1.13	.870762
.34	.633072	.74	.770350	1.14	.872857
.35	.636831	.75	.773373	1.15	.874928
.36	.640576	.76	.776373	1.16	.876976
.37	.644309	.77	.779350	1.17	.879000
.38	.648027	.78	.782305	1.18	.881000
.39	.651732	.79	.785236	1.19	.882977
.40	.655422	.80	.788145	1.20	.884930

z		z		z	
1.21	.886861	1.61	.946301	2.01	.977784
1.22	.888768	1.62	.947384	2.02	.978308
1.23	.890651	1.63	.948449	2.03	.978822
1.24	.892512	1.64	.949497	2.04	.979325
1.25	.894350	1.65	.950529	2.05	.979818
1.26	.896165	1.66	.951543	2.06	.980301
1.27	.897958	1.67	.952540	2.07	.980774
1.28	.899727	1.68	.953521	2.08	.981237
1.29	.901475	1.69	.954486	2.09	.981691
1.30	.903200	1.70	.955435	2.10	.982136
1.31	.904902	1.71	.956367	2.11	.982571
1.32	.906582	1.72	.957284	2.12	.982997
1.33	.908241	1.73	.958185	2.13	.983414
1.34	.909877	1.74	.959070	2.14	.983823
1.35	.911492	1.75	.959941	2.15	.984222
1.36	.913085	1.76	.960796	2.16	.984614
1.37	.914657	1.77	.961636	2.17	.984997
1.38	.916207	1.78	.962462	2.18	.985371
1.39	.917736	1.79	.963273	2.19	.985738
1.40	.919243	1.80	.964070	2.20	.986097
1.41	.920730	1.81	.964852	2.21	.986447
1.42	.922196	1.82	.965620	2.22	.986791
1.43	.923641	1.83	.966375	2.23	.987126
1.44	.925066	1.84	.967116	2.24	.987455
1.45	.926471	1.85	.967843	2.25	.987776
1.46	.927855	1.86	.968557	2.26	.988089
1.47	.929219	1.87	.969258	2.27	.988396
1.48	.930563	1.88	.969946	2.28	.988696
1.49	.931888	1.89	.970621	2.29	.988989
1.50	.933193	1.90	.971283	2.30	.989276
1.51	.934478	1.91	.971933	2.31	.989556
1.52	.935745	1.92	.972571	2.32	.989830
1.53	.936992	1.93	.973197	2.33	.990097
1.54	.938220	1.94	.973810	2.34	.990358
1.55	.939429	1.95	.974412	2.35	.990613
1.56	.940620	1.96	.975002	2.36	.990863
1.57	.941792	1.97	.975581	2.37	.991106
1.58	.942947	1.98	.976148	2.38	.991344
1.59	.944083	1.99	.976705	2.39	.991576
1.60	.945201	2.00	.977250	2.40	.991802

z		z		z	
2.41	.992024	2.81	.997523	3.21	.999336
2.42	.992240	2.82	.997599	3.22	.999359
2.43	.992451	2.83	.997673	3.23	.999381
2.44	.992656	2.84	.997744	3.24	.999402
2.45	.992857	2.85	.997814	3.25	.999423
2.46	.993053	2.86	.997882	3.26	.999443
2.47	.993244	2.87	.997948	3.27	.999462
2.48	.993431	2.88	.998012	3.28	.999481
2.49	.993613	2.89	.998074	3.29	.999499
2.50	.993790	2.90	.998134	3.30	.999517
2.51	.993963	2.91	.998193	3.31	.999534
2.52	.994132	2.92	.998250	3.32	.999550
2.53	.994297	2.93	.998305	3.33	.999566
2.54	.994457	2.94	.998359	3.34	.999581
2.55	.994614	2.95	.998411	3.35	.999596
2.56	.994766	2.96	.998462	3.36	.999610
2.57	.994915	2.97	.998511	3.37	.999624
2.58	.995060	2.98	.998559	3.38	.999638
2.59	.995201	2.99	.998605	3.39	.999651
2.60	.995339	3.00	.998650	3.40	.999663
2.61	.995473	3.01	.998694	3.41	.999675
2.62	.995604	3.02	.998736	3.42	.999687
2.63	.995731	3.03	.998777	3.43	.999698
2.64	.995855	3.04	.998817	3.44	.999709
2.65	.995975	3.05	.998856	3.45	.999720
2.66	.996093	3.06	.998893	3.46	.999730
2.67	.996207	3.07	.998930	3.47	.999740
2.68	.996319	3.08	.998965	3.48	.999749
2.69	.996427	3.09	.998999	3.49	.999758
2.70	.996533	3.10	.999032	3.50	.999767
2.71	.996636	3.11	.999065	3.51	.999776
2.72	.996736	3.12	.999096	3.52	.999784
2.73	.996833	3.13	.999126	3.53	.999792
2.74	.996928	3.14	.999155	3.54	.999800
2.75	.997020	3.15	.999184	3.55	.999807
2.76	.997110	3.16	.999211	3.56	.999815
2.77	.997197	3.17	.999238	3.57	.999822
2.78	.997282	3.18	.999264	3.58	.999828
2.79	.997365	3.19	.999289	3.59	.999835
2.80	.997445	3.20	.999313	3.60	.999841

z		z		z	
3.61	.999847	3.62	.999853	3.63	.999858
3.64	.999864	3.65	.999869	3.66	.999874
3.67	.999879	3.68	.999883	3.69	.999888
3.70	.999892	3.71	.999896	3.72	.999900
3.73	.999904	3.74	.999908	3.75	.999912
3.76	.999915	3.77	.999918	3.78	.999922
3.79	.999925	3.80	.999928	3.81	.999931
3.82	.999933	3.83	.999936	3.84	.999938
3.85	.999941	3.86	.999943	3.87	.999946
3.88	.999948	3.89	.999950	3.90	.999952
3.91	.999954	3.92	.999956	3.93	.999958
3.94	.999959	3.95	.999961	3.96	.999963

LESSON 6. SOME ALGEBRA OF RANDOM VARIABLES

As indicated before, we consider only certain highlights of measure theory in probability spaces. First we clarify the arithmetic of the extended real line $\overline{R}$: all real numbers and the two ideal points $+\infty$, $-\infty$. The introduction of these ideal points is for the mathematical convenience that every set of real numbers have a sup (least upper bound) and an inf (greatest lower bound); these will be pursued in lesson 7. On the other hand, the probability side, there is the real Galton–Watson process wherein we just might have positive probability that the time to extinction is infinite; the earliest example of this is the computation of the probabilities that the lineage of a certain British peer should die out.

Definition: *$(+\infty) + (+\infty) = +\infty$; $(-\infty) + (-\infty) = -\infty$.*

Let r be a fixed real (finite) number;

$$r + (+\infty) = +\infty = (+\infty) + r \; ; r + (-\infty) = -\infty = (-\infty) + r \; .$$

For $|r| < \infty$, $r/\pm\infty = 0$; for $r = 0$, $r(\pm\infty) = 0 = (\pm\infty)r$.

For $0 < r \le +\infty$, $r(+\infty) = +\infty = (+\infty)r$ and $r(-\infty) = -\infty = (-\infty)r$.

For $-\infty \le r < 0$, $r(+\infty) = -\infty = (+\infty)r$ and $r(-\infty) = +\infty = (+\infty)r$.

For s and t in R , the absolute value relations hold:

$$|s + t| \le |s| + |t| \; ; \; |st| = |s| \cdot |t| \; .$$

The expressions $+\infty - \infty$ and $-\infty + \infty$ remain undefined.

For $\overline{R}$ the corresponding Borel σ–field, $\overline{\mathscr{B}}$ can be generated by the intervals $[-\infty, a]$ or the intervals $(a, +\infty]$, or $\cdots$; every $\overline{B}$ in $\overline{\mathscr{B}}$ is a countable expression of unions and/or intersections and/or complementations of such intervals. Except for "ideal points", a measurable function is the same as a random variable; either term may be used here.

Definition: *Let $[\Omega, \mathscr{B}, P]$ be a probability space. Let X be a(n) extended) real valued function on Ω (symbolically, $X : \Omega \to \overline{R}$). Then X is measurable iff $\{\omega \mid X(\omega) \le x\} = X^{-1}[-\infty, x]$ is in $\mathscr{B}$ for* *each (all) x in Ř.*

We abbreviate $X^{-1}[-\infty, x]$ as $\{X \leq x\}$; we say that X is a measurable function. Because of the closure of $\mathscr{B}$ (and $\overline{\mathscr{B}}$ as σ–fields), the condition $\{X \leq x\} \in \mathscr{B}$ is equivalent to $\{X \geq x\}$ or $\{X > x\}$ or $\{X < x\} \in \mathscr{B}$. Moreover, since X^{-1} "preserves" unions, intersections, and complements (exercise 1 below), X is measurable iff $X^{-1}(\overline{B}) \in \mathscr{B}$ for all $\overline{B} \in \overline{\mathscr{B}}$.

Exercise 1: Complete the proof of the following; note that measurability is not needed for these results.

Lemma: *Consider the spaces Ω and Ř. Let $X : \Omega \to \check{R}$. Let B_α be sets in Ř. Then:*

a) $X^{-1}(\cup_\alpha B_\alpha) = \cup_\alpha X^{-1}(B_\alpha);$

Proof: $\omega \in X^{-1}(\cup_\alpha \overline{B}_\alpha)$ iff $X(\omega) \in \cup_\alpha \overline{B}_\alpha$ iff $X(\omega) \in \overline{B}_\alpha$ for some α iff $\omega \in X^{-1}(\overline{B}_\alpha)$ for some α iff $\omega \in \cup_\alpha X^{-1}(\overline{B}_\alpha)$.

b) $X^{-1}(\cap_\alpha B_\alpha) = \cap_\alpha X^{-1}(B_\alpha) ;$

c) $X^{-1}(B_\alpha{}^c) = (X^{-1}(B_\alpha))^c.$

Although the following example is artificial, the distinction between measurable and non–measurable cannot be missed; some people consider the corresponding examples in general measure theory "artificial".

Example: First recall that probability is defined on $\mathscr{B}$ not on Ω and, when Ω is a finite set, all σ–fields reduce to fields. Take
$$\Omega = \{a\ b\ c\ d\ e\}, \quad \mathscr{B} = \{\phi, \{a\ b\}, \{c\ d\ e\}, \Omega \} \text{ and}$$

$$P(\{a\ b\}) = 1/3 , P(\{c\ d\ e\}) = 2/3 .$$

a)   If $X(a) = X(b) = 2$ and $X(c) = X(d) = X(e) = 3$, then

$$X^{-1}[-\infty, x] = \phi \qquad \text{for } x < 2$$
$$= \{a \ b\} \qquad \text{for } 2 \le x < 3$$
$$= \Omega \qquad \text{for } 3 \le x .$$

Since each of these sets is in $\mathscr{B}$, X is measurable.

b)   If Y(a) = 1, Y(b) = 2, Y(c) = Y(d) = Y(e) = 3, then

$$Y^{-1}[-\infty, x] = \phi \qquad \text{for } x < 1$$
$$= \{a\} \qquad \text{for } 1 \le x < 2$$
$$= \{a \ b\} \qquad \text{for } 2 \le x < 3$$
$$= \Omega \qquad \text{for } 3 \le x .$$

Since {a} is not in $\mathscr{B}$, Y is not measurable.

c)   P(X = 2) = P(a b) = 1/3 but P(Y = 2) = P(b) is undefined.

Exercise 2: Continuing this example, show that

$$Z(a) = Z(c) = Z(d) = 3, \ Z(b) = Z(e) = 5$$

does not define a measurable function.

The point is that given $[\Omega, \mathscr{B}, P]$, it is not true that any real valued function will be measurable yet we need measurability to define its CDF, say P(X ≤ x). If, in the example, $\mathscr{B}$ were the power set, there would be no difficulty but it can be shown that in general measure theory this escape is not always available. Fortunately, there is another point too; almost any real valued function useful in an application will be measurable and its CDF will be describable even when not calculable explicitly. In particular, when X is a measurable function from Ω to $\bar{R}$, and g is a continuous function from $\bar{R}$ to $\bar{R}$, then g(X) will be a measurable function from Ω to $\bar{R}$.

The following definition begins the algebra of random variables with the two simplest functions; we have seen special cases in earlier lessons. The lemma following the definition is "the next logical step".

Definition: *Let X be a measurable function on Ω of the space $[\Omega, \mathscr{B}, P]$.*

a)   *X is a degenerate RV if for some (one) r in $\bar{R}$,*

$$P(\{\omega : X(\omega) = r\}) = 1 .$$

b)   *X is an indicator RV if for some (one) set A in $\mathscr{B}$,*

$X(\omega) = 1$ *for ω in A and $X(\omega) = 0$ for ω not in A.*

Note that for degeneracy, it is not necessary that $X(\omega) = r$ for all ω in Ω; we say that $X = r$ *almost surely* (abbreviated a.s.). For a continuous type RV X, $P(X = 136) = 0$ so that $P(X \neq 136) = 1$ and X is almost surely not 136; recall the discussion of the bees in lesson 4. In a similar way, for all the gamma type RVs, $P(X > 0) = 1$ so X is positive a.s.; this X is also a.s. non–negative. One could attach a.s. to most results in probability but only particularly important ones are so flaged. Degenerate RVs are usually just "constants". The indicator X is usually written as I_A .

Lemma: *If I_{A_1} and I_{A_2} are indicator random variables for $[\Omega, \mathscr{B}]$ and c_1, c_2 are real non-zero constants, then*

$$X = c_1 I_{A_1} + c_2 I_{A_2}$$

is a measurable function in $[\Omega, \mathscr{B}]$: for all $x \in \bar{R}$, $\{\omega : c_1 I_{A_1}(\omega) + c_2 I_{A_2}(\omega) \leq x\} \in \mathscr{B}$.

Partial proof: Consider the case $c_1 < 0 < c_1 + c_2 < c_2$. Then,

$$
\begin{aligned}
\{X(\omega) \leq x\} &= \phi & &\text{for } x < c_1 \\
&= A_1 \cap A_2{}^c & &\text{for } c_1 \leq x < 0 \\
&= A_2{}^c & &\text{for } 0 \leq x < c_1 + c_2 \\
&= A_1 \cup A_2{}^c & &\text{for } c_1 + c_2 \leq x < c_2 \\
&= \Omega & &\text{for } c_2 \leq x .
\end{aligned}
$$

Since each of these sets is in $\mathscr{B}$, X is measurable.

Exercise 3: Complete the proof of the lemma for the cases $0 < c_1 < c_2$ and $c_1 < c_2 < 0$.

Although we are not doing the formal induction, we conclude from this that any *finite linear combination* of indicator

RVs is another RV:

$$\sum_{i=1}^{k} c_i I_{A_1} \text{ is a measurable function when the } A_i \text{ are}$$

measurable sets and the c_i are (finite) constants.

Now we examine the lemma from a different perspective.

Let $B_1 = A_1 \cap A_2^c$, $B_2 = A_1^c \cap A_2$,

$B_3 = A_1 \cap A_2$ and $B_4 = A_1^c \cap A_2^c$ like

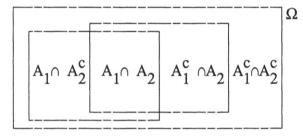

These B_i form a partition of Ω:

$$\text{for } i \neq j, B_i \cap B_j = \phi \text{ and } B_1 \cup B_2 \cup B_3 \cup B_4 = \Omega.$$

The function X in the lemma is the same as if we had written

$$X = c_1 I_{B_1} + c_2 I_{B_2} + (c_1 + c_2) I_{B_3} + 0 I_{B_4}.$$

This leads to the construction of another RV and the reader should note where properties of $\overline{R}$ may need to be invoked.

Definition: *a) A collection $\{B_i\}$ of subsets of a set Ω is a partition of Ω if   the B_i are pairwise disjoint*

($i \neq j$ implies $B_i \cap B_j = \phi$) and

their union is Ω .

If these B_i are also in an associated σ-field $\mathcal{B}$, then they form a measurable partition.

b)   If $A_1, A_2, \cdots, A_k \in \mathcal{B}$ is a finite partition of Ω in $[\Omega, \mathcal{B}, P]$ and $c_1, c_2, \cdots, c_k$ are constants, then

$$X = \sum_{i=1}^{k} c_i I_{A_i} \quad \text{is called a simple RV.}$$

c)   If $B_1, B_2, \cdots \in \mathscr{B}$ is a countable partition of Ω of
$[\Omega, \mathscr{B}, P]$ and $c_1, c_2, \cdots$ are constants, then $X = \sum_{i=1}^{\infty} c_i I_{B_i}$
is called an elementary RV.

Example: Consider the sample space for families with three single births: $\Omega = \{bbb \ bbg \ bgb \ gbb \ ggb \ bgb \ bgg \ ggg\}$. Let X be the number of girls in a sample point. Then X induces the partition

$$B_1 = X^{-1}(0) = \{bbb\} \ , \quad B_2 = X^{-1}(1) = \{bbg \ gbb \ bgb\} \ ,$$

$$B_3 = X^{-1}(2) = \{ggb \ bgg \ gbg\} \ , \quad B_4 = X^{-1}(3) = \{ggg\} \ .$$

For $c_1 = 0, c_2 = 1, c_3 = 2, c_4 = 3$,

$$X = c_1 I_{B_1} + c_2 I_{B_2} + c_3 I_{B_3} + c_4 I_{B_4} \ .$$

Inconveniently, such representations of simple (elementary) functions are not unique as you are now asked to demonstrate.

Exercise 4:   Consider the traditional space for the outcome of a toss of a pair of dice: $\Omega = \{11 \ \ 12 \ \cdots \ 66\}$. Let $X(\omega)$ be the sum of the components of ω :

$$X(11) = 2, \ X(12) = 3 = X(21), \ \cdots \ .$$

   a)   Find the partition $A_1, A_2, \cdots, A_{11}$ which makes
$$X = \sum_{j=1}^{11} (j + 1) I_{A_j} \ .$$
   b)   Find a partition $B_1, B_2, \cdots, B_{36}$ which makes
$$X = \sum_{i=1}^{6} \sum_{j=1}^{6} (i + j) I_{B_{ij}} \ .$$

Exercise 5:   For a probability space $[\Omega, \mathscr{B}, P]$, let

$$X(\omega) = \begin{array}{l} 1 \text{ for } \omega \in A_1 \\ 4 \text{ for } \omega \in A_2 \\ 2 \text{ for } \omega \in A_3 \\ 3 \text{ for } \omega \in A_4 \end{array} .$$

a)   What can you conclude about A_1, A_2, A_3, A_4 ?

b)   Write the CDF of X in terms of $\delta_1, \delta_2, \delta_3, \delta_4$ (lesson 4).

In classical analysis, attention focuses on continuous functions whereas here we need the algebra of measurable functions. We used a special case of the next theorem when dealing with the normal (lesson 5). Although the proof of this theorem is straightforward, it is long and cumbersome so we leave it as an INE–exercise; see, for example, Loève, 1963.

Theorem: *Let X and Y be measurable functions in $[\Omega, \mathscr{B}]$ and let r be a real number in R. Then:*
   a)   rX is measurable;
   b)   X + Y is measurable;
   c)   X·Y is measurable.

Since the details establishing measurability can become quite tricky, in the remainder, we shall assume that all the functions we deal with are measurable and that whenever such proofs are needed, they have been, or at least could be, carried out.

LESSON 7. CONVERGENCE OF SEQUENCES OF RANDOM VARIABLES

It has been seen that the inclusion of the ideal points creates worrisome details in measurability; we will be mollifying this somewhat. In this lesson and the next, we combine properties of measurable functions and sequences of real numbers; we include some specific results which will be needed later but we do not include all the interesting facts about sequences.

Definition: *Let S be a subset of $\overline{R}$. Any $b \in \overline{R}$ with $b \le s$ for all $s \in S$ is a lower bound of S; S is bounded below if S has a lower bound. The greatest lower bound $glb\ S = b$ iff*
b is a lower bound and
for each $\varepsilon > 0$, there is some s_ε in S such that $s_\varepsilon < b + \varepsilon$.
If $b = glb\ S$ is in S, then b is the min(imum) of S.

Note: glb   is also called   inf(inum); the set of positive integers is denoted by $\mathscr{J}^+$.

Example: a) Let S be the set of positive rational numbers. Then,
$0 = glb\ S$ does not belong to S.
b)   If S is the set of positive integral primes, $2 = glb\ S$ is in S.
c)   Let $S = \{(n+1)/n : n \in \mathscr{J}^+\}$; $1 = glb\ S$ is in S.
d)   Let $S = \{x^2/(x^2 + 1) : x \in R\}$; $0 = glb\ S$ is in S.
e) $\inf \mathscr{J}^+ = 1 \in \mathscr{J}^+$.

Exercise 1:   a) Formulate definitions of upper bound, least upper bound [lub = sup(remum)], and max(imum) of a subset S of $\overline{R}$.
b)   Prove the

Theorem: *If $S_1 \subset S_2 \subset \overline{R}$ and sup $S_2 = b$ is finite, then S_1 has a finite sup $\le b$.*

Definition: *Let $X, X_1, X_2, X_3, \cdots$ be extended real valued functions on a set A. Then the sequence of functions $\{X_n\}$ converges pointwise to the function X on A iff for each $\omega \in A$,* *the sequence of real numbers* $\{X_n(\omega)\}$ *converges to* $X(\omega)$. *We write* $A = \{X_n \to X\}$ *or* $X_n \to X$ *when A is understood; moreover, "pointwise" is often omitted.*

More symbolically: for each $\omega \in A$ and each $\varepsilon > 0$, there is an $N \ (= N_{\varepsilon\omega})$ depending on ε and ω such that

$n > N$ and $X(\omega)$ finite imply $|X_n(\omega) - X(\omega)| < \varepsilon$,

$n > N$ and $X(\omega) = +\infty$ imply $X_n(\omega) > \varepsilon$,

$n > N$ and $X(\omega) = -\infty$ imply $X_n(\omega) < -\varepsilon$.

When in fact $N \ (= N_\varepsilon)$ does not depend on ω, then the convergence is said to be uniform: $X_n \to X$ *uniformly* on A.

Examples: Let $\Omega = [0, 1]$.

a)   Consider $X_n : \Omega \to \Omega$ such that $X_n(\omega) = \omega^n$; $X_n(\omega) \to 0$ for $0 \leq \omega < 1$ and $X_n(1) \to 1$. More precisely, for $0 < \omega < 1$,

$$|X_n(\omega) - 0| = \omega^n < \varepsilon$$

when $n > (\log \varepsilon)/(\log \omega)$

while $|X_n(1) - 1| = |1 - 1| < \varepsilon$ for all n. Obviously, the value of n needed to make the remainder $|X_n(\omega) - X(\omega)|$ "small" does indeed depend on ω.

b)   Consider X_n as above but restricted to $B = [0, 1/2]$. Then $X_n(\omega) \to 0$ for all $\omega \in B$: $|X_n(\omega)| < \varepsilon$ iff

$\omega^n < \varepsilon$ iff $n > (\log \varepsilon)/(\log \omega)$ as before.

But $0 \leq \omega \leq 1/2$ implies $\omega^n < 1/2^n$ so that

$\omega^n < \varepsilon$ when $n > -(\log \varepsilon)/(\log 2) = N_\varepsilon$ no matter which $\omega \in B$ is observed. Hence,

$$X_n \to 0 \text{ uniformly in (on) } B.$$

c)   Consider the forms $Y_n(\omega) = n/|1 - n\omega|$ for $n \geq 2$.
This is undefined when $\omega = 1/n$ so, to have each
determine values of a function, the domain must be
restricted to $A = \{\omega \in \Omega \mid \omega \neq 1/2, 1/3, \cdots\}$. Then,

$$Y_n(0) = n \to \infty;$$

$$Y_n(1) = n/|1 - n| \to 1;$$

for $0 < \omega < 1$, $Y_n(\omega) \to 1/\omega$.

The convergence is not uniform.

Exercise 2: Show that:

a) $X_n(\omega) = n\omega^n(1 - \omega)$ converges pointwise but not
uniformly on $\Omega = [0,1]$;

b) $nxe^{-n^2x^2}$ converges uniformly on any interval not
containing $x = 0$.

When all X_n and X are finite, the set of convergence

$$\{X_n \to X\}$$

$$= \cap_{\varepsilon>0} \left[\cup_{N=1}^{\infty} \left[\cap_{m=1}^{\infty} \{\omega : |X_{N+m}(\omega) - X(\omega)| < \varepsilon\} \right] \right]$$

is arrived at by translating the verbal definition:

for all (each) $\varepsilon > 0$ becomes $\cap_{\varepsilon>0}$;

there is an N becomes $\cup_{N\geq 1}$;

$n > N$ becomes $\cap_{m\geq 1}$

and

$$B_{Nm\varepsilon} = \{\omega : |X_{N+m}(\omega) - X(\omega)| < \varepsilon\} = \{|X_{N+m} - X| < \varepsilon\}.$$

Exercise 3: Show that the set $\{X_n \to X\}$ in the above paragraph

is equivalent to $\cap_{k=1}^{\infty} \cup_{N=1}^{\infty} \cap_{m=1}^{\infty} B_{Nm}(k)$ where $B_{Nm}(k)$ is
the set $B_{Nm\varepsilon}$ with ε replaced by $1/k$.

When the X and X_n are measurable (with respect to a given Borel σ–field $\mathcal{B}$), so are the differences $X_n - X$ and then $B_{Nm}(k) \in \mathcal{B}$. Since $\mathcal{B}$ is closed under countable unions and intersections, $A = \{X_n \to X\} \in \mathcal{B}$.

Examples: Continued.

    a) $X_n(\omega) \to 0$ except for $\omega = 1$. Impose a Borel σ–field $\mathcal{B}$ generated by intervals and probability (Lebesgue) measure generated by $P(a, b] = b - a$. Then

$$P(\{1\}) = P(\,(1,1]\,) = 0.$$

    c)   Let A be the set where Y_n are defined:

$$A = \Omega - \cup_{n=2}^{\infty}\{1/n\}\,.$$

With the same $[\Omega, \mathcal{B}, P]$, "points" are measurable so that $P(\cup_{n=2}^{\infty}\{1/n\}) = \Sigma_{n=2}^{\infty}P(\{1/n\})$

$$\begin{aligned} &= \lim_{N\to\infty} \Sigma_{n=2}^{N}P(1/n) \\ &= \lim_{N\to\infty} 0 = 0\,. \end{aligned}$$

In both cases, the "exceptional set" has probability 0, leading to:

Definition: *Let A be an event of a probability space. If P(A) = 1, then A is said to occur almost surely (a.s.). In particular, if $P(\{X_n \to X\}) = 1$, then X_n converges to X a.s., $X_n \overset{as}{\to} X$.*

Examples: Recontinued.

    a)   As in the above, P is uniform on Ω. X_n converges for all ω but $X_n \to 0$ almost surely; we have

$$P(X_n \to X) = 1 = P(X_n \to 0)$$

even though the two sets are not equal.

    c)   Here, Y_n is defined almost surely since P(A) = 1.

Moreover, $P(Y_n \to \infty) = P(\{0\}) = 0$

$$= P(Y_n \to 1) = P(\{1\})$$

so that $Y_n \to 1/\omega$ almost surely.

For all Lebesgue measures, the corresponding phrase is almost everywhere (a.e.); for any measure μ, one might say "almost everywhere μ" or "except for a μ–null set N", or, "on N with $\mu(N) = 0$". Of course, if in the last of example c), we had assigned $\mu(1/n) = 6/(\pi n)^2$ then $\mu(A)$ would be 1; the measure on Ω would not be uniform probability and our conclusions would vary.

Exercise 4: See "An Elementary Proof of $\sum_{n=1}^{\infty} 1/n^2 = \pi^2/6$" by Choe, 1987.

In the rest of this lesson, we use one probability space $[\Omega, \mathscr{B}, P]$ and random variables which are finite a.s. When $P(A) = 0$, we ignore the points in A to calculate probabilities; in particular, when $P(|X| = +\infty) = 0$, we don't have to check those worrisome details. Here is another result from real analysis.

Lemma: *(Cauchy Criterion) The sequence $\{X_n\}$ converges to X on some set A iff $X_n - X_m \to 0$ on A as $n,m \to \infty$ in any mode whatsoever.*

Proof: Necessity– if $X_n(\omega) \to X(\omega)$, then

$$X_n(\omega) - X_m(\omega) = X_n(\omega) - X(\omega) + X(\omega) - X_m(\omega) \to 0 \text{ as } n,m \to \infty .$$

Sufficiency– If $X_n(\omega) - X_m(\omega) \to 0$ for all $n,m \to \infty$, then

for some N, when $n,m > N$, we have $|X_n(\omega) - X_m(\omega)| < 1$.

In particular, from $|X_n(\omega) - X_{N+1}(\omega)| < 1$ we get

$$|X_n(\omega)| < 1 + |X_{N+1}(\omega)| \text{ for } n > N .$$

Let   max $\{|X_1(\omega)|, \cdots, |X_N(\omega)|, |X_{N+1}(\omega)|\} = M$; then $|X_n(\omega)| < M+1$ for all n. But (Bolzano–Weierstrass theorem), a bounded sequence of real numbers has at least one limit (accumulation) point. Let the subsequence $\{X_{n'}(\omega)\}$ converge to $Y(\omega)$; that is,

for some $N_1 \geq N$, $n' > N_1$, implies $|X_{n'}(\omega) - Y(\omega)| < \varepsilon/2$.

Then for some $N_2 \geq N_1$, n and $n' > N_2$, imply

$$|X_n(\omega) - X_{n'}(\omega)| < \varepsilon/2$$

and

$$|X_n(\omega) - Y(\omega)| \leq |X_n(\omega) - X_{n'}(\omega)| + |X_{n'}(\omega) - Y(\omega)| < \varepsilon.$$

Thus, $X_n(\omega) \to Y(\omega)$.

If $P(A^c) = 0$, then this lemma says that $X_n \to X$ a.s. iff $X_n - X_m \to 0$ a.s. Uniqueness of limits is the equality: if $X_n \to X$ and $X_n \to Z$ on A, $X = Z$ on A. More generally, if $X_n \to X$ on A and $X_n \to Z$ on B, then

$$|X(\omega) - Z(\omega)| = |X(\omega) - X_n(\omega) + X_n(\omega) - Z(\omega)|$$

$$\leq |X(\omega) - X_n(\omega)| + |X_n(\omega) - Z(\omega)|$$

will be "small" when $\omega \in A \cap B$ and n is "large". Since the difference $X - Z$ does not contain ε or n, $X = Z$ on $A \cap B$. If it happens that $P(A) = P(B) = 1$, then $P(A \cap B) = 1$ (obvious?) and X equals Z a.s. A simple case appears in part a) of the examples: when $Z(\omega) = 0$ for all $\omega \in \Omega$, $X(\omega) = Z(\omega)$ a.s. This leads to the notion in the:

Definition: *Two RVs X and Y on [Ω, $\mathcal{B}$,P] are equivalent iff*

$$P(X = Y) = 1.$$

Write X ase Y.

Thus when we talk about convergence of sequences of RVs and the like, we are really talking about properties of equivalence classes of random variables. We can replace "finite" with "finite a.s.", "measurable" with "measurable a.s.", etc. We don't always do this formally.

Exercise 5: Show that "ase" is a true equivalence relation:

X ase X (reflexivity);

X ase Y implies Y ase X (symmetry);

X ase Y and Y ase Z imply X ase Z (transitivity).

Exercise 6: Let Ω, ϕ, Ψ have corresponding Borel σ–fields $\mathcal{A}$, $\mathcal{B}$, $\mathcal{C}$. Then the function $X : \Omega \to \phi$ is measurable with respect to $(\mathcal{A}, \mathcal{B})$ iff $X^{-1}(B) \in \mathcal{A}$ for all $B \in \mathcal{B}$. Suppose that the function $Y : \phi \to \Psi$ is measurable with repsect to $(\mathcal{B}, \mathcal{C})$. Show that the composition

$$Y \circ X : \Omega \to \Psi \text{ with } Y \circ X(\omega) = Y(X(\omega))$$

is measurable with respect to $(\mathcal{A}, \mathcal{C})$. Hint: first show that $(Y \circ X)^{-1}(C) = X^{-1}(Y^{-1}(C))$.

Exercise 7: For RVs X and Y, prove that

$$P(|X + Y| \geq \varepsilon) \leq P(|X| \geq \varepsilon/2) + P(|Y| \geq \varepsilon/2).$$

Exercise 8: For simplicity, assume that all the RVs involved are finite a.s. Prove that convergence almost surely obeys the usual laws of the algebra of limits: $X_n \to X$ and $Y_n \to Y$ imply

a) $X_n \pm Y_n \to X \pm Y$;

b) $X_n \cdot Y_n \to X \cdot Y$;

c) $X_n/Y_n \to X/Y$ when $P(Y = 0) = 0$.

INE–Exercises:
1.   Let Ω be the unit interval, $\mathcal{B}$ be the Borel σ–field, P be the uniform probability on Ω. Let Q be the set of rational numbers in Ω. Define $X_n(\omega) = I_{\Omega-Q}(\omega)$. Show that

$$X_n \overset{as}{\to} 1.$$

Does $X_n(\omega) \to 1$ for all $\omega \in \Omega$?

2.   Consider the equivalence relation
$$X \text{ ase } Y \text{ iff } P(X = Y) = 1.$$

Let $d(X,Y) = \inf \{\varepsilon > 0 : P(|X - Y| \geq \varepsilon) \leq \varepsilon\}$.

a)   Show that d is a "distance" on the equivalence classes of random variables:
$$d(X,Y) = 0 \text{ when X is equivalent to Y;}$$

.

$d(X,Y) > 0$ when it is not;

$d(X,Y) = d(Y,X); \quad d(X,Y) \leq d(X,Z) + d(Z,Y)$.

b)   Show that $X_n \overset{P}{\to} X$ iff for each $\varepsilon > 0$, there is an N_ε such that $n \geq N_\varepsilon$ implies $P(|X_n - X| \geq \varepsilon) \leq \varepsilon$.

c)   Show that $X_n \overset{P}{\to} X$ iff $d(X_n, X) \to 0$.

LESSON 8. CONVERGENCE ALMOST SURELY AND IN PROBABILITY

In this lesson we look at some theorems on convergence of sequences of numerical (extended real) valued functions. All will be used at some point later particularly in the lessons on integration. We keep a fixed $[\Omega, \mathscr{B}, P]$.

If $f_1, f_2, f_3, \cdots$ are numerical functions on Ω, then

$$(\sup_{n \geq 1} f_n)(\omega) = \sup\{f_1(\omega), f_2(\omega), \cdots\}.$$

It is easy to see that $\inf f_n = -\sup(-f_n)$. Then "limsup" is defined by $\limsup f_n = \inf_{n \geq 1} \sup_{k \geq n} f_k$ and it follows that

$$\liminf f_n = -\limsup (-f_n).$$

Hence the measurability of all these functions will follow from that of the sup.

Theorem: *If each numerical function in the sequence $\{f_n\}$ is measurable, then $f = \sup f_n$ is measurable.*

Proof: For r in $\overline{R}$, $\{\omega : f(\omega) \leq r\} = \cap_{n=1}^{\infty} \{\omega : f_n(\omega) \leq r\}$ is in $\mathscr{B}$ by its closure property for intersection.

Exercise 1: Prove the following:

Corollary: *Let $f, f_1, f_2, \cdots$ be measurable functions on Ω into R such that $f_n(\omega) \to f(\omega)$ for each $\omega \in \Omega$. Then f is measurable.* Hint: see exercise 2 lesson 7.

The following construction is another "classical" result. As we shall see, this is just the trick which will get us a general "expectation".

Let $X \geq 0$ be a measurable function on $[\Omega, \mathscr{B}]$. For each $n = 1(1)\infty$, define the function $X_n(\omega)$

$$= n \qquad \text{for} \qquad n \qquad\qquad \leq X(\omega)$$
$$= (n2^n-1)/2^n \quad \text{for } (n2^n-1)/2^n \qquad \leq X(\omega) < n$$
$$= (n2^n-2)/2^n \quad \text{for } (n2^n-2)/2^n \qquad \leq X(\omega) <(n2^n-1)/2^n$$
$$= (n2^n-3)/2^n \quad \text{for } (n2^n-3)/2^n \qquad \leq X(\omega) < (n2^n-2)/2^n$$
$$= \cdots$$
$$= 1/2^n \qquad \text{for} \qquad 1/2^n \qquad \leq X(\omega) < 2/2^n$$
$$= 0 \qquad \text{for} \qquad 0 \qquad \leq X(\omega) < 1/2^n .$$

Since X is measurable, the inverse image of each interval [a,b) is a measurable set, that is, a member of $\mathcal{B}$; this means that each indicator $I_{[a,b)}$ is a measurable function. Therefore, each X_n is simple and can be written as

$$X_n(\omega) = \sum_{i=1}^{n2^n} (i - 1)/2^n \cdot I_{[(i-1)/2^n,\ i/2^n)}(\omega)$$
$$+ nI_{[n,\ \infty]}\ (\omega). \qquad\qquad (*)$$

Also, $|X_n(\omega) - X(\omega)| < 1/2^n$ for $X(\omega) < n$ from which it follows that for such ω, $X_n(\omega) \to X(\omega)$. For $X(\omega) = +\infty$, $X_n(\omega) = n$ and again $X_n(\omega) \to X(\omega)$.

Example: Let $X(\omega) = -\omega \qquad$ for $-3 \leq \omega < 0$
$$= \omega \qquad \text{for } 0 \leq \omega < 2$$
$$= 4 - \omega \quad \text{for } 2 \leq \omega \leq 3 .$$

In both sketches, dots represent the values of the function X . In the first sketch, hyphens represent the values of X_1: 0, 1/2, 2/2, 3/2, $\cdots$.

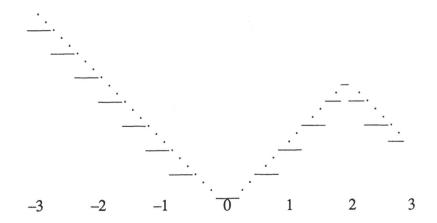

In the second sketch, the hyphens represent the values of
$$X_2: 0, 1/4, 2/4, 3/4, \cdots.$$

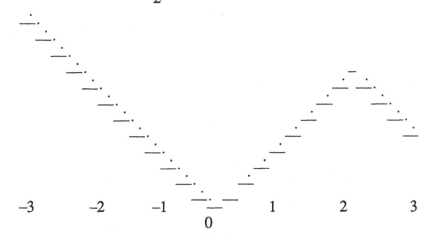

Exercise 2: Take $\Omega = [-1,6]$, sketch each X and the corresponding X_1, X_2. a) $X(\omega) = \omega^2$. b) $X(\omega) = \exp(-\omega)$.

Exercise 3: Show that the sequence of simple functions defined in (*) is montonically increasing: $X_{n+1} \leq X_n$.

Hint: $(k-1)/2^n = (2k-2)/2^{n+1}$.

These few paragraphs imply a proof of the following theorem. Of course, one need not divide the "function axis" in precisely this way so there are many sequences with these properties.

Theorem: *Let X be a non-negative measurable function for* $[\Omega, \mathscr{B}]$. *Then there exists a non-decreasing sequence of non-negative simple functions which converges (everywhere) to* X .

Next, we introduce another useful tool, the *positive* and *negative* parts (components) of a function.

Definition: *Let* $X : \Omega \rightarrow R;$ *then,*

$$X^+(\omega) = X(\omega) \qquad X^-(\omega) = 0 \qquad \textit{when } X(\omega) \geq 0$$

$$X^+(\omega) = 0 \qquad X^-(\omega) = -X(\omega) \qquad \textit{when } X(\omega) < 0 .$$

For each $\omega \in \Omega ,$ $\quad X^+(\omega) = max \{X(\omega), 0\} ;$

$$X^-(\omega) = - min \{X(\omega), 0\} ;$$

$$X(\omega) = X^+(\omega) - X^-(\omega) ;$$

$$|X(\omega)| = X^+(\omega) + X^-(\omega) .$$

Example: On $\Omega = (-2, 3)$, let

$$X(\omega) = -(2+\omega)I_{(-2,-1)}(\omega) + \omega I_{[-1,1]}(\omega) + (2-\omega)I_{(1,3)}(\omega) .$$

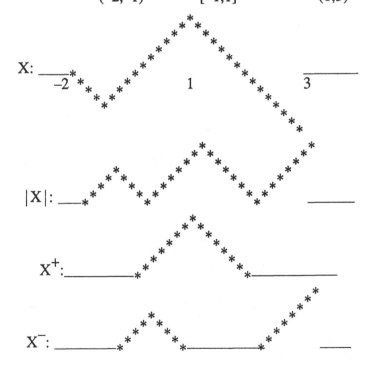

Exercise 4: Sketch $X^+, X^-, |X|$ in each case; take the domain as R. a) $X(\omega) = \tan \omega$ b) $X(\omega) = \omega^5 - 5\omega^3 + 4\omega$

Exercise 5: Make use of the corollary and the last theorem above to prove the:

Theorem: *Let $X : \Omega \to R$. If X is measurable in $[\Omega, \mathcal{B}]$, so are X^+ and X^-. If X^+ and X^- are measurable in $[\Omega, \mathcal{B}]$, so are X and $|X|$.*

For an arbitrary measurable function, we can find one sequence of non–negative simple functions converging to X^+, say $\{Y_n\}$, and another sequence of non–negative simple functions converging to X^-, say $\{Z_n\}$; then $\{Y_n - Z_n\}$ is a sequence of simple functions converging to X.

Exercise 6: Prove the last sentence in the paragraph above.

In the following, we discover some simplification of the condition for convergence a.s. The theorem has an immediate corollary which introduces another kind of convergence.

Theorem: *Let X, X_n be (finite a.s.) RVs for $[\Omega, \mathcal{B}, P]$;*

$$X_n \to X \text{ a.s. iff } P(\cup_{m=1}^{\infty}\{|X_{n+m} - X| \geq \varepsilon\}) \to 0 \text{ as } n \to \infty.$$

Proof: Let $B_{nmk} = \{|X_{n+m} - X| \geq 1/k\}$, $B_{nk} = \cup_{m=1}^{\infty} B_{nmk}$ and $B_k = \cap_{n=1}^{\infty} B_{nk}$. Note that each of these sequences of sets is increasing in k. As in Lesson 7, let $A = \{X_n \to X\}$; then,

$$A^c = \cup_{k=1}^{\infty} B_k.$$

When $X_n \to X$ a.s., $P(A) = 1$ and $P(A^c) = 0$. Now by continuity of P (Lesson 2) and monotonicity of $\{B_k\}$,

$$P(\cup_{k=1}^{\infty}B_k) = \lim_{K\to\infty} P(\cup_{k=1}^{K}B_k) = \lim_{K\to\infty} P(B_K).$$

It follows that $P(B_k) = 0$ for all $k \geq 1$. It also follows that

$\lim_{n\to\infty} P(B_{nk}) = P(B_k) = 0$. This is equivalent to the conclusion.

Corollary: *For the same hypotheses, $X_n \to X$ a.s. implies*

$$\lim_{n\to\infty} P(|X_n - X| \geq \varepsilon) = 0 \text{ for each } \varepsilon > 0.$$

Proof: $P(|X_{n+1} - X| \geq \varepsilon) \leq P(\cup_{m=1}^{\infty}\{|X_{n+m} - X| \geq \varepsilon\})$ tends to 0 as above.

Definition: *When $\lim_{n\to\infty} P(|X_n - X| > \varepsilon) = 0$ for each $\varepsilon > 0$, we say that X_n converges to X in probability.*

The corollary shows that convergence a.s. implies convergence in probability; symbolically, $X_n \xrightarrow{as} X$ implies $X_n \xrightarrow{P} X$. The following classical example shows that the converse is false. A "best possible" result is given in Lesson 8, Part III.

Example: Let $[\Omega, \mathscr{B}, P]$ be the uniform probability space on $[0,1]$. Let each Y be the indicator function of the corresponding interval:

$$Y_{11} = I_{(0, 1]},$$

$$Y_{21} = I_{(0, 1/2]}, \quad Y_{22} = I_{(1/2, 1]},$$

$$Y_{31} = I_{(0, 1/3]}, \quad Y_{32} = I_{(1/3, 2/3]}, \quad Y_{33} = I_{(2/3, 3/3]}, \quad \cdots$$

or,

$$Y_{i,j} = I_{((j-1)/i, \, j/i]}, \, j = 1(1)i, \, i = 1(1)\infty.$$

a)   Now relabel (taking Y in lexicographical order):

$X_1 = Y_{11}$, $X_2 = Y_{21}$, $X_3 = Y_{22}$, $X_4 = Y_{31}$, $X_5 = Y_{32}$, $X_6 = Y_{33}$, $\cdots$; (without worrying about the exact formula), $X_n = Y_{i_n, j_n}$.

Then, $P(|X_n - 0| \geq \varepsilon) = P(X_n = 1) = 1/i_n \to 0$ as $n \to \infty$. Therefore, $X_n \overset{P}{\to} 0$.

b)   (Looking at the sketch below will enhance the next few arguments.) For each $\omega \neq 0$, some $Y_{i,j}(\omega) = 1$ and in fact infinitely many $X_n(\omega) = 1$; at the same time, there are also infinitely many $X_n(\omega) = 0$. Any (sub)sequence of such zeroes and ones cannot converge so $\{X_n(\omega)\}$ does not converge for any $\omega \neq 0$; of course, the sequence $X_n(0) = 0$ converges to 0. Symbolically,

$\{\omega : X_n(\omega)$ converges$\} = \{0\}$ (is a one point set)

and $P(X_n$ converges$) = 0$ (is a number).

Example: The line segments represent the $Y_{ij} = 1$ "in lexicographical order"; 0 is not in any segment.

i = 1:
      0   1

i = 2:

i = 3:

i = 4:

i = 5:

$\cdots$.

At $\omega = 1/4$, $X_1 = 1$, $X_2 = 1$, $X_3 = 0$, $X_4 = 1$, $X_5 = X_6 = 0$,

$X_7 = 1$, $X_8 = 0 = X_9 = X_{10} = X_{11}$, $X_{12} = 1$, $X_{13} = 0 = X_{12} = X_{15}$, $\cdots$.

Exercise 6:   Prove that convergence in probability obeys the customary laws for algebra of limits:

a) $X_n \to X$ and $X_n \to Z$ imply $X = Z$ a.s.

b) $X_n \to X$ implies $X_n - X_m \to 0$

c) $X_n \to X$ and $Y_n \to Y$ imply $X_n + Y_n \to X + Y$,
$$X_n Y_n \to XY$$

d) $X_n \to X$, $Y_n \to Y$ and $P(Y \neq 0) = 1$ imply
$X_n/Y_n \to X/Y$

Hint: $|X - Z| > \varepsilon$ implies $|X - X_n| > \varepsilon/2$ or $|X_n - Z| > \varepsilon/2$.

LESSON 9. INTEGRATION — I

In this lesson, we begin Lebesgue(–Stieltjes) integration. This form of an integral is particularly suited to probability since all properties of expectation are under one guise and we are not limited to the discrete and continuous forms illustrated earlier. We consider a fixed measure space $[\Omega, \mathscr{B}, P]$; P is non–negative but, in general, $P(\Omega)$ may be different from 1, even infinite. Only measurable sets and (extended) real–valued measurable functions enter the discussion.

Definition I: *Let* $X = \sum_{j=1}^{J} x_j I_{A_j}$ *be a non-negative simple function. The integral of X ,* $\int_{\Omega} X(\omega)dP(\omega)$, *is defined to be*

$$\sum_{j=1}^{J} x_j P(A_j). \qquad (*)$$

The sum in the value (*) of the integral has only a finite number of non–negative terms so it is well–defined and non–negative; the value is $+\infty$ if one or more of the $x_j P(A_j) = +\infty$.

Some confusion could arise because the representation of a simple function is not unique (see Lesson 8). Once this is corrected (in the lemma below), the integral of a non–negative simple function becomes well–defined and, of course, also non–negative.

Lemma: *If the simple function in the definition above also has the representation* $X = \sum_{k=1}^{K} y_k I_{B_k}$ *where* $B_1, B_2, \cdots, B_K$ *is another partition, then* $\sum_{j=1}^{J} x_j P(A_j) = \sum_{k=1}^{K} y_k P(B_k).$

Proof: With $\Omega = \cup_{j=1}^{J} A_j = \cup_{k=1}^{K} B_k$, one has

$A_j = A_j \cap \Omega = \cup_k (A_j \cap B_k)$ and $B_k = \Omega \cap B_k = \cup_j (A_j \cap B_k)$.
Since every function is (by definition) single–valued,

$$X(\omega) = x_j = y_k \quad \text{when} \quad \omega \in A_j \cap B_k \neq \phi .$$

Then, $\Sigma_j \, x_j P(A_j) = \Sigma_j \, x_j P(\cup_k (A_j \cap B_k))$

$$= \Sigma_j \, x_j \, \Sigma_k P(A_j \cap B_k) = \Sigma_j \, \Sigma_k \, x_j P(A_j \cap B_k)$$

$$= \Sigma_j \, \Sigma_k \, y_k P(A_j \cap B_k) = \Sigma_k \, y_k \, \Sigma_j \, P(A_j \cap B_k)$$

$$= \Sigma_k \, y_k P(\cup_j (A_j \cap B_k)) = \Sigma_k \, y_k P(B_k) .$$

Exercise 1: Justify each of the equalities in the proof above.

Definition II: *If X is a non-negative simple function and A is a measurable set, then the integral of X on A , $\int_A X(\omega)dP(\omega)$, is defined to be $\int_\Omega X(\omega) \cdot I_A(\omega)dP(\omega)$.*

At various times, notation for integrals may be shortened:

$$\int_\Omega X \;\text{for}\; \int_\Omega X(\omega)dP(\omega) , \int_A X \;\text{for}\; \int_\Omega X(\omega) \cdot I_A(\omega)dP(\omega) ,$$

even $\int X$ when the context is clear.

The integral in Definiton II will be well–defined by Definition I as soon as we show that $X \cdot I_A$ is also non–negative and simple.

Lemma: *If A is a measurable set and X is a non-negative simple function, then $X \cdot I_A$ is non-negative and simple.*

Proof: Let $A_1, \cdots, A_J$ be the partition associated with X. Construct another partition: $B_j = A \cap A_j$ for $j = 1(1)J$, $B_{J+1} = A^c$; let $x_{J+1} = 0$. Then it is obvious that

$$X \cdot I_A = \sum_{j=1}^{J+1} x_j I_{B_j} \;\text{is non–negative and simple.}$$

Exercise 2: Assume the conditions of the lemma above.
a)   Explain the logical need of B_{J+1} and the reduction of

the sum to: $\sum_{j=1}^{J} x_j I_{A_j \cap A}.$

b)   Show that $\int_A X = \sum_{j=1}^{J} x_j P(A_j \cap A).$

c)   Show that for a measurable set A,

$$\int_\Omega I_A(\omega)dP(\omega) = \int I_A = \int_A 1 = P(A).$$

Exercise 3: Let X be a non–negative simple function; let r be a finite real number. Show that rX is also simple.

The next theorem (presented in four pieces) contains the basic properties of the integral of simple functions. Note that the measurable set A could be Ω itself and other symbolic results could be listed; for example, when $A = \Omega = R$, part iii) might

be written as $\int_{-\infty}^{+\infty} X \le \int_{-\infty}^{+\infty} Y$ or $\int_{(-\infty,+\infty)} X \le \int_{(-\infty,+\infty)} Y.$

Theorem: *Let* $X = \sum_{j=1}^{J} x_j I_{A_j}$ *and* $Y = \sum_{k=1}^{K} y_k I_{B_k}$ *be*

non-negative simple functions. Let A and B be measurable sets.

a)   *If* $0 \le r < +\infty$, $\int_A rX = r\int_A X.$

Proof: Note that $\sum_j (rx_j)P(A \cap A_j) = r\sum_j x_j P(A \cap A_j)$ is an old

property of summation which (in $\overline{R}$) is valid even when $r = 0$ and some $x_j P(A \cap A_j) = +\infty .$

Exercise 4: Continue with the non–negative simple X and Y in the theorem. Show that $X + Y$ is non–negative simple.

Theorem continued:
b) $\int_A (X + Y) = \int_A X + \int_A Y.$

Proof: From exercise 4, $X + Y = \sum_{j=1}^{J} \sum_{k=1}^{K} (x_j + y_k) I_{A_j \cap B_k}$ is

non–negative simple; therefore its integral is

$$\sum_{j=1}^{J} \sum_{k=1}^{K} (x_j + y_k) P(A \cap A_j \cap B_k) \; .$$

Again the finiteness of the number of terms and their non–negativity allow this double sum to be rearranged as

$$\sum_{j=1}^{J} \sum_{k=1}^{K} x_j P(A \cap A_j \cap B_k) + \sum_{j=1}^{J} \sum_{k=1}^{K} y_k P(A \cap A_j \cap B_k)$$

$$= \sum_{j=1}^{J} x_j \sum_{k=1}^{K} P(A \cap A_j \cap B_k) + \sum_{k=1}^{K} y_k \sum_{j=1}^{J} P(A \cap A_j \cap B_k)$$

$$= \sum_{j=1}^{J} x_j P(A \cap A_j \cap \{\cup_k B_k\}) + \sum_{k=1}^{K} y_k P(A \cap \{\cup_j A_j\} \cap B_k)$$

$$= \sum_{j=1}^{J} x_j P(A \cap A_j) + \sum_{k=1}^{K} y_k P(A \cap B_k); \qquad (**)$$

we have also used the fact that the disjoint unions $\cup_k B_k$, $\cup_j A_j$ both equal Ω. The sums in (**) are the integrals of X and Y, respectively, so that the conclusion follows.

c)   If $0 \le Y \le X$, then $\int_A Y \le \int_A X$.

Proof: Now $Y \le X$ means $Y(\omega) \le X(\omega)$ for all $\omega \in \Omega$ so that

$X - Y$ is non–negative simple and its integral exists. Then

$$\int_A X = \int_A \{(X - Y) + Y\} = \int_A (X - Y) + \int_A Y$$

by the additivity in ii). Since $\int_A (X - Y)$ is non–negative, the inequality follows.

d)   If $P(A \cap B) = 0$, in particular when $A \cap B = \phi$,

$$\int_{A \cup B} X = \int_A X + \int_B X .$$

Proof: Now $X \cdot I_{A \cup B} = X \cdot I_A + X \cdot I_{(A^c) \cap B}$ and both terms of this sum are non–negative simple functions. Hence, by the additivity proved in ii) (with that $A = \Omega$),

$$\int_{A \cup B} X = \int_\Omega X \cdot I_{A \cup B} = \int_\Omega (X \cdot I_A + X \cdot I_{(A^c) \cap B})$$

$$= \int_\Omega X \cdot I_A + \int_\Omega X \cdot I_{(A^c) \cap B} = \int_A X + \int_{(A^c) \cap B} X .$$

Since $P(B) = P(A \cap B) + P(A^c \cap B) = P(A^c \cap B)$, the last integral

is $\displaystyle\sum_{j=1}^J x_j P(A_j \cap A^c \cap B) = \sum_{j=1}^J x_j P(A_j \cap B) = \int_B X$. With the

arithemtic in $\overline{R}$, the conclusion follows.

Exercise 5: Suppose that $A \subset B$ and both are measurable. Let X be non– negative simple; show that $\displaystyle\int_A X \le \int_B X$.

Hint: $X \cdot I_A \le X \cdot I_B$.

The next step is to extend the definition and the properties in the theorem to non–negative functions.

Definition III: *Let a non-decreasing sequence of non-negative simple functions* X_n *converge to the non-negative measurable function* X *. Then the integral of* X *,* $\displaystyle\int_\Omega X(\omega) dP(\omega)$ *, is defined*

to be $\displaystyle\lim_{n \to \infty} \int_\Omega X_n(\omega) dP(\omega)$ *. For any measurable set* A *,*

$$\int_A X = \int X \cdot I_A \text{ and, briefly, } \int_A X_n \to \int_A X .$$

By part c) of the theorem,

$$0 \le X_n \le X_{n+1} \text{ implies } \int X_n \le \int X_{n+1}$$

and so the limit of the monotonic sequence of integrals in Definition III exists (but may be +∞) for each $\{X_n\}$. Here again, confusion may arise because the representation is not unique: there are many sequences of such functions converging to a given X (Lesson 8). However, the value of the integral is independent of that choice; we leave the long boring proof of such a lemma to books like Chung, 1974. Since such results hold when $A = \Omega$, the integral in Definition III becomes well–defined and non–negative.

It is now very easy to use the definitions and basic properties in the theorem to extend those very properties to the integral of non–negative measurable functions. For this discussion, $0 \le X_n \uparrow X$ denotes a non–decreasing sequence of non–negative simple functions converging to the non–negative measurable function X ; also, Y is a non–negative measurable function.

a') From $0 \le X_n < \uparrow X$ and the constant $0 \le r < +\infty$, we get $0 \le rX_n \uparrow rX$. Hence,

$$\int_A rX = \lim \int_A rX_n = \lim r \int_A X_n$$

$$= r \cdot \lim \int_A X_n = r \int_A X .$$

b') From $0 \le X_n \uparrow X$ and $0 \le Y_m \uparrow Y$, we get

$$0 \le (X_n + Y_n) \uparrow (X + Y) . \text{ Hence,}$$

$$\int_A (X + Y) = \lim \int_A (X_n + Y_n)$$

$$= \lim \int_A X_n + \lim \int_A Y_n = \int_A X + \int_A Y .$$

c') As before, $X \ge Y$ implies $X - Y \ge 0$ and
$$\int_A (X - Y) \ge 0 . \text{ Hence,}$$

$$\int_A X = \int_A \{(X - Y) + Y\}$$

$$= \int_A (X - Y) + \int_A Y \geq \int_A Y .$$

d')   If $P(A \cap B) = 0$ and $0 \leq X_n \uparrow X$, then

$$\int_{A \cup B} X = \lim \int_{A \cup B} X_n = \lim \{\int_A X_n + \int_B X_n\}$$

$$= \lim \int_A X_n + \lim \int_B X_n = \int_A X + \int_B X .$$

Exercise 6: Write out full hypotheses and conclusions corresponding to the four properties whose proofs are given above.

Exercise 7: By induction, extend the rule in part d') to any finite number of mutually exclusive sets.

The next lesson (INE) completes the formalities of integration but all the results we need for *expectation* are given in Lesson 11.

LESSON *10. INTEGRATION—II

In this lesson, we complete the fundamental theory of Lebesgue (–Stieltjes) integration of measurable functions on a fixed measure space $[\Omega, \mathcal{B}, P]$. As before, results valid on the arbitrary measurable set A are valid on Ω.

Definition IV: *Let X be a measurable (extended) real valued function on Ω with positive, negative parts X^+, X^-. The integral of X, $\int_{\Omega} X(\omega)dP(\omega)$, is defined to be*

$$\int_{\Omega} X^+(\omega)dP(\omega) - \int_{\Omega} X^-(\omega)dP(\omega)$$

whenever this sum exists. In abbreviated form, when A is a measurable set, $\int_A X = \int_A X^+ - \int_A X^-$ whenever the right hand side exists. If $\int_A X$ is finite, then X is integrable on A. The set of all functions integrable on A is denoted by LS_A.

Since X^+ and X^- are non–negative functions, each of their integrals is well–defined (lesson 9); consequently, the condition of existence in Definition IV, namely at least one of $\int_A X^+, \int_A X^-$ is finite, guarantees that the integral of X is well–defined. When both of these are finite, so are their sum and difference and the following is immediate for A or Ω.

Corollary: *The measurable function X is integrable iff X^+ and X^- are integrable iff $|X| = X^+ + X^-$ is integrable.*

In addition, another little consequence allows us to simplify some work by looking only at "finite functions":

Corollary: *If X is integrable on A, then X is finite almost everywhere (finite a.e.) on A. More precisely,*

$$\textit{if } B = \{\omega : |X(\omega)| = +\infty\} \cap A, \, P(B) = 0 \, .$$

Proof: Since $|X|$ is non–negative, $|X| \geq |X| \cdot I_B \geq r \cdot I_B$ for any postive integer r . By applying part iiia) of Lesson 9, we get

$$\int_A |X| \geq \int_A |X| \cdot I_B \geq \int_A r \cdot I_B = rP(B) \, .$$

If $P(B)$ were positive, the right–hand side would tend to infinity with r and this would contradict the finiteness of the integral on the left. Therefore, $P(B) = 0$.

Exercise 1: Prove: if X is integrable, then

$$\left| \int_A X \right| \leq \int_A |X| = \int_A X^+ + \int_A X^- \, .$$

Exercise 2: Suppose that X and Y are measurable, $Y \in LS_A$ and $|X| \leq Y$ on A . Prove that $X \in LS_A$.

In the following theorem, we illustrate a procedure that is used for many proofs (in particular some that we omit):
the proof for non–negative simple functions is "easy";
the proof for non–negative measurable functions comes via the limit;
finally, definition IV takes over.

Theorem 1: *If $P(A) = 0$ and X is measurable, then*

$$\int_A X = 0 \, .$$

Proof: a) Suppose that X is non–negative and simple, say $\Sigma_j \, x_j I_{A_j}$. Since $P(A \cap A_j) = 0$, and in $\overline{R}$ every $x_j P(A \cap A_j) = 0$,

$$\int_A X = \Sigma_j \, x_j P(A \cap A_j) = 0 \, .$$

b)   Suppose that X is non–negative. Then for some sequence of simple functions, $0 \leq X_n \uparrow X$, and

$$\int_A X = \lim \int_A X_n = 0 \text{ because each } \int_A X_n = 0 \, .$$

c) When X is arbitrary, X^+ and X^- are non–negative so that each of their integrals is 0 as in b). Hence,

$$\int_A X = 0 = \int_A |X| \ .$$

Note that in this instance, X is integrable although trivially; this does not contradict the last corollary since $P(B)$ is still 0 .

Theorem 2: *Let A and B be fixed measurable sets with $A \subset B$. If $X \in LS_B$, then $X \in LS_A$.*

Proof: From $0 \leq X^+ \cdot I_A \leq X^+ \cdot I_B$, we get $0 \leq \int_A X^+ \leq \int_B X^+$; since the right–hand side is finite, so is the left–hand side. Similarly, $\int_A X^-$ is finite; the conclusion follows.

The fact that the following theorem has a name attests to its fundamental importance and usefulness, some of which we shall see. We could have included this in Lesson 9 for we are merely dropping "simple" from Definition III.

Lebesgue Monotone Convergence Theorem: *Let $\{X_n\}$ be a non-decreasing sequence of non-negative measurable functions with measurable limit X a.e., that is,*

$$P(\{\omega : \lim_{n \to +\infty} X_n(\omega) \neq X(\omega)\}) = 0 \ .$$

Then, $\lim \int_A X_n = \int_A X$.

Proof: $\{\int_A X_n(\omega)\}$ is a monotone increasing sequence of (extended) real numbers which has a limit, say $M \ (\leq +\infty)$. Since $X_n \leq X$, $\int_A X_n \leq \int_A X$ whence $M \leq \int_A X$. Fix $0 < r < 1$; let Y be a simple function such that $0 \leq Y \leq X$. For each positive integer n, let $A_n = \{\omega : rY(\omega) \leq X_n(\omega)\} \cap A$. From

$$X_n \cdot I_A \geq X_n \cdot I_{A_n} \geq rY ,$$

we get

$$\int_A X_n \geq \int_{A_n} X_n \geq \int_{A_n} rY = r \int_{A_n} Y = r \int_\Omega Y \cdot I_{A_n} . \qquad (*)$$

(As you are to prove,) the sequence $A_n \subset A_{n+1}$ has a limit $B = \cup_n A_n$; note that $P(A - B) = 0$ by convergence of $\{X_n\}$. Then $0 \leq Y \cdot I_{A_n} \uparrow Y \cdot I_B$ so that

$$\lim_{n \to +\infty} r \int_\Omega Y \cdot I_{A_n} = r \int_\Omega Y \cdot I_B = r \int_\Omega Y \cdot I_A = r \int_A Y .$$

As $n \to +\infty$ in $(*)$, we get $M \geq r \int_A Y$ and letting $r \to 1$, yields $M \geq \int_A Y$ for all such simple functions Y . We may then take a sequence of such Y, say $\{Y_k\}$, increasing to X ; with one more limit operation, we get $M \geq \int_A X$. It follows that $\int_A X = M$.

Exercise 3: Prove the statement referred to after $(*)$.

Of course, we are heading for a general additivity of the integral but along the way, we have other results, some of which are of independent interest. For example, the following is a partial converse to Theorem 1.

Corollary: *If* $X \geq 0$ *and* $\int_A X = 0$ *, then* $X = 0$ *a.e. on* A .

Proof: The sequence $A_n = \{\omega : X(\omega) > 1/n\} \cap A$ is increasing with limit $\cup_n A_n = B = \{\omega : X(\omega) > 0\} \cap A$. For each n,

$$(1/n)P(A_n) \leq \int_{A_n} X \leq \int_A X = 0 .$$

Therefore, $P(A_n) = 0$ and $P(B) = \lim P(A_n) = 0$.

Theorem 3: *Let $\{A_n\}$ be a sequence of disjoint measurable sets*
with $\cup_n A_n = A$. If $X \in LS_A$, then $\int_A X = \sum_{n=1}^{\infty} \int_{A_n} X$.

Proof: Since $A \supset A_n$, $X \in LS_{A_n}$, that is, $\int_{A_n} X^+$ and $\int_{A_n} X^-$
are finite.

a)   If $X \geq 0$, let $B_n = \cup_{k=1}^{n} A_k$ and $X_n = X \cdot I_{B_n}$. Then
X_n is a sequence of non-negative functions
increasing to X and the LMC Theorem applies:

$$\int_A X = \lim \int X_n .$$

yise 8, Lesson 9,

$$\int_A X \cdot I_{B_n} = \int_{B_n} X = \int_{A_1 \cup \cdots \cup A_n} X = \sum_{k=1}^{n} \int_{A_k} X .$$

Hence,

$$\int_A X = \lim \sum_{k=1}^{n} \int_{A_k} X = \sum_{k=1}^{\infty} \int_{A_k} X .$$

b)   For arbitrary X, $\int_A X = \int_A X^+ - \int_A X^-$. Part a)
may be applied to each X^+ and X^-. Then

$$\int_A X = \sum_{k=1}^{\infty} \int_{A_k} X^+ - \sum_{k=1}^{\infty} \int_{A_k} X^-$$

$$= \sum_{k=1}^{\infty} (\int_{A_k} X^+ - \int_{A_k} X^-) = \sum_{k=1}^{\infty} \int_{A_k} X$$

because all terms and sums are finite valued.

Exercise 4: Let $X \in LS_\Omega$. Let $\varphi : \mathscr{B} \to \bar{R}$ have value $\varphi(A) = \displaystyle\int_A X$. Prove that φ is σ–additive on $\mathscr{B}$, that is, for the disjoint sequence $\{A_n\}$ in $\mathscr{B}$, $\varphi(\cup_n A_n) = \Sigma_n \varphi(A_n)$.

The following theorem is another "simplifier" allowing us to ignore sets of measure zero when integrating.

Theorem 4: *If* $X \in LS_A$ *and* $X = Y$ *a.e. (on A), then* $Y \in LS_A$ *and* $\displaystyle\int_A X = \int_A Y$.

Proof: Let $B = \{\omega : X(\omega) = Y(\omega)\} \cap A$; from the hypothesis a.e., we get $P(A - B) = 0$. Since X^+ is non–negative, part iva) of lesson 9 applies and

$$\int_A X^+ = \int_{A \cap B} X^+ + \int_{A \cap B^c} X^+ .$$

The last integral is 0 by Theorem 1 so that $\displaystyle\int_A X^+ = \int_B X^+$; by similar arguments, $\displaystyle\int_A Y^+ = \int_B Y^+$. Obviously, $\displaystyle\int_B X^+ = \int_B Y^+$.

The same logic may be applied to X^-, Y^- . This also shows that all the integrals are finite so that we may add the pieces to get the final conclusion.

Theorem 5: *If* $X \in LS_A$ *and* r *is any finite real number, then* $rX \in LS_A$ *and* $\displaystyle\int_A rX = r \int_A X$.

Proof: If $r = 0$, the conclusion is obvious. The case $r > 0$ and $X \geq 0$ is part ia) in Lesson 9. The general result is obtained by multiple applications of this to X^+ and X^- :

for $r > 0$,
$$\int_A rX = \int_A (rX)^+ - \int_A (rX)^-$$

$$= r\int_A X^+ - r\int_A X^- = r\int_A X \; ;$$

for $r < 0$, $(rX)^+ = (-r)X^-$ and $(rX)^- = (-r)X^+$ and

$$\int_A rX = \int_A (-r)X^- - \int_A (-r)X^+$$

$$= -r\int_A X^- - (-r)\int_A X^+$$

$$= r\int_A X^+ - r\int_A X^- = r\int_A X \; .$$

In proving a general form of additivity, we assume that $\int_\Omega X$, $\int_\Omega Y$, and $\int_\Omega X + \int_\Omega Y$ exist so that (at least) one of $\int_\Omega X$, $\int_\Omega Y$ is finite. If $\int_\Omega |Y|$ is the finite one so that Y is finite a.s., then $X + Y$ will be defined a.s. Now Ω can be decomposed into the following six sets:

$$A_1 = \{X \geq 0, Y \geq 0, X+Y \geq 0\}$$

$$A_2 = \{X \geq 0, Y < 0, X+Y \geq 0\}$$

$$A_3 = \{X < 0, Y \geq 0, X+Y \geq 0\}$$

$$A_4 = \{X < 0, Y \geq 0, X+Y < 0\}$$

$$A_5 = \{X \geq 0, Y < 0, X+Y < 0\}$$

$$A_6 = \{X < 0, Y < 0, X+Y < 0\} \; .$$

Since $(X+Y) \cdot I_{A_2}$ and $-Y \cdot I_{A_2}$ are nonnegative, by results in Lesson 9,

$$\int_{A_2} X = \int_{\Omega} (X + Y - Y) \cdot I_{A_2}$$

$$= \int_{\Omega} (X+Y) \cdot I_{A_2} + \int_{\Omega} (-Y) \cdot I_{A_2} .$$

By theorem 5, the last integral is $-\int_{\Omega} Y \cdot I_{A_2} = -\int_{A_2} Y$ which is

finite by application of exercise 2. Numerical addition yields

$$\int_{A_2} X + \int_{A_2} Y = \int_{A_2} (X+Y) .$$

Similarly, we obtain integrals for the other five A_i. By theorem 3, we may add each (finite) sequence of integrals to obtain

Theorem 6: $\int_A (X + Y) = \int_A X + \int_A Y$ *whenever the right- hand side is defined.*

Exercise 5: Prove the following corollaries:
   a)   If X and Y $\in$ LS$_A$ and $X \le Y$ a.e. on A ,

$$\int_A X \le \int_A Y .$$

   Hint: Note that $Y = (Y - X) + X$ and $Y - X \ge 0$.

   b)   Suppose that $P(A) < +\infty$ and $X \in$ LS$_A$. If there are constants m and M , such that $m \le X \le M$ on A , then $mP(A) \le \int_A X \le MP(A)$.

   A combination of theorems 3 and 6 make up the *linearity* of the integral: for finite real numbers r and s, and integrable functions X and Y , $\int_{\Omega} (rX + sY) = r\int_{\Omega} X + s\int_{\Omega} Y$.

We close this lesson with two more "named" theorems.

Fatou Theorem: *If* $\{X_n\}$ *is a sequence of non-negative* *measurable functions such that* $\liminf\limits_{n} X_n = X$ *a.e., then*

$$\int_A X \le \liminf \int_A X_n .$$

Proof: For a positive integer m, let $Y_m = \inf \{X_n : n \ge m\}$. Then $\{Y_m\}$ is an increasing sequence of non–negative functions with

$$\lim_{m \to \infty} Y_m = \liminf_{n \to \infty} X_n = X \text{ a.e.}$$

and we may apply the LMC Theorem to $\{Y_m\}$:

$$\int_A X = \lim_{m \to \infty} \int_A Y_m .$$

But $Y_m \le X_m$ implies $\liminf \int_A Y_m \le \liminf \int_A X_m$ so that

$$\int_A X \le \liminf \int_A X_m \text{ as desired.}$$

Lebesgue Dominated Convergence Theorem: *Let* $\{X_n\}$ *be a sequence of measurable functions with* $\lim\limits_{n} X_n = X$ *a.e. and* $|X_n| \le Y$ *where* $Y \in LS_A$. *Then* $\lim \int_A X_n = \int_A X$.

Proof: The limit on X_n implies $|X| \le Y$ a.e. so that $X, X_1, X_2,$ $\cdots$ are all integrable (exercise 2). Apply Fatou's theorem to $\liminf \{Y + X_n\} = Y + X$:

$$\int_A (Y + X) \le \liminf \int_A (Y + X_n) = \int_A Y + \liminf \int_A X_n .$$

Linearity of the integral yields

$$\int_A Y + \int_A X \le \int_A Y + \liminf \int_A X_n$$

from which it follows that $\int_A X \le \liminf \int_A X_n .$ (**))

Similarly, from the non–negative sequence $\{Y - X_n\}$, we get

$$\int_A (-X) \leq \liminf \int_A (-X_n)$$

which implies

$$\limsup \int_A X_n \leq \int_A X .$$

Since $\liminf \leq \limsup$, a combination of (**) and (***) yields equality.

LESSON 11. THEOREMS FOR EXPECTATION

In this lesson, we take a fixed probability space $[\Omega, \mathscr{B}, P]$ with events $A, B, \cdots$ and rephrase the properties of integration of random variables (measurable functions $X, Y, Z, X_1, X_2, \cdots$) in terms of the following definition; now "almost everywhere (a.e.)" is "almost surely (a.s.)". Only a few proofs are included, the details of lesson 10 not withstanding. Some results are obvious or at least like those in "sophomore calculus"; others involve substantially more theory. Since some writers restrict "expected value" to finite cases, the reader may notice small differences in phraseology.

Definition: *Let $[\Omega, \mathscr{B}, P]$ be a probability space and let $X : \Omega \to \hat{R}$ be measurable. If* $\displaystyle\int_\Omega X(\omega)^+ dP(\omega) - \int_\Omega X(\omega)^- dP(\omega)$ *exists, it is* $\displaystyle\int_\Omega x(\omega) dP(\omega)$ *, and called the mean of the distribution of the RV X or the expected value of X or the first moment (of the distribution) of X . Write* $\displaystyle \mu_X = E[X] = \int_\Omega X = \int X$.

Theorem:

1. *If $E[X] + E[Y]$ exists, then $E[X + Y] = E[X] + E[Y]$; loosely, the expected value of a sum is the sum of the expected values.*

2. *If $P(A \cap B) = 0$, $E[X \cdot I_{A \cup B}] = E[X \cdot I_A] + E[X \cdot I_B]$ when this sum exists; loosely, this is the sum of the integrals on "disjoint" sets.*

3. *For a finite real number r, $E[rX] = rE[X]$; loosley, constants factor outside the expectation (integral) sign.*

4. *$X \geq 0$ implies $E[X] \geq 0$.*

5. *$X \geq Y$ implies $E[X] \geq E[Y]$.*

6. *$X = Y$ a.s. implies $E[X] = E[Y]$.*

7. *$E[X]$ is finite iff $E[\,|X|\,]$ is finite; then X is finite a.s.*

8. *$|X| \leq Y$ and $E[Y]$ finite imply $E[X]$ finite.*

9.   *E[X] and E[Y] finite imply E[X + Y] finite.*

10.   *E[X] finite implies E[X·I$_A$] finite for all A ∈ ℬ.*

11.   *If {A$_n$} is a sequence of disjoint sets in ℬ, then*

$$E[X \cdot I_{\cup A_n}] = \Sigma_n E[X \cdot I_{A_n}] .$$

This is a good place (and about time) to look at an

Example: Let C be the set of non–negative integers in the real line Ω = R; take ℬ as the power set of Ω. Choose $q_n \geq 0$ such

that $\sum_{n=0}^{\infty} q_n = 1$. For A in ℬ, let $P(A) = \sum_{n \in A} q_n$ which is

zero when A ∩ C = ϕ . For singleton sets of non–negative

integers, $P(\{n\}) = q_n$; also, $C = \cup_{n=0}^{\infty} \{n\}$ and $P(\Omega - C) = 0$.

For $X : \Omega \to \overline{R}$, $\int_{\Omega} X(\omega)dP(\omega)$ $= \int_C X(\omega)dP(\omega)$

$$= \sum_{n=0}^{\infty} \int_{\{n\}} X(\omega)dP(\omega)$$

$$= \sum_{n=0}^{\infty} X(n)q_n.$$

Voilà, we have recovered expectation of discrete random

variables. In particular, when X(ω) = ω, $E[X] = \sum_{n=0}^{\infty} nq_n$. The

extra condition that we had to impose previously, we now get

from LS: $\sum_{n=0}^{\infty} X(n)q_n$ is finite iff $\sum_{n=0}^{\infty} |X(n)|q_n$ is finite. In

the case that $q_n = 0$ for n ≥ N (fixed),

$$E[X] = \sum_{n=0}^{N} X(n)q_n$$

like in the binomial: $\displaystyle\sum_{n=0}^{N} n^2 \binom{N}{n} \theta^n (1-\theta)^{N-n}$.

We resume the translation of results:

12.   *If a and b are fixed real numbers such that $a \le X \le b$ on A , then $aP(A) \le E[X \cdot I_A] \le bP(A)$.*

13. $|E[X]| \le E[|X|]$.

14.   *Lebesgue Monotone Convergence Theorem: If the sequence of non-negative RVs $\{X_n\}$ is non-decreasing with limit X , then $E[X_n]$ is non-decreasing with limit $E[X]$ ($\le +\infty$) .*

15.   *If $\{X_n\}$ is a sequence of non-negative RVs, then*

$$E[\sum_{n=1}^{\infty} X_n] = \sum_{n=1}^{\infty} E[X_n] .$$

Exercise 1: Prove #15 from #14 using the fact that

$$0 \le S_N = \sum_{n=1}^{N} X_n \text{ is non-decreasing.}$$

Again, resuming the list:

16.   *If X is integrable and $P(A) \to 0$ (in some mode, not necessarily by sequences), then $E[X \cdot I_A] \to 0$.*

17.   *Fatou Theorems: Let Y and Z be integrable RVs.*

   a) $\{Y \le X_n\}$ imply $E[\liminf X_n] \le \liminf E[X_n]$.

   b) $\{X_n \le Z\}$ imply $\limsup E[X_n] \le E[\limsup X_n]$.

   c) $\{Y \le X_n\}$ and $X_n \uparrow X$ imply $E[X_n] \uparrow E[X]$.

   d) $\{Y \le X_n \le Z\}$ and $X_n \to X$ a.s. imply $E[X_n] \to E[X]$.

18.   *Lebesgue Dominated Convergence Theorem: If $|X_n| < Y$ a.s., $E[Y]$ is finite, and $X_n \to X$ a.s. (or in probability),*

then $E[X_n] \to E[X]$.

18a) Indeed, $\int_A X_n - \int_A X \to 0$ *uniformly in* A ; *equivalently,*

$E[|X_n - X|] \to 0$.

Proof: $|\int_A (X_n - X)| \le \int_A |X_n - X|$

$$\le \int |X_n - X|$$

$$= \int (X_n - X)^+ + \int (X_n - X)^- .$$

Now $\int_A |X_n - X| \to 0$ iff $\int_A (X_n - X)^+$ and $\int_A (X_n - X)^-$

both tend to 0 whence it follows that $\int_A (X_n - X) \to 0$.

Uniformity comes about because this holds when $A = \Omega$.

19. *Bounded Convergence Theorem: Let M be a finite real number. If $X_n \to X$ a.s. (or in probability) on A and $|X_n| \le M$ a.s. on A , then* $\lim E[X_n \cdot I_A] = E[X \cdot I_A]$.

20. *Let X be integrable and let $A_r = \{\omega : |X(\omega)| \le r\}$. Then,*

$$\lim_{r \to \infty} \int_{A_r} |X| = E[|X|]$$

and

$$\lim_{r \to \infty} \int_{(A_r{}^c)} |X| = 0 = \lim_{r \to \infty} rP(A_r{}^c) .$$

Exercise 2: Use #18 to prove # 19.

Exercise 3: Prove #20.

Hints: $|X| \cdot I_{A_r}$ is non–negative, increasing with r to $|X|$;

also, $\int_{\Omega} |X| = \int_{A_r} |X| + \int_{(A_r^c)} |X|$.

Herein *random functions* are families of RVs $\{ X_t \}$ with the index set $T \subset \overline{R}$ not necesssarily a set of integers and perhaps not even countable. For example, the real–valued function $f(x,y) = \sin(\sqrt{x})y$ may be symbolized as $X_t(\omega) = \sin(\sqrt{t})\omega$ for $\omega \in \Omega = R$ and $t \in T = [0, +\infty)$.

We can translate all the theorems involving sequences to "random functions" by approximating convergences like:

$t \to 7$ has bounding sequences $\{7-1/n \le t \le 7+1/n\}$;

$t \to +\infty$ has bounding sequences $\{n \le t < n + 1\}$; etc.

The following three results are the most important. We use dX_t/dt for the function on Ω with value,

$$dX_t(\omega)/dt = \lim_{h \to 0} (X_{t+h}(\omega) - X_t(\omega))/h)$$

when this limit exists.

Theorem: *Let all RVs in the family $\{X_t\}$ be integrable:*
$$E[X_t] \text{ is finite for each } t \in T .$$
Let $t_o \in T$ and let Y be integrable.

21.   *(Taking a limit under the integral sign) If $|X_t| \le Y$ and $X_t \to X_{t_o}$ as $t \to t_o$, then $E[X_t] \to E[X_{t_o}]$ as $t \to t_o$.*

22.   *(Differentiation at a point) If dX_{t_o}/dt exists for each*

$\omega \in \Omega$ *and, for all $t \ne t_o \in T$, $|(X_t - X_{t_o})/(t - t_o)| \le Y$,*

then $dE[X_t]/dt = E[dX_t/dt]$ at $t = t_o$.

23.   *(Differentiation on an interval) If for each t in [a, b] both finite, dX_t/dt exists for each $\omega \in \Omega$, and $|dX_t/dt| \le Y$, then for each $t \in [a, b]$, $dE[X_t]/dt = E[dX_t/dt]$.*

The following lemma from analysis will be used to obtain an interesting result about expectation.

Lemma: *Let the sequence of non-negative numbers* $\{a_k\}$ *be such that* $\sum_{k=1}^{\infty} a_k < \infty$. *Then the sequence of nonnegative numbers* $\{ka_k\}$ *has limit 0.*

Proof: The sequence $\{ka_k\}$ has at least a limsup; say,

$$\limsup_k ka_k = b .$$

If $b = \infty$, some subsequence $\{k'a_{k'}\}$ has $k'a_{k'} \geq 100$ which implies $\sum a_{k'} \geq \sum 100/k' = \infty$ and this contradicts the hypothesis. Similarly, if $0 < b < \infty$, some subsequence has $k''a_{k''} \geq b/2$ and a like contradiction obtains. Therefore, $\limsup_k ka_k = b = 0$ and $\lim_k ka_k = 0$.

Theorem: *Let X be a real random variable for* $[\Omega, \mathscr{B}, P]$. *For* $n = 0(1)\infty$, *let* $B_n = A_n - A_{n+1}$ *with* $A_n = \{\omega : |X(\omega)| \geq n\}$. *Then,* $\sum_{n=0}^{\infty} nP(B_n) = \sum_{n=1}^{\infty} P(A_n)$; $E[|X|]$ *is finite iff these sums are finite.*

Proof: $\Omega = \cup_{n=0}^{\infty} B_n$ is a disjoint union and

$$E[|X|] = \int_\Omega |X| = \sum_{n=0}^{\infty} \int_{B_n} |X|$$

(#14 above). In B_n, $n \leq |X| < n + 1$ so

$$nP(B_n) \leq \int_{B_n} |X| \leq (n+1)P(B_n)$$

(#12 above). Hence,

$$\sum_{n=0}^{\infty} nP(B_n) \le E[\,|X|\,] \le \sum_{n=0}^{\infty} nP(B_n) + 1 . \qquad (*)$$

Now $\displaystyle\sum_{n=0}^{N} nP(B_n) = \sum_{n=1}^{N} n(P(A_n) - P(A_{n+1}))$

$$= P(A_1) - P(A_2) + 2P(A_2) - 2P(A_3) + 3P(A_3)$$

$$- 3P(A_4) + \cdots + NP(A_N) - NP(A_{N+1}) .$$

This collapses to:

$$P(A_1) + P(A_2) + P(A_3) + \cdots + P(A_N) - NP(A_{N+1})$$

whence $\displaystyle\sum_{n=0}^{N} nP(B_n) = \sum_{n=1}^{N} P(A_n) - NP(A_{N+1}) . \qquad (**)$

a)   If $\displaystyle\sum_{n=1}^{\infty} P(A_n)$ is finite, $(N+1)P(A_{N+1}) \to 0$ by the

lemma so that $NP(A_{N+1}) \to 0$. Finiteness of $\displaystyle\sum_{n=0}^{\infty} nP(B_n)$

follows from (**).

b)   If $\displaystyle\sum_{n=0}^{\infty} nP(B_n)$ is finite, the finiteness of $E[\,|X|\,]$ follows

from (*). By #20 above, $(N+1)P(A_{N+1}) \to 0$ so that

$NP(A_{N+1}) \to 0$; again finiteness of $\displaystyle\sum_{n=1}^{\infty} P(A_n)$ follows

from (**).

c)   This shows that the equality holds when either one of the series is finite. Their equality when they sum to infinity is trivial.

Exercise 4:   For $\Omega = \{1,2,3,\cdots\}$ with the power set as σ–field,

let $X(\omega) = \omega$ and $P(X(\omega) = n) = q_n$, n=1(1)∞ . Show that

$$E[X] = \sum_{n=1}^{\infty} P(X \geq n) .$$

Exercise 5: Find $E[X]$ and $E[X^2]$ when:

a) $0 < \theta < 1$ and $q_n = (1 - \theta)^{n-1}\theta$ (the geometric distribution);

b) $q_n = 6/(\pi n)^2$ (this $\pi = 3.141592\cdots$).

In the following we first present some inequalities for real numbers and then make an application to expectation.

Lemma: *If a and b are real numbers, then (it is obvious that)*

$$|a + b| \leq |a| + |b| \leq 2 \, max\{|a|, |b|\} .$$

For $r > 0$, $|a + b|^r \leq 2^r max\{|a|^r, |b|^r\} \{ \leq 2^r(|a|^r + |b|^r)\}$.
Replacing a and b by RVs X and Y, It follows that

$$E[\,|X + Y|^r] \leq 2^r(E[\,|X|^r] + E[\,|Y|^r]) .$$

When $|X|^r$ and $|Y|^r$ are both integrable, so is $|X + Y|^r$.

Lemma: *For $|a| \leq 1$ and $0 < r < s$, $|a|^r < 1 < 1 + |a|^s$;*

for $|a| > 1$ and $0 < r < s$, $|a|^r \leq |a|^s < 1 + |a|^s$.

With a replaced by the RV X , it follows that $E[\,|X|^r]$ is finite whenever $E[\,|X|^s]$ is finite, $0 < r < s$. Then $E[X^r]$ is finite for $0 < r \leq s$.

Exercise 6: Fix $p > 1$; take $1/p + 1/q = 1$, and $b > 0$. Consider $f(a) = a^p/p + b^q/q - ab$ for $a \in (0, \infty)$. Sketch f; in particular, show that f has minimum value 0 .

Lemma (Hölder): *It follows from exercise 6 with $p > 1$ and $1/p + 1/q = 1$ that for any real numbers a and b ,*

$$|ab| \leq |a|^p/p + |b|^q/q .$$

Now suppose that X^p and Y^q are integrable RVs. When a is replaced by $X/(E[\,|X|^p])^{1/p}$ and b by $Y/(E[\,|Y|^q])^{1/q}$, the resulting inequality can be arranged as

$$|E[XY]| \leq E[\,|XY|\,] \leq (E[\,|X|^p])^{1/p} \cdot (E[\,|Y|^q])^{1/q} .$$

Exercise 7: Fill in the missing steps of this last lemma.

When $p = q = 2$, Hölder's inequality becomes the Cauchy(–Bunyakovski–Schwartz) inequality:

$$(E[XY])^2 \leq E[X^2] \cdot E[Y^2] .$$

INE– In particular, when $\Omega = \{1,2,3, \cdots \}$ and P is as in the first example above, we get

$$\left(\sum X(n)Y(n)q_n\right)^2 \leq \sum X(n)^2 q_n \cdot \sum Y(n)^2 q_n .$$

If also $q_n = 0$ for $n > N$, this reduces to

$$\left(\sum_{n=1}^{N} X(n)Y(n)q_n\right)^2 \leq \sum_{n=1}^{N} X(n)^2 q_n \cdot \sum_{n=1}^{N} Y(n)^2 q_n .$$

When the distribution on $\{1,2,3,\cdots,N\}$ is uniform, the last inequality is equivalent to

$$\left(\sum_{n=1}^{N} X(n)Y(n)\right)^2 \leq \sum_{n=1}^{N} X(n)^2 \cdot \sum_{n=1}^{N} Y(n)^2 .$$

In a common notation for vectors in R^N, this is written as

$$\left(\sum_{n=1}^{N} x_n y_n\right)^2 \leq \sum_{n=1}^{N} x_n^2 \cdot \sum_{n=1}^{N} y_n^2$$

or $|x \cdot y|^2 \leq \|x\|^2 \cdot \|y\|^2 .$

We think it is interesting that this result for vectors can be found as a special case of a result for expectation.

LESSON 12. STIELTJES INTEGRALS

In this lesson, we give some theory for a Stieltjes integral in direct generalization of that of Riemann, the familiar integral of "sophomore calculus".

We use the following notations:

1) a closed interval of real numbers: [a,b];

an indicator function $I_{[a,b]}(x) = I\{a \le x \le b\}$;

a *partition* of [a,b]: the finite set of points $D = \{x_i\}$ with

$$a = x_0 < x_1 < x_2 < \cdots < x_n = b.$$

2) For a given a partition and $i = 1(1)n$: the domain increments are $\Delta x_i = x_i - x_{i-1}$; a bounded real valued function g has

$$m_i = \inf\{g(x) : x_{i-1} \le x \le x_i\}, \quad M_i = \sup\{g(x) : x_{i-1} \le x \le x_i\};$$

a real valued function F has increments

$$\Delta F(x_i) = F(x_i) - F(x_{i-1}) .$$

Each choice of ξ_i in $[x_{i-1}, x_i]$ defines an

$$S(D) = \sum_{i=1}^{n} g(\xi_i)\Delta F(x_i) , \quad \underline{S}(D) = \sum_{i=1}^{n} m_i\Delta F(x_i),$$

and $\overline{S}(D) = \sum_{i=1}^{n} M_i\Delta F(x_i) .$

The norm of the partition is

$$\Delta D = \max \{\Delta x_1, \Delta x_2, \cdots, \Delta x_n\} .$$

By the symbol $\lim_{\Delta D \to 0} S(D) = L$ (finite), we mean that for each $\varepsilon > 0$, there is a δ_ε such that for any partition D with $\Delta D < \delta_\varepsilon$, $|S(D) - L| < \varepsilon$. There is a "uniformity" involved in that it doesn't matter where the points of the partition are; it only matters that the differences $x_i - x_{i-1}$ be small enough. Note that this means that the limit must be independent of the way "ξ_i" are chosen. If $L = \pm\infty$, $\Delta D < \delta_\epsilon$ implies $|S(D)| > \epsilon$.

Definition: *When* $\lim\limits_{\Delta D \to 0} S(D)$ *exists, it is called the Stieltjes integral of g with respect to F on [a,b]; the value of the limit is symbolized by* $\int_a^b g(x) \, dF(x)$. *We say that g is integrable with respect to F (wrt F) when this is finite.*

Note that when $F(x) = x$, we are back to Riemann! As in that case, the "improper integral"

$$\int_{-\infty}^{\infty} g(x) \, dF(x) \text{ is } \lim\limits_{\substack{a \to -\infty \\ b \to \infty}} \int_a^b g(x) \, dF(x)$$

when this exists.

Example: Take $F(x) = I\{3 \le x < \infty\}$. A graph of F looks like

$$* \cdots\cdots\cdots\cdots\cdots\cdots\cdots \rightarrow$$

$$\leftarrow\cdots\cdots\cdots\cdots\cdots\cdots$$
$$x = 3$$

where the $*$ emphasizes the right hand continuity. From this, we see that $\Delta F(x_i) = 0$ except for an interval which contains 3; say, $x_{i_3-1} < 3 \le x_{i_3}$. For any function g on R, $S(D) = g(\xi_{i_3})$.

    a)   Let g be continuous at 3. Since ξ_{i_3} has limit 3 as n $\to \infty$, $S(D)$ will have the limit $g(3)$. Then

$$\int_{-\infty}^{\infty} g(x) \, dF(x) = g(3);$$

in fact, $\int_a^b g(x) \, dF(x) = g(3)$ for any of the intervals

$$a < 3 < b, \ a < 3 = b \text{ while } \int_3^b g(x) \, dF(x) = 0.$$

Of course, $\displaystyle\int_a^b g(x)\,dF(x) = 0$ for any $b < 3$.

b) On the other hand, if g is bounded but not continuous at 3, S(D) will have no limit and the first three integrals in a) will not exist. Of course,

$$\int_3^b g(x)\,dF(x) \text{ is still } 0.$$

The following is just a trifle more involved.

Example: Let $F(x) = 2 \cdot I\{3 \le x < \infty\} + 5 \cdot I\{7 \le x < \infty\}$. Let a partition have $x_{i_3 - 1} < 3 \le x_{i_3}, x_{i_7} < 7 \le x_{i_7}$.

*.............→

*..............

←..................
x = 3   x = 7

For each $a < 3$ and $b > 7$, $\displaystyle\int_a^b g(x)\,dF(x)$ will have all terms of $S(D) = 0$ except the corresponding $g(\xi_{i_3}) \cdot 2 + g(\xi_{i_7}) \cdot 5$. When g is continuous at 3 & 7, all these integrals will have the value

$$g(3) \cdot 2 + g(7) \cdot 5.$$

If g is not continuous at 3 or 7, none of these integrals will exist.

Exercise 1: Find, if they exist, integrals of F wrt F & g wrt F :
a) on $[1,2]$ with $F(x) = I\{1 \le x < \infty\}$, g continuous;
b) on $[0.9]$ with $F(x) = 2 \cdot I\{3 \le x < \infty\} + 5 \cdot I\{7 \le x < \infty\}$,

$g(x) = 4 \cdot I\{5 < x \le 8\}$. Hint: draw pictures!.

The long proof of the basic existence theorem following is not a detail to be retained; mutatis mutandis, the proof remains valid for F "decreasing". Recall that a real valued function F on the real line is non–decreasing (colloquially, increasing) if

$$\text{for } x_1 < x_2, \; F(x_1) \leq F(x_2) \, .$$

Theorem: *Let g be continuous and let F be non-decreasing on [a,b]. Then the Stieltjes integral of g wrt F on [a,b] exists.*

INE–Proof: a) In terms of the notation introduced earlier, since $m_i \leq M_i$ and $\Delta F(x_i) \geq 0$, it is obvious that $\underline{S}(D) \leq \overline{S}(D)$.

b)   Let the partition $\tilde{D}$ be refinement of D; that is, $\Delta \tilde{D}$ is made smaller than ΔD by the putting in of a finite number of additional points. Let $\tilde{x}$ be one of the new points, say $x_{i-1} < \tilde{x} < x_i$. Now $\tilde{m}_{i-i}, \tilde{m}_i$ are taken over the smaller sub–intervals, $[x_{i-1}, \tilde{x}]$ and $[\tilde{x}, x_i]$ respectively, so these glb are greater than or equal to the old m_i taken over $[x_{i-1}, x_i]$. Within $\underline{S}(\tilde{D})$, we have

$$\tilde{m}_{i-1}(F(\tilde{x}){-}F(x_{i-1})) + \tilde{m}_i(F(x_i){-}F(\tilde{x})) \geq m_i(F(x_i){-}F(x_{i-1})).$$

Since each new point can be treated in a like manner, it follows that $\underline{S}(\tilde{D}) \geq \underline{S}(D)$. Similarly, $\overline{S}(\tilde{D}) \leq \overline{S}(D)$.

c)   Now let D_1 and D_2 be any two partitions of [a,b] and let $\tilde{D}$ be their union, a partition made up of all points of either; $\tilde{D}$ is a refinement of both. By part b),

$$\underline{S}(D_1) \leq \underline{S}(\tilde{D}) \text{ and } \overline{S}(\tilde{D}) \leq \overline{S}(D_2) \, ;$$

by part a), $\underline{S}(\tilde{D}) \leq \overline{S}(\tilde{D})$. By transitivity, $\underline{S}(D_1) \leq \overline{S}(D_2)$.

Similarly, $\underline{S}(D_2) \leq \overline{S}(D_1)$.

d)   This monotonicity means that $\{\underline{S}(D) : D$ is any partition$\}$ is bounded above by any $\overline{S}(D)$ and so this set has a lub, say s. Similarly, $\{\overline{S}(D) : D$ is any partition$\}$ has a glb, say S.

Moreover, $s \leq S$.

e)   Since g is continuous on a closed bounded interval, it is uniformly continuous there:

for $\varepsilon > 0$ there is a $\delta_\varepsilon > 0$ such that any $|x' - x''| < \delta_\varepsilon$

has $|g(x') - g(x'')| < \varepsilon$.

Consider all partitions D with $\Delta D < \delta_\varepsilon$; in these, $M_i - m_i < \varepsilon$. Then,

$$0 \leq [\bar{S}(D) - S] + [S - s] + [s - \underline{S}(D)] \qquad (*)$$

$$= \bar{S}(D) - \underline{S}(D) \leq \sum_{i=1}^{n} (M_i - m_i)\Delta F(x_i)$$

$$\leq \varepsilon \sum_{i=1}^{n} (F(x_i) - F(X_{i-1})) = \varepsilon(F(b) - F(a)).$$

Since ε is arbitrary and each of the terms $[\cdots]$ in the first line of (*) is not negative, it follows that all three terms tend to 0 as ε and δ_ε tend to 0. In particular, $s = S$.

f)   Since $\underline{S}(D) \leq S(D) \leq \bar{S}(D)$ and the left and right hand sides of this inequality tend to $s = S$, so does $S(D)$. In other words, $\int_a^b g(x)\, dF(x) = \lim_{\Delta D \to 0} S(D)$ as desired.

INE–Exercise: Complete the arguments for "Similarly,$\cdots$" in parts b), c), d) of the proof above.

Example: $\int_0^1 x\, d(x^2)$ exists since $g(x) = x$ is continuous and $F(x) = x^2$ is increasing on [0,1] . Therefore, we can evaluate this integral by choosing any convenient partition and then taking the limit of $S(D)$.

Let $x_0 = 0$, $x_1 = \xi_1 = 1/n$,

$$x_2 = \xi_2 = 2/n, \cdots, x_n = \xi_n = n/n = 1.$$

$$S(D) = \sum_{i=1}^{n} \left[\frac{i}{n}\right]\left[\left[\frac{i}{n}\right]^2 - \left[\frac{i-1}{n}\right]^2\right] = \sum_{i=1}^{n} i(2i - 1)/n^3$$

$$= \sum_{i=1}^{n} \left[2\sum_{i=1}^{n} i^2 - \sum_{i=1}^{n} i\right]/n^3$$

$$= \left[2n(n+1)(2n+1)/6 - n(n+1)/2\right]/n^3 \to 4/6.$$

Exercise 3: Prove by induction:

$$\sum_{i=1}^{n} i = n(n+1)/2 \; ; \quad \sum_{i=1}^{n} i^2 = n(n+1)(2n+1)/6 \; ;$$

$$\sum_{i=1}^{n} i^3 = (n(n+1))^2/4 \;.$$

Example: Take $g(x) = x^2$ and $F(x) = (x + 1)^2$ for $x \in [-1,2]$. Let $a = -1$, $x_1 = -1 + 3/n$, $x_2 = -1 + 2\cdot3/n$, $x_3 = -1 + 3\cdot3/n$, $\cdots$,

$$x_i = -1 + i\cdot3/n, \cdots, x_n = -1 + n\cdot3/n = 2 \;.$$

Note that $\Delta D = 1/n \to 0$ as $n \to \infty$. Then,

$$\sum_{i=1}^{n} g(x_i)\Delta F(x_i) =$$

$$\sum_{i=1}^{n} (-1 + i\cdot3/n)^2\left[(i\cdot3/n)^2 - ((i-1)\cdot3/n)^2\right]$$

$$= \sum_{i=1}^{n} (1 - 6\cdot i/n + 9\cdot i^2/n^2)(9/n^2)(2\cdot i - 1)$$

$$= (9/n^2)\sum_{i=1}^{n} (2i - 12i^2/n + 18i^3/n^2 - 1 + 6i/n - 9i^2/n^2)$$

$$= (9/n^2)\left[2n(n+1)/2 - 12(n(n+1)(2n+1)/6n\right.$$

$$\left. + 18n^2(n+1)^2/4n^2 - n + 6n(n+1)/2n - 9n(n+1)(2n+1)/6n^2\right].$$

For $n \rightarrow \infty$, this is $9(1 - 24/6 + 18/4 - 0 + 0 - 0) = 13.5$.

Exercise 4: Verify the limit in the previous example.

Exercise 5: $\begin{aligned} F(x) &= 0 &&\text{for} && x < -1 \\ &= .1 &&\text{for} && -1 \le x < 0 \\ &= .3 &&\text{for} && 0 \le x < 1 \\ &= .6 &&\text{for} && 1 \le x < 2 \\ &= 1.0 &&\text{for} && 2 \le x \ . \end{aligned}$

a)   Investigate $\displaystyle\int_{-1}^{2} x\,dF(x)$ and $\displaystyle\int_{-1}^{2} x^2\,dF(x)$.

b)   Describe these in terms of expectation.

Now we consider a special case of the integral, omitting some details of the proof; the reader should recognize its usefulness immediately.

Theorem: *Let g and F be real valued functions on [a, b]. Let g be continuous; let F have a continuous derivative, F' = f on [a, b]. Then, the Stieltjes integral* $\displaystyle\int_{a}^{b} g\,dF$ *and the Riemann integral* $\displaystyle\int_{a}^{b} g(x)f(x)dx$ *both exist and are equal.*

Partial proof: When f is non–negative, F is increasing and g is integrable wrt F; the integral will be the limit of a sum

$$\sum g(\xi_i)[F(x_i) - F(x_{i-1})] \ .$$

Now F satisfies the hypotheses of the mean value theorem so that for some η_i in each (x_{i-1}, x_i) ,

$$F(x_i) - F(x_{i-1}) = f(\eta_i)(x_i - x_{i-1}).$$

Also, $g \cdot f$ is continuous, in fact, uniformly continuous. Hence,

a)   as $x_i - x_{i-1} \rightarrow 0$ implies $\xi_i - x_i \rightarrow 0$ and $\eta_i - x_i \rightarrow 0$, it follows that $\sum_i g(\xi_i)f(\eta_i)(x_i - x_{i-1})$ also approximates the integral.

b)   Moreover, $g \cdot f$ is Riemann integrable so that this last sum approximates that integral. Together these yield the conclusion.

Example: $\int_{-1}^{2} x^2 \, d(x+1)^2 = \int_{-1}^{2} x^2 \cdot 2(x+1) \, dx$

$$= 2\left[x^4/4 + x^3/3\right]_{-1}^{2} = 13.5.$$

Exercise 6: Evaluate $\int_{-1}^{2} g(x) \, dF(x)$ in each case.

a) $g(x) = x^3$, $F(x) = (\sqrt{(x+2)}) \cdot I\{-1 \le x \le 2\}$
b) $g(x) = x$,
$$F(x) = (1/3)x^2 \cdot I\{0 \le x \le 1\}$$
$$+ (1 - 2(x - 2)^3/3) \cdot I\{1 \le x \le 2\} .$$

The following theorem has no counterpart in the Riemann theory. As is evident, the result is useful when a CDF is decomposed into continuous and discrete parts (lesson 3).

Theorem: Let g be integrable wrt to each of F_1, F_2 . Then, g is integrable wrt $F = F_1 + F_2$ and

$$\int_{-\infty}^{\infty} g(x) \, d(F_1 + F_2)(x) = \int_{-\infty}^{\infty} g(x) dF_1(x) + \int_{-\infty}^{\infty} g(x) dF_2(x) .$$

Partial proof: For a given partition,
$$\Delta F(x_i) = F(x_i) - F(x_{i-1}) =$$
$$\{F_1(x_i) + F_2(x_i)\} - \{F_1(x_{i-1}) + F_2(x_{i-1})\} =$$
$$\Delta F_1(x_i) + \Delta F_2(x_i) .$$

Hence, $\Sigma_i m_i \Delta F(x_i) = \Sigma_i m_i \Delta F_1(x_i) + \Sigma_i m_i \Delta F_2(x_i)$. Since each of the latter two sums converges to its corresponding integral, the conclusion follows.

Example: Let $g(x) = x^2 \cdot I\{-1 \le x \le 2\}$,

$$F_1(x) = (x+1)^2 \cdot I\{-1 \le x \le 2\} + 9I\{2 < x < \infty\},$$

$$F_2(x) = (1/3) \cdot I\{-1 \le x < 0\} + (1/2) \cdot I\{0 \le x < 1\}$$

$$+ I\{1 \le x < \infty\} .$$

a) $\displaystyle\int_{-1}^{2} g(x) d(F_1(x) + F_2(x))$

$$= \int_{-1}^{2} x^2 \, dF_1(x) + \int_{-1}^{2} x^2 \, dF_2(x)$$

$$= \int_{-1}^{2} x^2 \, d(x+1)^2 + g(0)(1/2 - 1/3) + g(1)(1 - 1/2)$$

$$= \int_{-1}^{2} x^2 \cdot 2(x+1) \, dx + 0(1/6) + 1(1/2)$$

$$= 13.5 + .5 = 14 .$$

b) $\displaystyle\int_{-\infty}^{\infty} x^2 \, dF_1(x)$

$$= \int_{-\infty}^{-1} x^2 \, dF_1(x) + \int_{-1}^{2} x^2 \, dF_1(x) + \int_{2}^{\infty} x^2 dF_1(x).$$

The first and last integrals are zero since $\Delta F_1(x_i) = 0$ outside $[-1, 2]$ and the middle integral is 13.5 .

c) $\displaystyle\int_{-\infty}^{\infty} x^2 \, dF_2(x)$

$$= (-1)^2(1/3 - 0) + (0)^2(1/2 - 1/3) + (1)^2(1 - 1/2)$$

$$= 1.5.$$

d) For $F = F_1 + F_2$, $\displaystyle\int_{-\infty}^{\infty} x^2 \, dF(x) = 15$.

Exercise 7: Continue the example; evaluate, (when they exist):

$$\int_{-1}^{2} F(x)\ dF(x) \quad \text{and} \quad \int_{-\infty}^{\infty} F(x)\ dF(x)\ .$$

The following is true and proved in more advanced courses.

Consider $\Omega = R$ and the corresponding Borel σ–field $\mathscr{B}$ Let P be any measure on $\mathscr{B}$ assigning a finite value to each finite interval. Then the relation

$$P((a,b]) = F(b) - F(a) \tag{*}$$

defines F as a finite non–decreasing right–hand continuous function on R. If g is integrable wrt P ala Lesson 10, then that "Lebesgue–Stieltjes" integral, $\int g\ dP$, has the same value as the "Riemann–Stieltjes" integral $\int g\ dF$ in this lesson. As noted earlier, given an F with these properties, (*) defines a measure P.

LESSON 13. PRODUCT MEASURES AND INTEGRALS

The material in this lesson is needed to complete the properties associated with "independence" in probability spaces but here we focus on measure spaces: $[\Omega_1, \mathscr{B}_1, \Lambda_1]$, $[\Omega_2, \mathscr{B}_2, \Lambda_2]$.

Definition: *I)* If $A_i \subset \Omega_i$, the cartesian product of these two subsets is $A_1 \times A_2 = \{(\omega_1, \omega_2) : \omega_1 \in A_1, \omega_2 \in A_2\}$. In particular, $\Omega_1 \times \Omega_2$ is the collection of all ordered pairs (ω_1, ω_2), $\omega_i \in \Omega_i$.

Example: Let $A_1 = \{a, b, c\}$, $A_2 = \{3, 5\}$. Then,

$$A_1 \times A_2 = \{(a,3)\ (a,5)\ (b,3)\ (b,5)\ (c,3)\ (c,5)\}\ \text{ but}$$

$$A_2 \times A_1 = \{(3,a)\ (3,b)\ (3,c)\ (5,a)\ (5,b)\ (5,c)\}.$$

Exercise 1: a) Describe $\Omega_1 \times \Omega_2$ if $\Omega_i = R$.

b)   Describe $A_1 \times A_2$ if $A_1 = (-1, 2)$ and $A_2 = (1/4, 2)$.

Definition: *II)* $\mathscr{B} = \mathscr{B}_1 \times \mathscr{B}_2$ is the minimal σ-field generated by the "rectangles" $A_1 \times A_2$, $A_i \in \mathscr{B}_i$.

Note: $\mathscr{B}_1 \times \mathscr{B}_2$ is really just a symbol; the σ-field is built up by countable unions/intersections/complements of the true products $A_1 \times A_2$; traditionally, the same symbol $\times$ is used throughout. The same spirit prevails in the next part on product measure where the last $\times$ is ordinary multiplication in $\overline{R}$. The term "generated" hides many constructive details.

Definition: *III)* $\Lambda = \Lambda_1 \times \Lambda_2$ is the product measure on $\mathscr{B}$ generated by $\Lambda(A_1 \times A_2) = \Lambda_1(A_1) \times \Lambda_2(A_2)$, $A_i \in \mathscr{B}_i$.

Example: If Λ_i is simple Lebesgue measure,

$$\Lambda([1,3]\times[3,6]) = \Lambda_1([1,3])\times\Lambda_2([3,6]) = 2\cdot 3 = 6,$$

which is the area of the rectangle $[1,3]\times[3,6]$.

It is possible that for a given Λ_1,Λ_2 pair, one could not construct a corresponding Λ but, in the context of probability, such difficulties do not arise. Moreover, we will assume that we have σ–*finite* measures: there is a sequence of sets $B_n \in \mathcal{B}$ such that $\Lambda(B_n) < \infty$, $\Omega = \cup_{n=1}^{\infty} B_n$. Since $P(\Omega) = 1$, probability is (trivially) σ–finite; so is any bounded measure. For Lebesgue measure, $R = \cup_{n=-\infty}^{\infty}(n, n+1]$ with $\Lambda(n,n+1] = 1$.

Definition: *IV) Let $\Omega = \Omega_1\times \Omega_2$; let X be a function on Ω; let B be a subset of Ω. The section of B at ω_1 is $B(\omega_1;) = \{\omega_2 : (\omega_1,\omega_2) \in B\}$; the section of B at ω_2 is $B(;\omega_2) = \{\omega_1 : (\omega_1,\omega_2) \in B\}$. The section of X at ω_1 [at ω_2] is the function $X(\omega_1;)$ on Ω_2 [$X(;\omega_2)$ on Ω_1] with values $X(\omega_1;)(\omega_2) = X(\omega_1,\omega_2) = X(;\omega_2)(\omega_1)$.*

In a "section", one variable is fixed and the surrounding discussion focuses on the other variable.

Example: Let $\Omega = R\times R$, $B = [1,3]\times(3,6)$,

$$C = \{(\omega_1,\omega_2) : \omega_1 + \omega_2 < 6\} \text{ and } X(\omega_1,\omega_2) = \omega_1^2 + \omega_1\omega_2.$$

    a)   All along the line segment
$$B(2;) = \{\omega_2: (2,\omega_2) \in B\} = (3,6),$$

    ω_1 is fixed at 2 and $X(2;)(\omega_2) = 4 + 2\omega_2$.

    b)   All along the line segment
$$B(;5) = \{\omega_1 : (\omega_1,5) \in B\} = [1,3],$$
$$X(;5)(\omega_1) = \omega_1^2 + 5\omega_1.$$

c) $C(2;) = \{\omega_2: (2,\omega_2) \ \varepsilon \ C\} = (-\infty,4);$

$C(;5) = (-\infty,1)$ in Ω_1.

Exercise 2: Let X have domain $B \subset R \times R$. Find the sections of each X at $\omega_1 = 1, \omega_2 = 2$.

a) $B = [-1,4] \times [2,6]; \quad X(\omega_1,\omega_2) = |\omega_1| + |\omega_2|$.

b) $B = \{(\omega_1,\omega_2) : \omega_1^2 + \omega_2^2 \le 9\};$

$X(\omega_1,\omega_1) = \omega_1/(1 + |\omega_2|)$

c) $B = \{(\omega_1,\omega_2) : \omega_1^2 + \omega_2^2 = 9\}; \quad X$ as in b) .

Now we are ready to state the product measure theorem. The proof is based on the Lebesgue monotone convergence theorem and can be found in books like Loève, 1963. What we use most often is the corollary following the theorem which justifies the "double integral".

Theorem: *Let Λ_1,Λ_2 in the measure spaces $[\Omega_1,\mathscr{B}_1,\Lambda_1]$, $[\Omega_2,\mathscr{B}_2,\Lambda_2]$ be σ-finite; let $B \in \mathscr{B}_1 \times \mathscr{B}_2$. Then,*

a)   *the sections $B(\omega_1;)$ and $B(;\omega_2)$ are measurable;*

b)   *the functions Q_1,Q_2 with values $\Lambda_1(B(;\omega_2))$, $\Lambda_2(B(\omega_1:))$ are measures on $\mathscr{B}_1$, $\mathscr{B}_2$ respectively;*

c)   *The function Λ on $\mathscr{B}_1 \times \mathscr{B}_2$ with values*

$$\Lambda(B) = \int_{\Omega_2} \Lambda_1(B(;\omega_2))d\Lambda_2(\omega_2)$$

$$= \int_{\Omega_1} \Lambda_2(B(\omega_1;))d\Lambda_1(\omega_1)$$

is a σ-finite measure uniquely determined by

$$\Lambda(B_1 \times B_2) = \Lambda_1(B_1) \times \Lambda_2(B_2) .$$

Corollary: *(Fubini's Theorem) a) If $X : \Omega = \Omega_1 \times \Omega_2 \to R$ is* *actually non-negative and measurable wrt* $\Lambda = \Lambda_1 \times \Lambda_2$, *then*

$$\int_\Omega X \, d\Lambda = \int_{\Omega_1 \times \Omega_2} X \, d(\Lambda_1 \times \Lambda_2) = \int_{\Omega_1} \left[\int_{\Omega_2} X(\omega_1;) d\Lambda_2 \right] d\Lambda_1$$

$$= \int_{\Omega_2} \left[\int_{\Omega_1} X(;\omega_2) d\Lambda_1 \right] d\Lambda_2 \; .$$

b)   *If X is integrable wrt* Λ, *this integral equality holds and almost all* $X(\omega_1;)$ $[X(;\omega_2)]$ *are integrable wrt* Λ_2 $[\Lambda_1]$.

Partial proof: a) Let $X = I_B$ for $B \in \mathscr{B}_1 \times \mathscr{B}_2$. Then

$$X(\omega_1;) = I_{B(\omega_1;)} \; [X(;\omega_2) = I_{B(;\omega_2)}]$$

and the conclusion of the corollary is just a special case of the theorem. As outlined in an earlier lesson, it follows that the conclusion holds for non–negative simple functions. By the LMCT, the conclusion will hold then for non–negative measurable functions.

b)   If $X \geq 0$ is integrable, so is $\int_{\Omega_2} X(\omega_1;) d\Lambda_2$ wrt Λ_1. But

then this integral itself is finite a.e. so that almost all $X(\omega_1;)$ are integrable wrt Λ_2. If $X = X^+ - X^-$ is integrable, these remarks apply to X^+, X^- and the conclusion follows.

In noting the following special cases which are used repeatedly, we also introduce more common notations. The manipulations are valid when all the integrals involved exist.

Lebesgue: $\Omega_1 = \Omega_2 = R$; $\Lambda_1 = \Lambda_2$ is Lebesgue measure on $[R, \mathscr{B}_1]$; $g : \Omega \to R$.

$$\int_\Omega g \, d\Lambda = \int_{R \times R} g(x_1, x_2) dx_1 dx_2 = \int_{-\infty}^{\infty} \left[\int_{-\infty}^{\infty} g(x_1, x_2) dx_1 \right] dx_2$$

$$= \int_{-\infty}^{\infty} \left[\int_{-\infty}^{\infty} g(x_1, x_2) dx_2 \right] dx_1.$$

Counting: $\Omega_1 = \Omega_2 = \{0,1,2,3,\cdots\}$, say W. $\Lambda_1 = \Lambda_2$ is defined by $\Lambda_i(A) =$ the number of elements in W which belong to A; $\Lambda_i(j) = \begin{bmatrix} 1 & \text{if } j \in \Omega_i \\ 0 & \text{otherwise} \end{bmatrix}$ imparts a uniform distribution (not probability) on Ω_i. For $g : \Omega \to R$,

$$\int_\Omega g d\Lambda = \int_{W \times W} g(\omega_1, \omega_2) \, d\Lambda_1(\omega_1) d\Lambda_2(\omega_2) =$$

$$\sum_{\omega_1} \sum_{\omega_2} g(\omega_1, \omega_2) = \sum_{i=0}^{\infty} \sum_{j=0}^{\infty} g(i,j).$$

Note that any countable set could be labeled as in W so that this result is more general than may first appear.

Stieltjes: Let P_F be a measure with the corresponding df F on Ω $= R \times R$. Suppose that $F(x_1, x_2) = F_1(x_1) \times F_2(x_2)$ where each F_i is also a df on R with corresponding measure P_{F_i}.

$$\int_\Omega g \, dP_F = \int_{R^2} g \, dF = \int_{-\infty}^{\infty} \int_{-\infty}^{\infty} g(x_1, x_2) dF(x_1, x_2) =$$

$$\int_{-\infty}^{\infty} \left[\int_{-\infty}^{\infty} g \, dF_2 \right] dF_1 = \int_{-\infty}^{\infty} \left[\int_{-\infty}^{\infty} g \, dF_1 \right] dF_2 .$$

Example: a) Take $F_i(x) = \int_{-\infty}^{x} f(t) \, d\Lambda_i(t)$ where Λ_i is Lebesgue measure, $F_i'(x) = f(x)$ is non–negative and continuous a.e. Then this integral is Riemann type and can be written as $\int_{-\infty}^{x} f(t) dt$.

b) For $F = F_1 \times F_2$, $\Lambda = \Lambda_1 \times \Lambda_2$, and g integrable wrt F [Λ],

$$\int_{R^2} g\ dF = \int_{-\infty}^{\infty}\left[\int_{-\infty}^{\infty}g(x_1,x_2)dF_1(x_1)\right]dF_2(x_2)$$

$$= \int_{-\infty}^{\infty}\left[\int_{-\infty}^{\infty}g(x_1,x_2)f(x_1)dx_1\right]f(x_2)dx_2$$

$$= \int_{-\infty}^{\infty}\int_{-\infty}^{\infty}g(x_1,x_2)f(x_1)f(x_2)dx_1dx_2$$

$$= \int_{R^2} g(x_1,x_2)f(x_1)f(x_2)\ d\Lambda$$

$$= \int_{-\infty}^{\infty}\left[\int_{-\infty}^{\infty}g(x_1,x_2)f(x_2)dx_2\right]f(x_1)dx_1$$

$$= \int_{-\infty}^{\infty}\left[\int_{-\infty}^{\infty}g(x_1,x_2)dF_1(x_1)\right]dF_2(x_2)\ .$$

ɔ)   Let $F_i(x) = \sum_{j=0}^{[x]}(1/2^{j+1})I\{0 \le x < \infty\}$ where $[x]$ is the greatest integer $\le x$. Let $F = F_1\times F_2$. Then,

$$\int_{R^2} g\ dF = \int_{-\infty}^{\infty}\left[\int_{-\infty}^{\infty}g(x_1,x_2)dF_1(x_1)\right]dF_2(x_2)$$

$$= \int_{-\infty}^{\infty}\left[\sum_{i=0}^{\infty}g(i,x_2)/2^{i+1}\right]dF_2(x_2)$$

$$= \sum_{j=0}^{\infty}\left[\sum_{i=0}^{\infty}g(i,j)/2^{i+1}\right]/2^{j+1}.$$

The theorem says that we can interchange the order of summation if $\sum_{i+j=0}^{\infty}g(i,j)/2^{i+j+2}$ converges absolutely.

Exercise 3: Let $g(x_1, x_2) = x_1 x_2$. Evaluate the integrals in parts b) and c) of the last example.

Exercise 4: Let $g(x_1, x_2) = \exp(tx_1 + tx_2)$. Evaluate the integrals in parts b) and c) of the last example by choosing sets of values for t which make the integrals finite.

Exercise 5: Let $F_1(x_1) = (1 - \exp(-x_1))I\{0 \le x_1 < \infty\}$ and let

$$F_2(x_2) = x_2 I\{0 \le x_2 \le 1\} + I\{1 < X_2 < \infty\} .$$

Let $g(x_1, x_2) = \exp(tx_1 + tx_2)$. Choose t appropriately so as to

make $\int_{R^2} g \, d(F_1 \times F_2)$ finite.

It is an easy induction to extend these results to any finite number of integrals, in particular, to random vectors

$$X = (X_1, \cdots, X_n) \text{ with CDF } F_{X_i} \text{ and } F = F_1 \times \cdots \times F_n .$$

Exercise 6: Write out the hypotheses and conclusions for the Fubini Theorem when $n = 3$. Hint: there are six integrals to compare.

We close this lesson with the following addition to the Stieltjes case. Let $h_1 : X_1(\Omega) \to R$; let $h_2 : X_2(\Omega) \to R$; we say that h_1 is a function of X_1 alone and h_2 is a function of X_2 alone. Suppose that h_1 and h_2 are both bounded and measurable wrt $(\mathscr{B}_1, \mathscr{B}_1)$. Then h_1 is integrable wrt F_1 and h_2 is integrable wrt F_2. It follows that $g = h_1 \cdot h_2$ is bounded, measurable, and integrable wrt $F_1 \times F_2$. Hence,

$$\int_{R^2} h_1 \cdot h_2 \, d(F_1 \times F_2) = \int_R h_1 \left[\int_R h_2 dF_2 \right] dF_1 = \int_R h_1 dF_1 \cdot \int_R h_2 dF_2 .$$

Induction leads to

$$\int_{R^n} h_1 \cdot h_2 \cdots h_n \, d(F_1 \times F_2 \times \cdots \times F_n) = \Pi_{k=1}^n \int_R h_k dF_k .$$

We shall refer to this as the *product rule* for expectation.

PART III: LIMITING DISTRIBUTIONS

Overview

Part III contains most of the properties of probability that are the day to day tools of mathematical statisticians. These lessons lead almost in a straight line from distributions of several RVs thru independence of RVs to the most famous result in the subject: the central limit theorem.

Lesson 1 contains the general definition of joint distributions, that is, the simultaneous distribution of more than one RV, and then focuses on the discrete case. Some of this is very like that of Part I; in fact, in lesson 2, conditional distributions and independent RVs are defined in terms of simple conditional probabilities and independent events. This leads to familiar formulas like

$$P(X = x, Y = y) = P(X = x) \cdot P(Y = y) .$$

In lessons 3 and 4, these notions are defined for continuous type RVs, meaning those having densities with respect to Lebesgue measure. The corresponding result for independence is $f(x,y) = f_1(x) \cdot f_2(y)$ so that

$$\int_{-\infty}^{a} \int_{-\infty}^{b} f(x,y) \, dx \, dy = \int_{-\infty}^{a} f_1(x) \, dx \cdot \int_{-\infty}^{b} f_2(y) \, dy .$$

As its title says, lesson 5 contains useful examples.

The first notion of a limit distribution appears in lesson 6:

$$P(X_n \le x) \to P(X \le x) .$$

Lesson 7 contains the inter–relations of the modes of convergence: almost surely, in probability, in distribution, in mean.

"Everybody believes in the law of large numbers; the "physicists" think it is a mathematical theorem and the "mathematicians" think it is an empirical fact." This refers to the use of a sample mean to approximate a population mean (an analogue of the spirit of relative frequency) and is explained properly in lesson 8; there are also some variations.

In lesson 9, the convergence of CDFs is seen as a special case of the convergence of DFs which are bounded montonic increasing functions on the real line R . These results are used in lesson 10 to justify convergence of sequences of integrals like

$$\int_R g(x)\, dF_n(x) \rightarrow \int_R g(x)\, dF(x)$$

as opposed to the earlier theory for

$$\int_R g_n(x)\, dF(x) \rightarrow \int_R g(x)\, dF(x).$$

Lesson 11 contains general formulas and examples of expectation.

Lessons 12 and 13 contain theory and examples of one of the most useful tools of mathematics – the Fourier transform – which statisticians call the characteristic function (CF); one reason for this utility is that there is a one–to–one correspondence between the CFs $\varphi_X(t) = E[e^{itX}]$, $t \in R$, and the CDFs $F(x) = P(X \le x)$, $x \in R$. Lesson 13 also contains some relations for the moment generating function $E[e^{tX}]$ and a probability generating function $E[t^X]$.

In lesson 14, the principal theorems of lessons 9 and 12 are arranged to allow limits of sequences of distribution functions to be studied as limits of sequences of characteristic functions; in a sense, the one–to–one correspondence is preserved.

The profit of lessons 9 thru 14 is an almost complete proof of the central limit theorem in lesson 15. Loosely speaking, this theorem allows probabilties for the sum of RVs,

$$X_1 + \cdots + X_n,$$

to be approximated by use of the normal distribution. Of course, there are extensions to multidimensional random vectors.

LESSON 1. JOINT DISTRIBUTIONS: DISCRETE

The work in Part II was concerned mostly with one random variable. Actual experiments usually involve more than one outcome; an obvious case is the yield of a corn field which depends on (the random) rainfall, ground conditions, sunlight even though (the non–random) seed, fertilizer, irrigation might be used. Here we continue to explain some probabilistic concepts for multiple or joint random variables (random vectors) concentrating, in this lesson and the next, on the discrete case. Following that we shall treat a continuous case.

First we extend without proofs the notion of measurable space. For the real line R, the Borel σ–field is generated by countable set operations beginning with intervals (a,b). For the Euclidean space, $R^n = R \times R \times \cdots \times R$, of n–tuples of real numbers, the Borel σ–field is generated by countable set operations from rectangles $(a,b) \times (c,d) \times \cdots \times (y,z)$. The measurable spaces are the pairs denoted by $[R^n, \mathscr{B}_n]$.

Definition: *Let $X_1, X_2, \cdots, X_n$ be real-valued RVs on the same probability space $[\Omega, \mathscr{B}, P]$. The random vector $X = (X_1, X_2, \cdots, X_n)$ is a mapping from $[\Omega, \mathscr{B}]$ to $[R^n, \mathscr{B}_n]$ satisfying the condition that for all n-tuples of real numbers $x = (x_1, x_2, \cdots, x_n)$,*

$$\{\omega : X_1(\omega) \le x_1, X_2(\omega) \le x_2, \cdots, X_n(\omega) \le x_n\}$$
$$= \{\omega : X_1(\omega) \le x_1\} \cap \{\omega : X_2(\omega) \le x_2\} \cap \cdots$$
$$\cap \{\omega : X_n(\omega) \le x_n\} \in \mathscr{B}.$$

The joint probability cumulative distribution function (CDF) is defined as the probability of these sets and is symbolized by

$$F_X(x) = P(X \le x) = P(X_1 \le x_1, X_2 \le x_2, \cdots, X_n \le x_n).$$

Note that the extension of "inequality" to vectors is taken to mean that the same inequality be applied in each component:

$(X_1, X_2) \le (x_1, x_2)$ iff $X_1 \le x_1$ and $X_2 \le x_2$; $(1,2) \le (2,2)$;

etc. Some of the proof of the following exercise is done more easily by formal verbal argument than by formal symbolic argument; visualizing the case n = 2 is helpful.

Exercise 1: From the definition, verify that:
   a)   for all x ε R^n, $F_X(x)$ ε $[0,1]$;
   b) F_X is non–decreasing in each component:
$$F_X(x) \leq F_X(y) \text{ if } x \leq y;$$
   c) $F_X(x) \to 0$ if at least one of the $\{x_i\} \to -\infty$;
   d) $F_X(x) \to 1$ if all of the $\{x_i\} \to +\infty$;
   e) F_X is right–hand continuous; that is, $F_X(x)$ "decreases" to $F_X(y)$ if all x_i "decrease" to y_i.

Definition: *If the components $X_1, X_2, \cdots, X_n$ of the random vector X are all discrete RVs, then X is a discrete random vector and the joint probability density function (PDF) is*

$$f_X(x) = P(X = x) = P(X_1 = x_1, X_2 = x_2, \cdots, X_n = x_n) \geq 0.$$

The support of f_X is the collection of tuples $(x_1, x_2, \cdots, x_n)$ at which f_X is actually positive.

As with the examples in Part I, when our subject appears to be too artificial, you are invited to make substitutions; in the following, "even" could be females or smart cats or honest senators or $\cdots$ and "odd" could be males or not so smart cats or not–so–honest senators or $\cdots$.

Example: Consider the selection of a handful of three chips from an urn containing six chips labeled 1, 2, 3, 4, 5, 6. Let $X_1[X_2]$ be the number of even [odd] numbers among the three selected. The sample space Ω and the values of $X = (X_1, X_2)$ can be tabulated:

Ω	X_1	X_2	Ω	X_1	X_2
1,2,3	1	2	2,3,4	2	1
1,2,4	2	1	2,3,5	1	2
1,2,5	1	2	2,3,6	2	1
1,2,6	2	1	2,4,5	2	1

1,3,4	1	2	2,4,6	3	0
1,3,5	0	3	2,5,6	2	1
1,3,6	1	2	3,4,5	1	2
1,4,5	1	2	3,4,6	2	1
1,4,6	2	1	3,5,6	1	2
1,5,6	1	2	4,5,6	2	1

If the selection is made "at random", every triple will have probability 1/20. After we list the values of $X = (X_1, X_2)$, we can obtain their probabilities by counting:

$$P(X_1 = 0, X_2 = 3) = 1/20 \quad P(X_1 = 1, X_2 = 2) = 9/20$$

$$P(X_1 = 2, X_2 = 1) = 9/20 \quad P(X_1 = 3, X_2 = 0) = 1/20.$$

Or, we can use a more sophisticated derivation. Since $\begin{bmatrix} 6 \\ 3 \end{bmatrix} = 20$, we know that there will be twenty triples. Since there are 3 even integers, $\begin{bmatrix} 3 \\ x_1 \end{bmatrix}$ is the number of ways in which we obtain x_1 even integers. Similarly, $\begin{bmatrix} 3 \\ x_2 \end{bmatrix}$ is the number of ways in which we obtain $x_2 = 3 - x_1$ odd integers. Therefore, the probability is

$$f_X(x_1, x_2) = \begin{bmatrix} 3 \\ x_1 \end{bmatrix} \cdot \begin{bmatrix} 3 \\ x_2 \end{bmatrix} \div \begin{bmatrix} 6 \\ 3 \end{bmatrix}$$

for $x = (x_1, x_2) \in \{(0,3)\ (1,2)\ (2,1)\ (3,0)\}$. This is the formula for a hypergeometric distribution.

Exercise 2: Consider the selection at random of a handful of six chips from an urn containing chips labeled 1,2,3,4,5,6,7,8,9; let X_1 [X_2] be the number of even [odd] integers among the six selected. Find the PDF of $X = (X_1, X_2)$.

In the previous example and exercise, we were able to find a "simple" formula for the PDF; that is not always the case. True, when there are only a finite numbers of points involved, one can find a formula by the method of Lagrange but these are not necessarily "nice".

Example: The following array represents the joint probability distribution of some

$$X = (X_1, X_2): P(X_1 = -5, X_2 = 1) = 1/24,$$

$$P(X_1 = -5, X_2 = 2) = 3/24, \text{ etc.}$$

	3	2/24	1/24	1/24	4/24
X_2	2	3/24	1/24	5/24	2/24
	1	1/24	2/24	1/24	1/24
		-5	-3	-1	0
			X_1		

Given such a distribution, we can answer probability questions about it.

a) $P(X_1 < -2, X_2 = 2)$

$$= P(X_1 = -3, X_2 = 2) + P(X_1 = -4, X_2 = 2) +$$

$$P(X_1 = -5, X_2 = 2) = 1/24 + 0 + 3/24 = 4/24.$$

b) $P(X_1 < -2, X_2 \geq 2)$

$$= P(X_1 = -5, X_2 = 2) + P(X_1 = -5, X_2 = 3)$$

$$+ P(X_1 = -3, X_2 = 2) + P(X_1 = -3, X_2 = 3)$$

$$= 3/24 + 2/24 + 1/24 + 1/24 = 7/24.$$

c)   Now the statement $P(X_2 > 1)$ does not contain an explicit condition on X_1; the implicit condition is that X_1 can be anything. Here of course, X_1 is in $\{0, -1, -3, -5\}$ so that

$$P(X_2 > 1) = P(X_1 = -5, X_2 = 2) + P(X_1 = -5, X_2 = 3)$$

$$+ P(X_1 = -3, X_2 = 2) + P(X_1 = -3, X_2 = 3)$$

$$+ P(X_1 = -1, X_2 = 2) + P(X_1 = -1, X_2 = 3)$$

$$+ P(X_1 = 0, X_2 = 2) + P(X_1 = 0, X_2 = 3)$$

$$= 19/24.$$

Exercise 3: Continuing the example above, find

   a) $P(X_1 = 5, X_2 > 1)$ b) $P(X_1 \le -1)$ c) $P(X_1 = -X_2)$

These examples also point out another aspect of applications. Many times we will be able to suggest a distribution for some RVs without formalizing a derivation from a probability space; in effect, we use the σ–field generated from rectangles in the range $X(\Omega)$ and $P(-\infty,x] = F(x)$. For discrete X, this CDF has the symbolic form

$$F(x) = \sum_{y \le x} f_X(y)$$

which looks like that for one RV in R^1. In R^2, this is the sum of the probabilities in the corner southwest of (x_1,x_2) including its boundaries:

The other way around,

$$F(b_1,c_2) - F(b_1,a_2) - F(a_1,c_2) + F(a_1,a_2) \qquad (*)$$

is the probability of the points in the rectangle including those on the north and east boundaries but not on the south and west boundaries:

$$(a_1,c_2)\ast\!\ast\!\ast\!\ast\!\ast\!\ast\!\ast\!\ast\!\ast\!\ast\!\ast\!\ast\!\ast\!\ast\!\ast\!\ast\!\ast\!\ast\!\ast\!\ast\ (b_1,c_2)$$

$$(a_1,a_2)\cdots\cdots\cdots\cdots\cdots\cdots\ast\ (b_1,a_2)$$

This corresponds to the one–variable form

$$P(a_1 < X_1 \le b_1) = F_{X_1}(b_1) - F_{X_1}(a_1)$$

way back in Lesson 11, Part I. As (a_1, a_2) is moved closer and closer to (b_1, c_2), the "second order difference" (*) representing the probability in the rectangle decreases to

$$f_X(b_1, c_2) = P(X_1 = b_1, X_2 = c_2).$$

This illustrates the one–to–one relationship of the CDF and PDF in the discrete case in R^2; we shall not write out a general proof.

Of course, seeing what goes on in higher dimensions is more difficult but algebra and calculus carry us through. For example, a *hypergeometric* joint PDF is

$$h_X(x_1, x_2, \cdots, x_k) = \begin{bmatrix} n_1 \\ x_1 \end{bmatrix} \begin{bmatrix} n_2 \\ x_2 \end{bmatrix} \cdots \begin{bmatrix} n_k \\ x_k \end{bmatrix} \div \begin{bmatrix} n_1 + n_2 + \cdots + n_k \\ x_1 + x_2 + \cdots + x_k \end{bmatrix}$$

where the n_j are fixed positive integers and the x_j are non–negative integers with sum equal to the fixed sample size s.

Exercise 4: Take $k = 5$; describe a sampling problem for which this PDF would be appropriate.

We symbolize this hypergeometric by thinking of $X_1, X_2, \cdots, X_k$ and their joint distribution. However, it should be recognized that the support is restricted to some "lattice points in a hyperplane" of dimension $k-1$. In fancier terms which we will not pursue, these variables are not functionally independent; indeed, they are not even linearly independent since the value of $X_1 + X_2 + \cdots + X_k = s$ is fixed.

Example: In ordinary Euclidean three–space,

$$x_1 + x_2 + x_3 = 4$$

represents a plane of two dimensions; for a hypergeometric, the support is limited to coordinates in this plane which are non–negative integers:

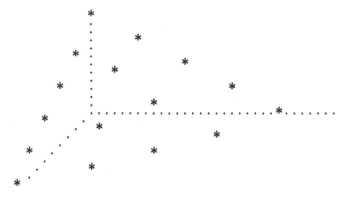

The range of each of X_1,X_2,X_3 is A = {0,1,2,3,4}. The range of (X_1,X_2,X_3) is {(4,0,0) (0,4,0) (0,0,4) (1,3,0) (1,0,3) (0,1,3) (0,3,1) (3,1,0) (3,0,1) (1,1.2) (1,2,1) (2,1,1) (2,0,2) (2,2,0) (0,2,2)} but this not equal to A×A×A .

Similar remarks apply to the *multinomial* distribution, which generalizes the binomial distribution from 2 to k cells, and has joint PDF

$$m_X(x) = P(X_1 = x_1, \cdots, X_k = x_k)$$

$$= \frac{s!}{n_1! \, n_2! \, \cdots \, n_k!} \, p_1^{n_1} p_2^{n_2} \cdots p_k^{n_k}$$

where for j = 1(1)k , the cell probabilities p_j are in [0,1] and the cell frequencies n_j are non–negative integers; of course,

$$p_1 + p_2 + \cdots + p_k = 1 \text{ and } n_1 + n_2 + \cdots + n_k = s$$

is the fixed sample size. In other words, there are s independent and identically distributed "trials" each of which has probability p_j of "falling in" cell j; the PDF is the probability of getting exactly n_1 in cell 1, n_2 in cell 2, $\cdots$, n_k in cell k.

Exercise 5:   For example, one might be interested in the distribution of grades in the statistics classes; one multinomial model for "s" students enrolled this semester has

$$p_1 = P(A), p_2 = P(B), p_3 = P(C), p_4 = P(D), p_5 = P(E).$$

Suppose that in the past, the distribution has been 10% A, 20% B, 40% C, 20% D, 10% E (failure).

    a)   If this distribution is valid for this semester, what is the probability that at least 10% of the class of s = 125 will get an E? Hint: what do you know about "E" and "not E"?

    b)   What is the probability that the distribution of grades will be, correspondingly, 15 25 45 25 15?

In the last exercise, you used the fact that the joint distribution of E and E^c is really biniomial. If we ask for the joint distribution of D's and E's, we get a three cell multinomial with PDF:

$$\frac{125!}{t!\ d!\ f!}\ (.70)^t(.20)^d(.10)^f \ \text{ for } t + d + f = 125 .$$

Although here, this reduced model is obvious (t is the number above D), a general form for reduction needs to be spelled out. We illustrate using a simpler example given earlier.

Example: Consider the array used for exercise 3. Then,

$$P(X_1 = 0) = P(X_1 = 0, X_2 = 1)$$

$$+ P(X_1 = 0, X_2 = 2) + P(X_1 = 0, X_2 = 3)$$

$$= 1/24 + 2/24 + 4/24 .$$

Similarly, $P(X_1 = -1) = 1/24 + 5/24 + 1/24 ;$

$$P(X_1 = -3) = 2/24 + 1/24 + 1/24 ;$$

$$P(X_1 = -5) = 1/24 + 3/24 + 2/24 .$$

These values are just the sums of the corresponding columns; in the same way, the distribution for X_2 is obtained as the sums of the corresponding rows. Loosely speaking, we just sum out the other variable(s). Older writers displayed all this in a different

array:

X_1:	-5	-3	-1	0	
1	1/24	2/24	1/24	1/24	5/24
X_2: 2	3/24	1/24	5/24	2/24	11/24
3	2/24	1/24	1/24	4/24	8/24
	6/24	4/24	7/24	7/24	

You see the distributions of X_1 and X_2 placed neatly in the margins; therefore, they are referred to as the marginal distributions.

The following is just a slight generalization of the arithmetic of the last example. Let the random vector $X = (X_1, X_2)$ have discrete components with joint PDF f_{X_1, X_2}; let the ranges of X_1, X_2 be A_1, A_2 respectively. The *marginal PDF of X_1* is the function on A_1 with value given by

$$P(X_1 = a_1) = f_{X_1}(a_1) = \sum_{a_2 \in A_2} f_{X_1, X_2}(a_1, a_2);$$

the *marginal PDF of X_2* is the function on A_2 with value given by

$$P(X_2 = a_2) = f_{X_2}(a_2) = \sum_{a_1 \in A_1} f_{X_1, X_2}(a_1, a_2).$$

Exercise 6: A box contains 100 light bulbs, 40 red, 35, white, 25 blue. Suppose that 12 bulbs are to be selected at random in one bunch. Let X_1, X_2, $X_3 = 12 - X_1 - X_2$ be the number of red, white, blue bulbs selected.
  a)   Find the joint distribution of (X_1, X_2).
  b)   Find the marginal distributions of X_1, X_2, X_3.

Exercise 7: The weight of a box of cereal is W (ounces). A box is termed light if $W < 12$, good if $12 \le W \le 12.08$, heavy if $W > 12.08$. The filling machine operates in a way such that $P(W < 12) = .01$, $P(W > 12.08) = .05$. Let X be the number of light boxes and Y be the number of heavy boxes in a random sample of n = 50 boxes. Find the joint and marginal PDFs of X and Y.

LESSON 2. CONDITIONAL DISTRIBUTIONS: DISCRETE

We are continuing directly from lesson 1. In our definition for marginal distributions, we use only n = 3; it is left to the reader's imagination to describe all the cases for arbitrary n.

Definition: *Let the random vector* $X = (X_1, X_2, X_3)$ *have discrete components with joint PDF* $f = f_{X_1, X_2, X_3}$*; let* X_i *have range* A_i.
The marginal PDFs are given by:

$$f_{X_2, X_3}(x_2, x_3) = \sum_{x_1 \in A_1} f(x_1, x_2, x_3) = \sum_{-\infty < x_1 < \infty} f(x_1, x_2, x_3);$$

$$f_{X_1, X_3}(x_1, x_3) = \sum_{x_2 \in A_2} f(x_1, x_2, x_3) = \sum_{-\infty < x_2 < \infty} f(x_1, x_2, x_3);$$

$$f_{X_1, X_2}(x_1, x_2) = \sum_{x_3 \in A_3} f(x_1, x_2, x_3) = \sum_{-\infty < x_3 < \infty} f(x_1, x_2, x_3);$$

$$f_{X_1}(x_1) = \sum_{x_2 \in A_2} \sum_{x_3 \in A_3} f(x_1, x_2, x_3)$$

$$= \sum_{x_2 < \infty} \sum_{x_3 < \infty} f(x_1, x_2, x_3);$$

$$f_{X_2}(x_2) = \sum_{x_1 \in A_1} \sum_{x_3 \in A_3} f(x_1, x_2, x_3)$$

$$= \sum_{x_1 < \infty} \sum_{x_3 < \infty} f(x_1, x_2, x_3);$$

$$f_{X_3}(x_3) = \sum_{x_1 \in A_1} \sum_{x_2 \in A_2} f(x_1, x_2, x_3)$$

$$= \sum_{x_1 < \infty} \sum_{x_2 < \infty} f(x_1, x_2, x_3).$$

The following illustrates arithmetical details, in particular double sums, without any hint of an application. We merely apply the definitions.

Example: Suppose the PDF

$$f_{X_1,X_2,X_3}(x_1,x_2,x_3) = c(x_1+x_2x_3)$$

has support $\{0,1\}\times\{1,2\}\times\{2,3,4\}$ and c is constant. Even though we don't know the value of c, yet, let us do some algebra:

$$f_{X_1}(x_1) = \sum_{x_2=1,2} \sum_{x_3=2,3,4} c(x_1+x_2x_3)$$

$$= \sum_{x_2=1,2} c\{(x_1+x_2\cdot 2) + (x_1+x_2\cdot 3) + (x_1+x_2\cdot 4)\}$$

$$= \sum_{x_2=1,2} c\{3x_1 + 9x_2\} = c\{3x_1 +9\} + c\{3x_1 + 18\}$$

$$= c\{6x_1 + 27\}.$$

This must hold for x_1 = 0 or 1 so (the total probability)

$$1 = f_{X_1}(0) + f_{X_1}(1) = c\{27\} + c\{33\}$$

whence c = 1/60 and

$$f_{X_1}(x_1) = (2x_1 + 9)/20.$$

$$f_{X_2,X_3}(x_2,x_3) = \sum_{x_1=0,1} (x_1+x_2x_3)/60$$

$$= \{(x_2x_3) + (1+x_2x_3)\}/60 = (1 + 2x_2x_3)/60.$$

A partial check is that the double sum of this function is 1:

$$f_{X_2,X_3}(1,2) + f_{X_2,X_3}(1,3) + f_{X_2,X_3}(1,4)$$

$$+ f_{X_2,X_3}(2,2) + f_{X_2,X_3}(2,3) + f_{X_2,X_3}(2,4)$$

$$= \{(1+4) + (1+6) + (1+8)$$

$$+ (1+8) + (1+12) + (1+16)\}/60 = 1.$$

Exercise 1: Continue with the example just above.
  a)   Find f_{X_2} and f_{X_1,X_3}.
  b)   Find $F_{X_1}(x) = P(X_1 \leq x)$

Example: (continued) You have just obtained the marginal CDF of X_1. Making use of the discussion of CDF in lesson 1, we can find other marginal distributions. The marginal CDF for X_1, X_2 is

$$F_{X_1,X_2}(a_1,a_2) = P(X_1 \leq a_1, X_2 \leq a_2, X_3 < \infty)$$

$$= F_{X_1,X_2,X_3}(a_1,a_2,\infty)$$

$$= \sum_{x_1 \leq a_1} \sum_{x_2 \leq a_2} \sum_{x_3 < \infty} f_{X_1,X_2,X_3}(x_1,x_2,x_3)$$

$$= \sum_{x_1 \leq a_1} \sum_{x_2 \leq a_2} f_{X_1,X_2}(x_1,x_2)$$

$$= \sum_{x_1 \leq a_1} \sum_{x_2 \leq a_2} (x_1 + 3x_2)/20 .$$

The four points in the support of f_{X_1,X_2} induce a partition of the plane into 9 pieces. Remember that in 2 dimensions, the CDF adds the probabilities to the "southwest" of its argument, here (a_1,a_2).

If (a_1, a_2) is in one of the positions like A,D,H,I,J, then

$$a_1 < 0 \text{ or } a_2 < 1 \text{ and } F_{X_1, X_2}(a_1, a_2) = 0;$$

if (a_1, a_2) is at a position like E, $0 \le a_1 < 1$, $1 \le a_2 < 2$, and

$$F_{X_1, X_2}(a_1, a_2) = P(X_1 \le 0, X_2 \le 1) = 3/20;$$

at G, $1 \le a_1$, $1 \le a_2 < 2$, and

$$F_{X_1, X_2}(a_1, a_2) = P(X_1 \le 1, X_2 \le 1) = 7/20;$$

at B, $0 \le a_1 < 1$, $1 \le a_2 < 2$, so

$$F_{X_1, X_2}(a_1, a_2) = P(X_1 \le 0, X_2 \le 1) = 9/20;$$

at C, $1 \le a_1$, $2 \le a_2$, and

$$F_{X_1, X_2}(a_1, a_2) = P(X_1 \le 1, X_2 \le 2) = 1.$$

Exercise 2: Continue the example. Find the marginal CDF of $Y = (X_1, X_3)$ and of $Z = X_3$ by itself.

Since in the discrete case PDFs are probabilities, it is natural to use conditional probabilities as conditional distributions for such RVs: if

$A = \{\omega : X_1(\omega) = a_1\}$ and $B = \{\omega : X_2(\omega) = a_2\}$ with $P(B) \ne 0$,

then $P(A|B) = P(A \cap B)/P(B)$ becomes

$$P(X_1 = a_1 \mid X_2 = a_2) = P(X_1 = a_1, X_2 = a_2)/P(X_2 = a_2).$$

For each value a_2 in the support for X_2, we get a distribution for X_1. The multiplication rule

$$P(X_1 = a_1, X_2 = a_2) = P(X_1 = a_1 \mid X_2 = a_2) \cdot P(X_2 = a_2)$$

may also be used.

Again, we cheat a bit by dealing with only the case $n = 3$. Need we remind you that not all notation is standard? Sometimes a simpler symbolization (even your own) helps to emphasize the meaning.

Definition: *Let the discrete RV X have PDF* f_{X_1,X_2,X_3}. *The conditional densities are defined only for those values of* x_1 *and/or* x_2 *and/or* x_3 *in the support of the density which appears in the denominator. The conditional density of:*

a) X_1 *given* (X_2,X_3) *is*

$$f_{1|2,3}(x_1 \mid x_2,x_3) = f_{X_1,X_2,X_3}(x_1,x_2,x_3)/f_{X_2,X_3}(x_2,x_3);$$

b) X_1 *given* X_3 *is*

$$f_{1|3}(x_1 \mid x_3) = f_{X_1,X_3}(x_1,x_3)/f_{X_3}(x_3)$$

c) (X_2,X_3) *given* X_1 *is*

$$f_{2,3|1}(x_2,x_3 \mid x_1) = f_{X_1,X_2,X_3}(x_1,x_2,x_3)/f_{X_1}(x_1)$$

d)   *etc.*

Example: a) Recall the array for exercise 3 of Lesson 1. The lack of a formula forces us to write out the terms individually:

$$P(X_1 = -5 \mid X_2 = 1) = (1/24)/(5/24) = 1/5$$

$$P(X_1 = -3 \mid X_2 = 1) = (2/24)/(5/24) = 2/5$$

$$P(X_1 = -1 \mid X_2 = 1) = (1/24)/(5/24) = 1/5$$

$$P(X_1 = 0 \mid X_2 = 1) = (1/24)/(5/24) = 1/5.$$

For $X_2 = 3$, $f_{1|2}(-5 \mid 3) = (2/24)/(8/24) = 1/4$

$$f_{1|2}(-3 \mid 3) = (1/24)/(8/24) = 1/8$$

$$f_{1|2}(-1 \mid 3) = (1/24)/(8/24) = 1/8$$

$$f_{1|2}(0 \mid 3) = (4/24)/(8/24) = 1/2.$$

b)   Let $f(x_1, x_2, x_3) = (x_1 + x_2 x_3)/60$ as in the first example of this lesson. Although it is very easy to see that

i) $f(x_2 \mid x_1) = (x_1 + 3x_2)/(2x_1 + 9)$,

ii) $f(x_1 \mid x_2, x_3) = (x_1 + x_2 x_3)/(1 + 2x_2 x_3)$,

and so on, it is important to note that these formulas represent very different distributions. In case i),

$$f(x_2 \mid 0) = x_2/3 \quad \text{but} \quad f(x_2 \mid 1) = (1+3x_2)/11;$$

in case ii), $f(x_1 \mid 1,2) = (x_1 + 2)/5$ but

$$f(x_1 \mid 2,3) = (x_1 + 6)/13,$$

and so on.

Exercise 3: Continue with part b) of the example above. First find the corresponding conditional PDFs and then evaluate the probabilities:

$$P(X_1 < 1 \mid X_2 = 2, X_3 = 2); \quad P(X_2 = 2 \mid X_1 = 0, X_3 = 3);$$

$$P(X_2 = 2 \mid X_1 = 0); \quad P(X_3 > 2 \mid X_1 = 1, X_2 = 1).$$

The following traditional example is attributed to Bernstein. The probability of $X = (X_1, X_2, X_3)$ is uniform on its support $S = \{(0,0,1)\ (0,1,0)\ (1,0,0)\ (1,1,1)\}$.

$$f_{X_1, X_2}(1,0) = P(X_1 = 1, X_2 = 0) = P((1,0,0)) = 1/4$$

$$= f_{X_1, X_2}(0,1) = f_{X_1, X_2}(0,0) = f_{X_1, X_2}(1,1).$$

Similarly, (X_1,X_3) and (X_2,X_3) have uniform distributions with the same support. The marginal distribution of X_1 or X_2 or X_3 is uniform on $\{0,1\}$. Now,

$$f_{X_1|X_2} = (1/4)/(1/2) = (1/2) = f_{X_1} .$$

Similarly,

$$f_{X_1|X_3} = f_{X_1}; \quad f_{X_2|X_1} = f_{X_2}; \quad f_{X_2|X_3} = f_{X_2};$$

$$f_{X_3|X_1} = f_{X_3}; \quad f_{X_3|X_2} = f_{X_3}.$$

These results are written more neatly in multiplicative form:

$$f_{X_1,X_2} = f_{X_1} \cdot f_{X_2}; \quad f_{X_1,X_3} = f_{X_1} \cdot f_{X_3}; \quad f_{X_2,X_3} = f_{X_2} \cdot f_{X_3}.$$

However, $f_{X_1,X_2,X_3} = 1/4 \neq 1/8 = f_{X_1} \cdot f_{X_2} \cdot f_{X_3}$.

The following is the most general definition of "independence" of random variables that we can invoke.

Definition: *Let $X_1,X_2, X_3, \cdots$ be real random variables on the same probability space $[\Omega, \mathscr{B}, P]$. Then these RVs are mutually stochastically independent (msi) iff*
for each positive integer $n \geq 2$,
for all selections $i_1,i_2,\cdots,i_n$ in $\{1,2,3,\cdots\}$,
for all selections $A_1,A_2,\cdots,A_n \in \mathscr{B}_1$ (the σ-field of R),
the product rule holds:

$$P(X_{i_1} \in A_1, X_{i_2} \in A_2, \cdots, X_{i_n} \in A_n) =$$

$$P(X_{i_1} \in A_1) \cdot P(X_{i_2} \in A_2) \cdots P(X_{i_n} \in A_n).$$

Corollary: *If $X_1, X_2, X_3, \cdots$ are msi, then conditional probabilities are equal to marginal probabilities.*

Partial proof:

$$P(X_1 \in A_1 \mid X_2 \in A_2) = P(X_1 \in A_1, X_2 \in A_2)/P(X_2 \in A_2)$$

$$= P(X_1 \in A_1) \cdot P(X_2 \in A_2)/P(X_2 \in A_2)$$

$$= P(X_1 \in A_1);$$

$$P(X_1 \in A_1, X_3 \in A_3 \mid X_2 \in A_2) = P(X_1 \in A_1, X_3 \in A_3); \text{ etc.}$$

Since in all dimensions, the probabilities are determined by the CDF, this msi rule is equivalent to

$$F_{X_{i_1}, \cdots, X_{i_n}}(x_{i_1}, \cdots, x_{i_n}) = F_{X_{i_1}}(x_{i_1}) \cdots F_{X_{i_n}}(x_{i_n})$$

for all $n \geq 2$, $i_1, \cdots, i_n$ as before and $x_{i_1}, \cdots, x_{i_n} \in R$. "Selection" in the definition allows us to consider permuted forms like $F_{X_1, X_2}(a, b) = F_{X_2, X_1}(b, a)$ as equivalent. The following illustrates how independence is "inherited".

Exercise 4: Suppose that the CDF F_{X_1, X_2, X_3} is the product of the marginal CDFs $F_{X_1} \cdot F_{X_2} \cdot F_{X_3}$. Show that for $i \neq j$,

$$F_{X_i, X_j} = F_{X_i} \cdot F_{X_j}.$$

Example: Let (X_1, X_2) have CDF F; the probability in the rectangle $(a, b] \times (c, d]$ is (Lesson 1)

$$P(a < X_1 \leq b, c < X_2 \leq d) = F(b, d) - F(b, c) - F(a, d) + F(a, c).$$

Suppose now that $F(x_1, x_2) = F_1(x_1) \cdot F_2(x_2)$ where the F_i are CDFs on R (as in Lesson 13, Part II). Then,

$$P(a < X_1 \leq b, c < X_2 \leq d)$$

$$= F_1(b)F_2(d) - F_1(b)F_2(c) - F_1(a)F_2(d) + F_1(a)F_2(c)$$

$$= \{F_1(b) - F_1(a)\} \cdot \{F_2(d) - F_2(c)\}$$

$$= P(a < X_1 \leq b) \cdot P(c < X_2 \leq d).$$

If we allow a to approach b, c to approach d (from below),

$P(a < X_1 \leq b, c < X_2 \leq d)$ decreases to

$$P(X_1 = b, X_2 = d);$$

$P(a < X_1 \leq b) \cdot P(c < X_2 \leq d)$ decreases to

$$P(X_1 = b) \cdot P(X_2 = d).$$

In other words, the corresponding PDFs satisfy

$$f(x_1, x_2) = f_1(x_1) \cdot f_2(x_2).$$

A derivation of the general factorization equations is obviously more involved (and will be omitted). It should suffice to state a:

Corollary: X_1, X_2, X_3, $\cdots$ *are msi, or briefly independent, iff the joint densities all factor: for $i \neq j \neq k \neq m \neq \cdots$*

$$f_{X_i X_j} = f_{X_i} \cdot f_{X_j}, \quad (*)$$

$$f_{X_i X_j X_k} = f_{X_i} \cdot f_{X_j} \cdot f_{X_k}, \quad (**)$$

$$f_{X_i X_j X_k X_m} = f_{X_i} \cdot f_{X_j} \cdot f_{X_k} \cdot f_{X_m}, \quad etc.$$

Exercise 5: Let the PDF $f(x_1, x_2, x_3) = dx_1 x_2 x_3$ (d is a constant) have support $\{1,2\} \times \{1,2,3\} \times \{1,2,3,4\}$. Show that X_1, X_2, X_3 are msi; that is, verify that all the equations in (*) and (**) do hold.

Going back to the Bernstein example after exercise 3, we now see that X_1 and X_2 are independent, X_1 and X_3 are independent, X_2 and X_3 are independent but X_1, X_2, X_3 are not independent because

$$f_{X_1, X_2, X_3} \neq f_{X_1} \cdot f_{X_2} \cdot f_{X_3}.$$

We call X_1, X_2, X_3 *pairwise independent* but not mutually independent.

Exercise 6: For c constant, let the PDF be

$$f(x_1,x_2,x_3) = cx_1x_2x_3$$

but suppose that the support is

$$\{(1,2,3) \ (1,2,4) \ (1,2,5) \ (1,3,4) \ (1,3,5) \ .$$
$$(1,4,5) \ (2,3,4) \ (2,3,5) \ (2,4,5) \ (3,4,5)\}$$

Show that X_1,X_2,X_3 are not msi.

LESSON 3. JOINT DISTRIBUTIONS: CONTINUOUS

In Lesson 4, Part II, we saw traditional examples of continuous CDFs for one real valued random variable restricted to those cases for which the CDF was the Riemann integral of its PDF. Extending this fundamental theorem of integral calculus to Lebesgue integration involves a long array of details which we leave to such texts. The following is a summary for R; recall that in R, Lebesgue measure Λ is generated from the lengths of intervals: $\Lambda(a,b] = b - a$, etc.

Let F and f be real valued functions on a set containing $[a,b] \subset R$. For each set of non–overlapping intervals
$$D: \ x_0 \le x_1 \le \cdots \le x_n,$$
let $L(D) = \sum_{i=1}^{n}(x_i - x_{i-1})$ and $V(D) = \sum_{i=1}^{n}|F(x_i)-F(x_{i-1})|$.

1) Let $\mathscr{D}$ be the set of all D with $x_0 = a$ and $x_n = b$. F is of *bounded variation* on [a,b] iff $\{V(D) : D \in \mathscr{D}\}$ is bounded.

Then, the derivative F'(x) exists and is finite a.e. in [a,b].

2) If f is integrable on [a,b] & $F(x) = \int_{[a,x]} f \, d\Lambda$, then

$F'(x) = f(x)$ a.e.

3) If F is of bounded variation on [a,b], then

$F(x) - F(a) = \int_{[a,x]} F' \, d\Lambda$ iff F is *absolutely continuous*,

that is, for each $\varepsilon > 0$, there is a $\delta_\varepsilon > 0$ such that

for $L(D) < \delta_\varepsilon$, $V(D) < \varepsilon$.

Exercise 1: a) Let F be a CDF with F(a) = 0 and F(b) = 1. Show that F is of bounded variation.
Hint: for $x_{i-1} \le x_i$, $F(x_i)-F(x_{i-1}) \ge 0$.

b) Show that when F is absolutely continuous, F is continuous. Hint: n = 1.

The same kind of inter–relationship of derivatives and integrals exists in the multidimensional case which we begin by particularizing the first definition in Lesson 1.

Definition: *The real RVs in* $X = (X_1, X_2, \cdots, X_n)$ *have the joint probability cumulative distribution function (CDF)*

$$F_X(x) = P(X \leq x) = P(X_1 \leq x_1, \cdots, X_n \leq x_n)$$

for all n-tuples $x = (x_1, \cdots, x_n)$ *of real numbers. Then X is (absolutely) continuous iff there is a non-negative function* $f_X : R^n \to R$ *such that* $F_X(x) = \int_{(-\infty, x]} f_X \, d\Lambda$ *where* Λ *is the product Lebesgue measure on* R^n; f_X *is the probability density function (PDF). The support is the set in* R^n *where* $f_X > 0$.

Some confusion arises because "absolutely" is often omitted when discussing the RV X even though this adverb does distinguish the RVs with "density" from those whose CDF is merely continuous (see the corollary below). From time to time it is convenient to use Λ_1 for the measure on $R = R^1$, Λ_2 on R^2, etc. We leave the writing out of the general forms of the following definition to the reader; note that the notation for "partials" has not been standardized.

Definition: *Let* $F : R^2 \to R$ *and let* $x = (x_1, x_2)$ *be a generic point.*

a) *F is continuous at* $x_o = (b, c)$ *iff for each* $\varepsilon > 0$, *there is a* $\delta = \delta(b, c, \varepsilon)$ *such that*

$$\|x - x_o\| = \{(x_1 - b)^2 + (x_2 - c)^2\}^{1/2} < \delta \text{ implies}$$

$$|F(x_1, x_2) - F(b, c)| < \varepsilon.$$

b) *When the limit as* $h \to 0$ *exists, the indicated partial derivative of F at* $x = (x_1, x_2)$ *is:*

"first"- $F_1(x_1, x_2) = \partial F / \partial x_1$

$$= \lim (F(x_1 + h, x_2) - F(x_1, x_2)) / h \, ;$$

$$F_2(x_1, x_2) = \partial F / \partial x_2$$

$$= \lim (F(x_1, x_2 + h) - F(x_1, x_2)) / h \, ;$$

$$\text{"second"- } F_{11}(x_1,x_2) = \partial^2 F/\partial^2 x_1$$
$$= \lim \ (F_1(x_1+h,x_2)-F_1(x_1,x_2))/h \ ;$$

$$F_{21}(x_1,x_2) = \partial^2 F/\partial x_2 \partial x_1$$
$$= \lim (F_1(x_1,x_2+h)-F_1(x_1,x_2))/h \ ;$$

$$F_{22}(x_1,x_2) = \partial^2 F/\partial^2 x_2$$
$$= \lim (F_2(x_1,x_2+h)-F_2(x_1,x_2))/h \ ;$$

$$\text{"third"- } F_{121}(x_1,x_2) = \partial^3 F/\partial x_1 \partial x_2 \partial x_1$$
$$= \lim \ (F \quad (x_1+h,x_2)-F_{21}(x_1,x_2))/h \ ;$$

$$F_{222}(x_1,x_2) = \partial^3 F/\partial x_2 \partial x_2 \partial x_2 = \partial^3 F/\partial^3 x_2$$
$$= \lim(F_{22}(x_1,x_2+h)-F_{22}(x_1,x_2))/h \ ;$$

Exercise 2: $F(x_1,x_2) = (x_1 - x_2)/(x_1 + x_2)$ for $x_1 \neq -x_2$
$$= 1 \text{ for } x_1 = -x_2 \ .$$

Evaluate those limits which exist:

a) $\lim\limits_{(x_1,x_2)\to(0,0)} F(x_1,x_2)$; b) $\lim\limits_{x_1\to0} (\lim\limits_{x_2\to0} F(x_1,x_2))$;

c) $\lim\limits_{x_2\to0} (\lim\limits_{x_1\to0} F(x_1,x_2))$; d) $\lim\limits_{x_2\to x_1} F(x_1,x_2)$;

e) $\lim\limits_{x_2\to1} (\lim\limits_{x_1\to-1} F(x_1,x_2))$.

Exercise 3: (Traditional)

$F(x_1,x_2) = x_1 x_2 (x_1^2 - x_2^2)/(x_1^2 + x_2^2)$ for $(x_1,x_2) \neq (0,0)$;
$$= 0 \text{ for } (x_1,x_2) = (0,0) \ .$$

Apply the definition to find F_1, F_2, F_{12}, F_{21} at $(0,0)$, at $(x_1,0)$, at $(0,x_2)$ and compare the results.

In exercise 3, you should have noted that $F_{12} \neq F_{21}$ because neither is continuous at $(0,0)$. The proof in the following eliminates this disparity thru the Fubini theorem; note that except for the careful specification of the set where the relation fails, this is ordinary multivariate calculus.

Corollary: *Under the conditions of the definition of an absolutely continuous CDF* F_X *, for almost all* $(\Lambda = \Lambda_n)$ *points in* R^n,

$\partial^n F_X(x)/\partial x_1 \cdots \partial x_n$ *exists and equals* $f_X(x)$.

Proof: Note that f_X is integrable in each dimension since $F_X(\infty) = 1$. By a generalization of the corollary of the Fubini Theorem in Lesson 13, Part II, we rearrange

$$F_{X_1,\cdots,X_n}(x_1,\cdots,x_n) = \int_{(-\infty,x_1]\times\cdots\times(-\infty,x_n]} f_X \, d\Lambda_n$$

as

$$\int_{-\infty}^{x_1}\left[\int_{-\infty}^{x_2}\cdots\left[\int_{-\infty}^{x_n} f_{X_1,\cdots,X_n}(a_1,\cdots,a_n)da_n\right]\cdots da_2\right]da_1. \quad (*)$$

For $k = n = 1$, the conclusion is essentially part 2) in the discussion at the beginning of this lesson. Assume that the result holds for $k \leq n-1$. Let $G(a_1,w)$ denote the inner integrals of (*) with $w = (x_2,\cdots x_n)$. By the induction hypothesis,

$$\partial^{k-1} G(x_1,w)/\partial x_2 \cdots \partial x_n = f_{X_1,\cdots,X_n}(x_1,w)$$

for all (x_1,w) such that $w \in R^{k-1} - A(x_1)$ where
$$\Lambda_{k-1}(A(x_1)) = 0 \text{ for each } x_1 \in R.$$

Now (*) can also be written as

$$F_X(x_1, w) = \int_{-\infty}^{x_1} G(a_1, w) da_1$$

so that by 2) again

$$\partial F_X(x_1, w)/\partial x_1 = G(x_1, w) = f_X(x) \qquad (**)$$

for (x_1, w) such that $x_1 \in R - B(w)$ where $\Lambda_1(B(w)) = 0$ for each $w \in R^{k-1}$. The set where (*) and (**) fail is contained in a set M such that $\Lambda(M) = (\Lambda_1 \times \Lambda_{n-1})(M) = 0$. This completes the induction and the proof.

Example: a) In terms of the CDF F for (X_1, X_2), the probability for an R^2 rectangle is the same here as in Lesson 1 for a discrete RV:

$$P(a_1 < X_1 \le b_1, a_2 < X_2 \le c_2)$$

$$= F(b_1, c_2) - F(a_1, c_2) - F(b_1, a_2) + F(a_1, a_2) \qquad (***)$$

$$= F(x_1, c_2) \Big]_{x_1 = a_1}^{x_1 = b_1} - F(x_1, a_2) \Big]_{x_1 = a_1}^{x_1 = b_1}.$$

But now this evaluation is

$$\int_{(a_1, b_1]} \partial F(x_1, c_2)/\partial x_1 \, dx_1 - \int_{(a_1, b_1]} \partial F(x_1, a_2)/\partial x_1 \, dx_1$$

$$= \int_{(a_1, b_1]} \partial F(x_1, x_2)/\partial x_1 \Big]_{x_2 = a_2}^{x_2 = c_2} dx_1$$

$$= \int_{(a_1, b_1]} \left[\int_{(a_2, c_2]} \partial^2 F(x_1, x_2)/\partial x_1 \partial x_2 \, dx_2 \right] dx_1$$

$$= \int_{(a_1, b_1] \times (a_2, c_2]} f(x_1, x_2) \, d\Lambda_2 \quad \text{where f is the density.}$$

A more commmon notation for this last integral is

$$\int_{a_1}^{b_1} \int_{a_2}^{c_2} f(x_1,x_2)dx_2dx_1 \text{ meaning}$$

$$\int_{a_1}^{b_1} \left[\int_{a_2}^{c_2} f(x_1,x_2)dx_2 \right] dx_1.$$

b) Let (b_1,c_2) be a point of continuity for F. When $a_1 \rightarrow b_1$ and $a_2 \rightarrow c_2$, the limit of the integral in a), which is the difference (***), is 0 unlike the discrete case for which the limit was $P(X_1 = b_1, X_2 = c_2)$.

c) Consider instead the "iterated second–order difference"

$$\frac{F(b_1,c_2)-F(a_1,c_2)-F(b_1,a_2)+F(a_1,a_2)}{(b_1-a_1)(c_2-a_2)}$$

$$= \frac{\dfrac{F(b_1,c_2)-F(a_1,c_2)}{(b_1-a_1)} - \dfrac{F(b_1,a_2)-F(a_1,a_2)}{(b_1-a_1)}}{(c_2-a_2)}$$

Now take the iterated limit:
first as $a_1 \rightarrow b_1$ to get

$$\frac{\partial F(b_1,c_2)/\partial x_1 - \partial F(b_1,a_2)/\partial x_1}{(c_2-a_2)},$$

then as $a_2 \rightarrow c_2$ to get

$$\partial^2 F(x_1,x_2)/\partial x_2 \partial x_1 \Big|_{(b_1,c_2)} = f(b_1,c_2).$$

The (second partial) derivative of the distribution function is the density function.

Probabilities for X are computed by integrating the density

function over the appropriate region. As in the one–dimensional case, we may ignore pieces where the density is zero.

Example: Suppose that mathematics and reading scores for certain college students are both scaled to the interval $[0,1]$ and the density is to be approximated by

$$f(m,r) = c(m + r) \cdot I\{0 \leq m \leq 1, 0 \leq r \leq 1\}.$$

a)   The value of the "normalizing constant" c is determined from

$$1 = \int_0^1 \int_0^1 c(m + r)\, dm\, dr = c \int_0^1 (m^2/2 + rm) \Big]_0^1 dr$$

$$= c(r/2 + r^2/2) \Big]_0^1 = c.$$

b)   The proportion of such students who obtain a math score above $.9$ is

$$P(M > .9) = P(.9 < M < 1, 0 < R < 1)$$

$$= \int_{.9}^1 \left[\int_0^1 (m + r)dr \right] dm = \int_{.9}^1 (m + 1/2)dm$$

$$= (m^2/2 + m/2) \Big]_{.9}^1 = .55.$$

c)   The proportion of students whose reading score is less than their math score is

$$P(R < M) = \int_0^1 \left[\int_r^1 (m + r)\, dm \right] dr$$

$$= \int_0^1 (1/2 + r - 3r^2/2)dr = 1/2.$$

The region of integration is a triangle:

```
r ↑                    *
:                    *  *
:                 *     *
:              *        *
:           *           *
: *                     *
* * * * * *............ . .→
                    m
```

Exercise 4: Suppose that the joint PDF of (X_1, X_2) is given by

$$f(x_1, x_2) = c(4 - x_1 - x_2) \cdot I\{0 < x_1 < 2, 0 < x_2 < 2\} .$$

Find:

    a) $P(X_2 < 1)$,

    b) $P(X_1 > 2X_2)$,

    c) $P(X_1 < .5, X_2 < .5)$.

       It should be apparent from the examples and exercises that marginal distributions do come into play but without a neat natural array as in the discrete case. As we have done with other definitions, we leave the general forms of the following to the reader.

Definition: *Let the random vector (X_1, X_2, X_3) have a CDF F_{X_1, X_2, X_3}; then marginal CDFs are given by:*

$$F_{X_1, X_2}(x_1, x_2) = F_{X_1, X_2, X_3}(x_1, x_2, \infty) ;$$

$$F_{X_1}(x_1) = F_{X_1, X_2}(x_1, \infty) = F_{X_1, X_2, X_3}(x_1, \infty, \infty) ;$$

$$F_{X_2, X_3}(x_2, x_3) = F_{X_1, X_2, X_3}(\infty, x_2, x_3) ;$$

$$F_{X_3}(x_3) = F_{X_1, X_3}(\infty, x_3) = F_{X_2, X_3}(\infty, x_3) =$$

$$F_{X_1, X_2, X_3}(\infty, \infty, x_3) ; etc.$$

If the RVs are continuous with PDF f_{X_1, X_2, X_3}, then marginal PDFs are:

$$f_{X_1,X_2}(x_1,x_2) = \int_{-\infty}^{\infty} f_{X_1,X_2,X_3}(x_1,x_2,x_3) dx_3 \;;$$

$$f_{X_1}(x_1) = \int_{-\infty}^{\infty} f_{X_1,X_2}(x_1,x_2) \, dx_2$$

$$= \int_{-\infty}^{\infty} \left[\int_{-\infty}^{\infty} f_{X_1,X_2,X_3}(x_1,x_2,x_3) dx_3 \right] dx_2 \;;$$

$$f_{X_2}(x_2) = \int_{-\infty}^{\infty} f_{X_1,X_2}(x_1,x_2) dx_1$$

$$= \int_{-\infty}^{\infty} \left[\int_{-\infty}^{\infty} f_{X_1,X_2,X_3}(x_1,x_2,x_3) dx_3 \right] dx_1 \;;$$

etc. As before, the support of the density or distribution is the set of points for which the PDF is actually positive.

In words, to get the marginal density of one or more variables, we "integrate out" the other variables in the joint density. The general formulas are easy enough; the tricky part is in the actual evaluation of the integrals particularly when the support is not "rectangular".

We end this lesson with a final note on notation:

$$F_{X_1}(-\infty) = 0 = F_{X_1,X_2}(-\infty,x_2) = F_{X_1,X_2,X_3}(-\infty,x_2,x_3)$$

for all x_2,x_3 and similarly for any other "$-\infty$" point.

LESSON 4. CONDITIONAL DISTRIBUTIONS: CONTINUOUS

We are continuing directly from lesson 3.

Example: Let the CDF F_X have PDF

$$f(x_1, x_2) = 3x_1 I\{0 \le x_1 \le 1, 0 \le x_2 \le x_1\}.$$

The support is a triangular region which induces a partition of the plane:

a) For $a < 0$, $F_{X_1}(a) = 0$.

For $0 \le a \le 1$, $F_{X_1}(a) = F_X(a, \infty)$

$$= \int_{-\infty}^{a} \int_{-\infty}^{\infty} f(x_1, x_2) dx_2 \, dx_1.$$

But since f is zero outside the triangle (of stars), this integral immediately reduces to

$$\int_0^a \left[\int_0^{x_1} 3x_1 \, dx_2 \right] dx_1$$

which is equal to $\int_0^a 3x_1^2 \, dx_1 = a^3$.

For $1 < a$, $F_{X_1}(a) = 1$.

b) As with X_1, the interesting part of the CDF of X_2 is for $0 \le b \le 1$:

$$F_{X_2}(b) = F_X(\infty, b) = \int_0^b \left[\int_{x_2}^1 3x_1 \, dx_1 \right] dx_2$$

$$= \int_0^b 3(1 - x_2^2/2) dx_2 = (3b - b^3)/2 .$$

c)   If (a,b) is in a position like A or E or G or H, either
      $a < 0$ or $b < 0$ so that $F_X(a,b) = 0$.

d)   For (a,b) at K inside the triangle,

$$F_X(a,b) = \int_0^b \left[\int_{x_2}^a 3x_1 \, dx_1 \right] dx_2 = (3a^2 b - b^3)/2.$$

Here $b < a$ so when $a \to 1$,

$$F_X(a,b) \to F_X(1,b) = F_{X_2}(b) .$$

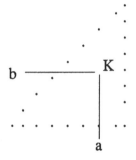

e)   For (a,b) at B, $F_X(a,b) = \int_0^a \left[\int_{x_2}^a 3x_1 \, dx_1 \right] dx_2 = a^3$

as it should since here $b > a > X_1$.

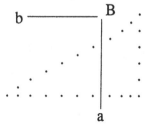

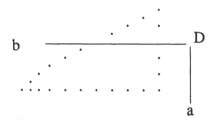

f)   For (a,b) at D,

$$F_X(a,b) = \int_0^b \left[\int_{x_2}^1 3x_1\, dx_1 \right] dx_2 = (3b - b^3)/2$$

as it should since here $a \geq 1$.

g)   For (a,b) at C, $F_X(a,b) = 1$ since $a \geq 1$ and $b \geq 1$.

In continuous cases, once we have the CDF, we can differentiate (at almost all points) to obtain the PDF. In the example above:

$$f_{X_1}(a) = \partial F_{X_1}(a)/\partial a = \begin{cases} 0 & \text{for } a < 0 \\ 3a^2 & \text{for } 0 \leq a < 1, \\ 0 & \text{for } 1 < a \end{cases}$$

$$f_{X_2}(b) = \partial F_{X_2}(b)/\partial b = \begin{cases} 0 & \text{for } b < 0 \\ 3(1 - b^2)/2 & \text{for } 0 < b < 1, \\ 0 & \text{for } 1 \leq b \end{cases}$$

$$f_X(a,b) = \partial^2 F_X(a,b)/\partial b \partial a = 3a \text{ for } 0 < b < a < 1.$$

Of course when we start with the joint density as in this example, we automatically find the marginal PDF while obtaining the CDF:

$$f_{X_1}(x_1) = \int_{-\infty}^\infty f(x_1,x_2)dx_2 = 0 \text{ if } x_1 < 0 \text{ or } x_1 > 1$$

$$= \int_0^{x_1} 3x_1\, dx_2 = 3x_1^2 \text{ otherwise;}$$

$$f_{X_2}(x_2) = \int_{-\infty}^{\infty} f(x_1, x_2) dx_1 = 0 \text{ if } x_2 < 0 \text{ or } x_2 > 1$$

$$= \int_{x_2}^{1} 3x_1 \, dx_1 = 3(1 - x_2^2)/2 \text{ otherwise.}$$

The differences in the "answers" at 0 and 1 are due to the fact that derivatives and integrals are "unique" only almost everywhere. (This does not affect the probabilities. Why?)

Exercise 1: Find all joint and marginal CDFs and PDFs.

a) $f_X(x_1, x_2) = I\{0 < x_2 \le x_1 \le 2\}/2x_1$.

b) $f_X(x_1, x_2) = [1 - \exp(-x_1)] \cdot [1 - \exp(-2(x_2-1))] \cdot$
$$I\{0 \le x_1, 1 \le x_2\}.$$

c) For c constant, $f_X(x_1, x_2) = c \cdot I\{x_1^2 + x_2^2 \le 4\}$.

d) For c constant and support $x_i \ge 0$, the non–zero part of the PDF is

$$f_X(x_1, x_2, x_3) = \exp(c - x_1 - 2x_2 - 3x_3).$$

e) For a constant c, the support is $(0,2) \times (1,3) \times (0,\infty)$, the non–zero part of the PDF is
$$f_X(x_1, x_2, x_3) = c(x_1 + x_2)\exp(-2x_3).$$

In lesson 2, the definition of mutually stochastically independent (msi) was given for arbitrary RVs so that it certainly applies to the continuous type; the following is a special case from which general rules can be inferred..

Example: Let $A \in \mathcal{B}_3$; for $i = 1,2,3$, let $A_i = X_i^{-1}(A) \in \mathcal{B}_1$.
The components of $X = (X_1, X_2, X_3)$ are msi iff

$$P(X^{-1}(A)) = P(X_1^{-1}(A) \cap X_2^{-1}(A) \cap X_3^{-1}(A))$$

$$= P(X_1^{-1}(A)) \cdot P(X_2^{-1}(A)) \cdot P(X_3^{-1}(A)).$$

For $A = (-\infty, x) = (-\infty, x_1) \times (-\infty, x_2) \times (-\infty, x_3)$, this is equivalent to

$$F_X(x) = F_{X_1, X_2, X_3}(x_1, x_2, x_3)$$

$$= F_{X_1}(x_1) \cdot F_{X_2}(x_2) \cdot F_{X_3}(x_3). \quad (*)$$

By properties of the derivative shown in Lesson 3, this implies

$$f_X(x) = f_{X_1, X_2, X_3}(x_1, x_2, x_3) = \partial^3 F_X / \partial x_1 \partial x_2 \partial x_3$$

$$= \partial F_{X_1} / \partial x_1 \cdot \partial F_{X_2} / \partial x_2 \cdot \partial F_{X_3} / \partial x_3$$

$$= f_{X_1}(x_1) . f_{X_2}(x_2) \cdot f_{X_3}(x_3).$$

By properties of the integral suggested at the end of Lesson 3, this in turn implies (*) again.

Exercise 2: Write out the general definitions and theorems inferred by this example.

It is possible that, as in the discrete case, X_1, X_2, X_3 are pairwise independent but not mutually independent. In either case, we could have other *partial independencies*:

$$f_{X_1, X_2, X_3, X_4} = f_{X_1, X_3} \cdot f_{X_2, X_4}$$

indicates that the pair (X_1, X_3) is independent of the pair (X_2, X_4); etc.

The first lemma below shows that a certain geometric condition is necessary for independence. Though the proof is for the discrete case, the principle carries over to the continuous case and, in both instances, is implicit in what is meant by factoring a function of several variables into the product of functions of the individual variables. The second lemma below shows that this kind of factorization is also sufficient for independence.

Lemma: *Let $X = (X_1, X_2, X_3)$ be a random vector with msi discrete components; let A, A_1, A_2, A_3 be the supports for the RVs X, X_1, X_2, X_3, respectively. Then $A = A_1 \times A_2 \times A_3$.*

Proof: Let $(a_1, a_2, a_3) \in A$. Then $P(X = (a_1, a_2, a_3)) > 0$ and independence imply $P(X_i = a_i) > 0$ for $i = 1, 2, 3$. Therefore, $A \subset A_1 \times A_2 \times A_3$. On the other hand, if for some $a_i \in A_i$, $(a_1, a_2, a_3) \notin A$, then

$$0 = P(X = (a_1, a_2, a_3))$$

$$= P(X_1 = a_1) \cdot P(X_2 = a_2) \cdot P(X_3 = a_3)$$

implies that at least one of $P(X_i = a_i)$ is 0. This contradiction means that all such $(a_1, a_2, a_3) \in A$ and the equality holds.

By an informal induction, we see that when the components in $X = (X_1, \cdots, X_n)$, are msi, the support of X must be the Cartesian product of the supports of the X_i; we call this product a "rectangle". The other way around, if the support of X is not a "rectangle", then the components of X cannot be msi.

Lemma: *For $(a,b) \subset R$ and $(c,d) \subset R$, let $g : (a,b) \to R$ and $h : (c,d) \to R$; then $f = g \cdot h : (a,b) \times (c,d) \to R$. If f is a PDF for (X_1, X_2), then X_1 and X_2 are independent.*

Partial proof: Suppose that X_1 and X_2 are discrete; let their supports be $A_1 \subset (a,b)$ and $A_2 \subset (c,d)$ respectively.. Then by Fubini's Theorem,

$$1 = \sum_{(x_1, x_2) \in A_1 \times A_2} f(x_1, x_2) = \sum_{x_1 \in A_1} \sum_{x_2 \in A_2} f(x_1, x_2)$$

$$= \sum_{x_1 \in A_1} \sum_{x_2 \in A_2} g(x_1) h(x_2)$$

$$= \sum_{x_1 \in A_1} g(x_1) \sum_{x_2 \in A_2} h(x_2) \, . \qquad (**)$$

On the other hand,

$$f_{X_1}(x_1) = \sum_{x_2 \in A_2} g(x_1)h(x_2) = g(x_1)\sum_{x_2 \in A_2} h(x_2)$$

and

$$f_{X_2}(x_2) = h(x_2)\sum_{x_1 \in A_1} g(x_1) \, .$$

Multiplying these last two equations together yields:

$$f_{X_1}(x_1)f_{X_2}(x_2) = g(x_1)h(x_2)\sum_{x_2 \in A_2} h(x_2) \cdot \sum_{x_1 \in A_1} g(x_1) \, .$$

By (**), the right–hand side of this equality reduces to

$$g(x_1)h(x_2) \cdot 1 \quad \text{so that} \quad f_{X_1}(x_1)f_{X_2}(x_2) = f(x_1,x_2)$$

which is the condition for independence.

Exercise 3:   a) Write out the proof of the last lemma for the continuous case.
b)   What modifications are necessary when a and/or c $= -\infty$, b and/or d $= +\infty$?
c)   State a general theorem embodying this principle.

When X_2 is continuous, $P(X_2 = b) = 0$ for all real b so that we cannot consider $P(X_1 = a, X_2 = b)/ P(X_2 = b)$ as we did in the discrete case. However, we can consider

$$P(X_1 \le a, b - \delta < X_2 \le b)/P(b - \delta < X_2 \le b)(***)$$

whenever $F_{X_2}(b) - F_{X_2}(b - \delta) > 0$. For continuous $X = (X_1, X_2)$,
(***) is

$$\int_{-\infty}^{a}\left[\int_{b-\delta}^{b} f(x_1,x_2)dx_2\right]dx_1 \div \int_{-\infty}^{\infty}\left[\int_{b-\delta}^{b} f(x_1,x_2)dx_2\right]dx_1 .$$

When f is uniformly continuous (a severe restriction) and

$f_{X_2}(b) \neq 0$, the limit of this ratio as $\delta \to 0$ is

$$\int_{-\infty}^{a} f(x_1,b)dx_1 + \int_{-\infty}^{\infty} f(x_1,b)dx_1 = \int_{-\infty}^{a} f(x_1,b)dx_1 + f_{X_2}(b) .$$

This may be taken as the conditional CDF of X_1 given $X_2 = b$.

The derivative of this function wrt a is $f(a,b)/f_{X_2}(b)$.

In other words, we now examine conditional probabilities for such X in terms of a "density ratio" just as in the discrete case. Although the formalism of these paragraphs applies to a particular case, the definition of each conditional density is translated directly from that of discrete RVs. For example:

when $f_{X_3}(x_3) \neq 0$,

$$f_{1|3}(x_1|x_3) = f(x_1|x_3) = f_{X_1,X_3}(x_1,x_3)/f_{X_3}(x_3);$$

when $f_{X_1,X_3}(x_1,x_3) \neq 0$,

$$f_{2|1,3}(x_2|x_1,x_3) = f(x_2|x_1,x_3) = f_X(x_1,x_2,x_3)/f_{X_1,X_3}(x_1,x_3);$$

etc.

In all cases, when some RVs are independent, certain conditional densities will be equivalent to marginal densities. For example: when X_1,X_2,X_3 are msi,

$$f_{X_1|X_3}(x_1|x_3) = f_{X_1,X_3}(x_1,x_3)/f_{X_3}(x_3)$$

$$= f_{X_1}(x_1) \cdot f_{X_3}(x_3)/f_{X_3}(x_3) = f_{X_1}(x_1);$$

when X_2 and (X_1,X_3) are independent,

$$f_{X_2|X_1,X_3}(x_2|x_1,x_3) = f_{X_1,X_2,X_3}(x_1,x_2,x_3)/f_{X_1,X_3}(x_1,x_3)$$

$$= f_{X_2}(x_2) \cdot f_{X_1,X_3}(x_1,x_3)/f_{X_1,X_3}(x_1,x_3) = f_{X_2}(x_2).$$

Example: With $f_X(x_1,x_2) = 3x_1$ for $0 \le x_2 \le x_1 \le 1$,

$$f_{X_1}(x_1) = 3x_1^2 \text{ for } 0 < x_1 \leq 1,$$

$$f_{X_2}(x_2) = 3(1 - x_2^2)/2 \text{ for } 0 < x_2 \leq 1,$$

For each $0 \leq x_2 < 1$, the conditional density

$$f(x_1 | x_2) = 2x_1/(1 - x_2^2) \text{ for } x_2 \leq x_1 < 1.$$

This is not constant as a function of x_2 so X_1 and X_2 are not independent. Note that for each x_2 in $(0,1)$. $f(x_1 | x_2)$ does satisfy the requirements for a density: it is non–negative and

$$\int_{-\infty}^{\infty} f(x_1 | x_2) dx_1 = \int_{x_2}^{1} 2x_1/(1 - x_2^2) dx_1 = x_1^2/(1 - x_2^2)\Big]_{x_2}^{1} = 1 .$$

Example: Consider the PDF in part e) of exercise 1. You should have found $c = 1/6$. Since f is factored into a function of (x_1, x_2) times another function of x_3 and the support is rectangular, the second lemma above (or rather an extension) guarantees that the vector (X_1, X_2) is independent of the singleton X_3. To check this directly, we find the non–zero portions of the PDFs :

$$f_{X_1,X_2}(x_1,x_2) = \int_0^{\infty} (x_1 + x_2)[\exp(-2x_3)] \, dx_3/6 = (x_1 + x_2)/12,$$

$$f_{X_3}(x_3) = \int_0^1 \left[\int_1^3 (x_1 + x_2)[\exp(-2x_3)] \, dx_2 \right] dx_1/6 = 2\exp(-2x_3).$$

Obviously, the product rule holds.

Exercise 4: a) Find all the conditional PDFs in exercise 1; indicate the domains precisely.
b)   In each part, determine which variables or vectors are "independent".

LESSON 5. EXPECTATION–EXAMPLES

The general theory of expectation was outlined in Lessons 11 and 12 Part II. In this lesson, we spell out more particular properties of expectation needed later. For simplicity, we assume that all RVs are defined on a given probability space $[\Omega, \mathcal{B}, P]$ and, unless otherwise indicated, that all the expectations are finite. The basic form is:

$X : \Omega \to R^n$ is measurable wrt $(\mathcal{B}, \mathcal{B}_n)$ and has CDF F_X;

$g : X(\Omega) \to R$ is measurable wrt $(\mathcal{B}_n, \mathcal{B}_1)$;

$$E[g(X)] = \int_\Omega g(X(\omega))dP(\omega) = \int_{R^n} g(x) \, dF_X(x).$$

The following particular cases are used in practice.

a)   For $n = 1$ with $a < b$, both finite such that $P(a \le X \le b) = 1$,

$$E[g(X)] = \int_a^b g(x)dF_X(x).$$

b)   If $n = 1$ but X is unbounded,

$$E[g(X)] = \lim_{a, b \to \infty} \int_{-a}^b g(x) \, dF_X(x).$$

c)   If $n = 1$ and $X(\Omega)$ can be well–ordered as $x_0 < x_1 < \cdots$, at most countable,

$$E[g(x)] = \sum_{i=0}^\infty g(x_i)(F_X(x_i) - F_X(x_{i-1}))$$

$$= \sum_{i=0}^\infty g(x_i)P(X = x_i).$$

d)   If the support of X is finite like $\{y_1, y_2, \cdots, y_m\}$ (for any $n \ge 1$), $E[g(X)]$ is the finite sum

$$g(y_1)P(X = y_1) + \cdots + g(y_m)P(X = y_m).$$

e)   If g is continuous and the density f_X is continuous a.e., then

$$E[g(X)] = \int_{R^n} g(x)f_X(x) \, d\Lambda_n =$$

$$\int_{-\infty}^{\infty} \cdots \int_{-\infty}^{\infty} g(x_1, \cdots, x_n) \dot{f}_X(x_1, \cdots, x_n) dx_1 \cdots dx_n.$$

Definition: *The mean or expected value of X is the vector of means:*

$$E[X] = (E[X_1], E[X_2], \cdots, E[X_n]).$$

Corollary: *Let F_X have density f_X wrt Lebesgue measure. For*

$$i = 1(1)n, \quad E[X_i] = \int_{-\infty}^{\infty} \cdots \int_{-\infty}^{\infty} x_i f_X(x_1, x_2, ..., x_n) \, dx_1 dx_2 \cdots dx_n$$

$$= \int_{-\infty}^{\infty} x_i f_{X_i}(x_i) \, dx_i.$$

Partial proof: Take $n = 3$ and $i = 2$. We are assuming the existence of the integrals; therefore, by Fubini's Theorem, we can arrange the multiple integral in any order.

$$E[X_2] = \int_{-\infty}^{\infty} x_2 \left[\int_{-\infty}^{\infty} \left[\int_{-\infty}^{\infty} f_X(x_1, x_2, x_3) \, dx_1 \right] dx_3 \right] dx_2$$

$$= \int_{-\infty}^{\infty} x_2 f_{X_2}(x_2) \, dx_2.$$

Exercise 1: State and prove (partially) a version of the corollary for X discrete.

As in the calculation of probabilities, the limits of the integrals or summations are determined by the support and not by the values of the integrand or summand.

Example: a) $f_X(x_1, x_2) = 3x_1 \cdot I\{0 \le x_2 \le x_1 \le 1\}$.

$$E[X_1 X_2] = \int_0^1 \left[\int_0^{x_1} x_1 x_2 \cdot 3x_1 \, dx_2 \right] dx_1$$

$$= \int_0^1 3x_1^2 \, x_2^2/2 \Big]_0^{x_1} dx_1 = 3/10.$$

$$E[X_1 + X_2] = \iint_{\{0 \le x_2 \le x_1 \le 1\}} (x_1 + x_2) \cdot 3x_1 \, d\Lambda_2$$

$$= \int_0^1 \int_0^{x_1} x_1 \cdot 3x_1 \, dx_2 dx_1 + \int_0^1 \int_{x_2}^1 x_2 \cdot 3x_1 \, dx_1 dx_2$$

$$= \int_0^1 x_1 \cdot 3x_1^2 \, dx_1 + \int_0^1 x_2 \cdot (3/2)(1 - x_2^2) \, dx_2$$

$$= \quad 3/4 \quad + \quad 1/4$$

which is, of course, $E[X_1] + E[X_2]$.

b)   The distribution of $X = (X_1, X_2)$ is given by the array:

X_1	−5	−3	−1	0
1	1/24	2/24	1/24	1/24
2	3/24	1/24	5/24	2/24
3	2/24	1/24	1/24	4/24

(where the left column is X_2 with values 1, 2, 3) ;

that is, $f_X(-3,2) = P_X(X_1 = -3, X_2 = 2) = 1/24$, etc. Let
$A_1 = \{-5 \ -3 \ -1 \ 0\}$ be the support of X_1; let
$A_2 = \{1 \ 2 \ 3\}$ be the support of X_2. Then the support of X
is $A = A_1 \times A_2$.

$$E[X_1 X_2] = \sum\sum_A x_1 x_2 \cdot f_X(x_1, x_2)$$

$$= (-5 \cdot 1) \cdot 1/24 + (-3 \cdot 1) \cdot 2/24 + (-1 \cdot 1) \cdot 1/24 + 0$$
$$+ (-5 \cdot 2) \cdot 3/24 + (-3 \cdot 2) \cdot 1/24 + (-1 \cdot 2) \cdot 5/24 + 0$$
$$+ (-5 \cdot 3) \cdot 2/24 + (-3 \cdot 3) \cdot 1/24 + (-1 \cdot 3) \cdot 1/24 + 0$$
$$= -109/24.$$

$$E[X_1 - 2X_2] = \sum\sum_A (x_1 - 2x_2) \cdot f_X(x_1, x_2)$$

$$= \sum\sum_A x_1 \cdot f_X(x_1, x_2) - 2 \cdot \sum\sum_A x_2 \cdot f_X(x_1, x_2)$$

$$= \sum_{A_1} x_1 \sum_{x_2=1}^{3} f_X(x_1, x_2) - 2 \cdot \sum_{x_2=1}^{3} x_2 \sum_{A_1} f_X(x_1, x_2).$$

The general theory tells us that this must be $E[X_1] - 2E[X_2]$ and, up to this point, the evaluation has been analogous to that for the double integral above. But without some general formulas for sums, we are forced to evaluate term by term.

$$E[X_1] = (-5)\sum_{x_2=1}^{3} f_X(-5, x_2) + (-3)\sum_{x_2=1}^{3} f_X(-3, x_2)$$

$$+ (-1)\sum_{x_2=1}^{3} f_X(-1, x_2) + (0)\sum_{x_2=1}^{3} f_X(0, x_2)$$

$$= (-5)(6/24) + (-3)(4/24) + (-1)(7/24) +$$
$$(0)(7/24) = -49/24.$$

$$E[X_2] = (1)\sum_{A_1} f_X(x_1, 1) + (2)\sum_{A_1} f_X(x_1, 2)$$

$$+ (3)\sum_{A_1} f_X(x_1, 3)$$

$$= (1)(5/24) + (2)(11/24) + (3)(8/24) = 51/24.$$
$$E[X_1 - 2X_2] = (-49/24) - 102/24 = -151/24.$$

Exercise 2: For the two densities in the example above, compute $E[g(X)]$ when $g(x)$ equals

a) $x_1^2 x_2$

b) $(x_1 + 1)(x_2 - 2)$.

The linearity of expectation (integration) used in these examples and exercises has a general form which can be proved by induction from Theorem 1 in Lesson 11, Part II.

Theorem: *Let* $X_1, X_2, \cdots, X_n$ *have finite expectation and let* $c_1, c_2, \cdots, c_n$ *be constants. Then* $E[\sum_{i=1}^{n} c_i X_i] = \sum_{i=1}^{n} c_i E[X_i]$.

Why has more attention has been paid to integrals of sums than to integrals of products? One reason is that the theorem above is true while, correspondingly,

$$E[X_1 X_2] = E[X_1] \cdot E[X_2]$$

is generally false. In example a) above,

$$E[X_1 X_2] = 3/10, \; E[X_1] = 3/4, \; E[X_2] = 1/4.$$

In addition, there are no other necessary and sufficient conditions characterizing such an equality. This is illustrated by the following theorem and example.

Theorem: *If* X_1 *and* X_2 *are independent with finite expectations, then*

$$E[X_1 \cdot X_2] = E[X_1] \cdot E[X_2].$$

Proof: By independence, we have $F_X = F_{X_1} \cdot F_{X_2}$. Then by Fubini's Theorem,

$$[X_1 X_2] = \int_{R^2} x_1 x_2 \, dF_X(x_1, x_2)$$

$$= \int_R x_2 \left[\int_R x_1 dF_{X_1}(x_1) \right] dF_{X_2}(x_2)$$

$$= \int_R x_2 E[X_1] dF_{X_2}(x_2) = E[X_1] \cdot E[X_2].$$

Example: The distribution is given by the array:

X_1		-1	0	1
X_2	0	p_4	p_1	p_4
	1	p_2	p_3	p_2

Even though the support is "rectangular", X_1, X_2 will not be independent if

$$P(X_2 = 0 \mid X_1 = 1) = p_4/(p_2 + p_4) \neq p_1/(p_1 + p_3)$$

$$= P(X_2 = 0 \mid X_1 = 0).$$

There are many choices for such p's but in all cases,

$$E[X_1] = -1(p_4 + p_2) + 0 + 1(p_4 + p_2) = 0$$

and

$$E[X_1 X_2] = 0 + 0 + 0 - p_2 + 0 + p_2 = 0.$$

It follows that

$$E[X_1 X_2] = E[X_1] \cdot E[X_2].$$

This has led to $E[X_1 X_2] - E[X_1] \cdot E[X_2]$ being taken as a measure of the "dependence" of X_1 and X_2. More precisely, (and including some notation),

Definition: *Let the RVs $X_1, \cdots, X_n$ have means $\mu_i = E[X_i]$. These RVs have covariances*

$$\sigma_{ij} = Cov(X_i, X_j) = E[(X_i - \mu_i)(X_j - \mu_j)]$$

when these expectations exist. For $i = j$, the covariance is called the variance:

$$\sigma_{ii} = \sigma_i^2 = E[(X_i - \mu_i)^2] \text{ and } \sqrt{\sigma_{ii}} = \sigma_i$$

is the standard deviation. When $\sigma_i \cdot \sigma_j > 0$, the correlation coefficients are $\rho_{ij} = \sigma_{ij}/\sigma_i \sigma_j$ with $\rho_{ii} = 1$.

The correlation coefficient is unit–less and may therefore be used to compare "dependencies" of different pairs of RVs.

Now consider the CBS inequality for real RVs (Lesson 11, Part II):

$$|E[XY]|^2 \leq E[X^2] \cdot E[Y^2].$$

For $i \neq j$, let $X = X_i - \mu_i$, $Y = X_j - \mu_j$. Then, CBS becomes

$$\sigma_{ij}^2 \leq \sigma_{ii}\sigma_{jj} \text{ whence } \rho_{ij}^2 \leq 1 \text{ or } -1 \leq \rho \leq 1.$$

It is also necessary to comment on the possibility of "dividing by 0" in ρ. The corollary to Theorem 2, Lesson 10, Part II, can be interpreted as:

$$X \geq 0 \text{ and } E[X] = 0 \text{ imply } X = 0 \text{ a.s.}$$

Hence, if $\sigma_i = 0$, $X_i - \mu_i = 0$ a.s. which means that X_i is μ_i for all practical purposes; it no longer contributes to the randomness of the problem. In fact:

Exercise 3: Prove that a "constant" or "degenerate RV" is independent of any other RV.

It is convenient to display the covariances of a vector of RVs in a matrix "Σ". For example, with $n = 3$,

$$\Sigma = (\sigma_{ij}) = \begin{bmatrix} \sigma_{11} & \sigma_{12} & \sigma_{13} \\ \sigma_{21} & \sigma_{22} & \sigma_{23} \\ \sigma_{31} & \sigma_{23} & \sigma_{33} \end{bmatrix}.$$

Since $\sigma_{ij} = \sigma_{ji}$, Σ is *symmetric*: interchanging rows and columns yields the same matrix; symbolically, $\Sigma' = \Sigma$, sigma transpose equals sigma.

Exercise 4: Find the mean vector and covariance matrix for the RVs X_1, X_2 in each part of the first example of this lesson.

Exercise 5: RVs X_i, X_j ($i \neq j$) are *uncorrelated* if $\rho_{ij} = 0$ or $\text{Cov}(X_i, X_j) = 0$. Show that the corresponding covariance matrix is *diagonal*.

The threorem and example above show that independent

RVs are uncorrelated but uncorrelated RVs need not be independent. But, there is one family of distributions (the multivariate normal) for which "uncorrelated" and "independent" are equivalent. We illustrate with n = 2.

Example: The *Bivariate Standard Normal* (family indexed by the correlation coefficient $-1 < \rho < 1$) has density

$$f(x_1, x_2) = \{1/2\pi\sqrt{(1-\rho^2)}\}\exp\{-(x_1^2 - 2\rho x_1 x_2 + x_2^2)/2(1-\rho^2)\}.$$

We see that f factors into a function of x_1 times a function of x_2 iff $\rho = 0$ whence iff X_1 and X_2 are independent. We need to verify that ρ is indeed the correlation coefficient. We begin with the first moment: $E[X_1]$ is equal to

$$\int_R x_1 \left[\int_R e^{-(x_1^2 - 2\rho x_1 x_2 + x_2^2)/2(1-\rho^2)} dx_2\right] dx_1 /2\pi\sqrt{(1-\rho^2)}. \quad (*)$$

In the inner integral, complete the square in x_2; then, substitute $z = (x_2 - \rho x_1)/\sqrt{(1-\rho^2)}$. This makes that inner integral equal to

$$e^{-x_1^2/2(1-\rho^2)} \cdot \int_{-\infty}^{\infty} e^{-(x_2 - \rho x_1)^2/2(1-\rho^2)} e^{\rho^2 x_1^2/2(1-\rho^2)} dx_2$$

or

$$e^{-x_1^2/2} \int_{-\infty}^{\infty} e^{-z^2/2} \{\sqrt{(1-\rho^2)}\} \, dz = e^{-x_1^2/2} \{\sqrt{(1-\rho^2)}\}\sqrt{(2\pi)}.$$

Then (*) becomes

$$E[X_1] = \int_R x_1 \left[e^{-x_1^2/2}\sqrt{\{(1-\rho^2)2\pi\}}\right] dx_1 /2\pi\sqrt{(1-\rho^2)}$$

$$= \int_{-\infty}^{\infty} x_1 \cdot e^{-x_1^2/2} \, dx_1 /\sqrt{2\pi} = 0.$$

(This also shows that the marginal distribution of X_1 is the univariate standard normal!) The same two "tricks" are used to reduce

$$E[X_1 X_2] = \int_{-\infty}^{\infty} \int_{-\infty}^{\infty} x_1 x_2 \cdot f(x_1, x_2) dx_1 dx_2$$

to $\int_{-\infty}^{\infty} \int_{-\infty}^{\infty} x_1 (z\sqrt{(1-\rho^2)} + \rho x_1) e^{-x_1^2/2} \cdot e^{-z^2/2} \, dx_1 \, dz/2\pi$

which equals ρ.

Exercise 6: Continue with the example above.

    a)   Verify that $E[X_1] = 0$ and $E[X_1^2] = 1$.

    b)   Fill in the missing details for $E[X_1 X_2]$.

LESSON 6. CONVERGENCE IN MEAN, IN DISTRIBUTION

We begin this lesson by reviewing the convergences introduced in Lessons 7 and 8, Part II; then we define and illustrate two new types. All together we will have the following "modes" of convergence:

pointwise convergence, convergence almost surely,
convergence a.s. completely, convergence in probability
convergence in a mean, convergence in distribution.

Perhaps the last mode is the one most used in statistics. Additional properties, including interdependencies, will be discussed in the next two lessons. To simplify matters, we take all random variables to be real valued, finite a.s., defined on a common probability space $[\Omega, \mathscr{B}, P]$.

Definition: *For each* $n = 1(1)\infty$, *let* X_n *be a real valued measurable function on* Ω; *the family of* X_n's *is a sequence of random variables.*

a) *The sequence* $\{X_n\}$ *converges at the point* $\omega \varepsilon \Omega$ *iff the sequence of real numbers* $\{X_n(\omega)\}$ *converges; if* $\{X_n(\omega)\}$ *converges to the RV* $X(\omega)$ *at each point of a set A, then* $\{X_n\}$ *converges pointwise on A (which could be* Ω*).*

b) *The sequence* $\{X_n\}$ *converges in probability to X iff*

for each $\varepsilon > 0$, $\lim_{n \to \infty} P(\{\omega : |X_n(\omega) - X(\omega)| \geq \varepsilon\}) = 0$.

There is no special shorthand for pointwise convergence but there are several forms for "in probability":

$$\lim_{n \to \infty} P(|X_n - X| \geq \varepsilon) = 0, \quad P(|X_n - X| \geq \varepsilon) \to 0,$$

$$X_n - X \xrightarrow{P} 0, \quad X_n \xrightarrow{P} X.$$

By complementation, these are all equivalent to

for each $\varepsilon > 0$, $\lim_{n \to \infty} P(|X_n - X| < \varepsilon) = 1$.

The simplest, and a very common case, is when X is a degenerate RV, a constant c: $P(X = c) = 1$; we write $X_n \overset{P}{\to} c$.

Example: a) For each $n = 1(1)\infty$, let
$$P(X_n = n) = 1/n \text{ and } P(X_n = 0) = 1 - 1/n.$$
Then, $P(|X_n - 0| \geq \varepsilon) = P(X_n \geq \varepsilon) = P(X_n = n)$
$$= 1/n \to 0 \text{ so } X_n \overset{P}{\to} 0.$$

b)   For a given RV X with CDF F, define a *truncate* of X by
$$X_n = -n \cdot I\{-\infty < X < -n\} + X \cdot I\{-n \leq X \leq n\}$$
$$+ n \cdot I\{n < X < \infty\}.$$
Then, $P(|X_n - X| \geq \varepsilon)$

$$= P(\{|-n - X| \geq \varepsilon\} \cap \{X < -n\})$$
$$+ P(\{|0| \geq \varepsilon\} \cap \{-n \leq X \leq n\})$$
$$+ P(\{|n - X| \geq \varepsilon\} \cap \{X > n\})$$
$$= P(\{-n - X \geq \varepsilon\} \cap \{X < -n\})$$
$$+ P(\{-n - X \leq -\varepsilon\} \cap \{X < -n\}) + 0$$
$$+ P(\{n - X \geq \varepsilon\} \cap \{X > n\})$$
$$+ P(\{n - X \leq -\varepsilon\} \cap \{X > n\})$$
$$= P(X \leq -n-\varepsilon) + P(X \geq n+\varepsilon)$$
$$= P(X \leq -n-\varepsilon) + (1 - P(X < n+\varepsilon)).$$
This last sum converges to $F(-\infty) + 1 - F(\infty) = 0$ and so $X_n \overset{P}{\to} X$.

Exercise 1: Show that each $\{X_n\}$ converges in probability:

a)   the other *truncate* $X_n(\omega) = X(\omega) \cdot I\{|X(\omega)| \leq n\}$;

b) $P(X_n = 2^n) = 1/n$, $P(X_n = 0) = 1 - 1/n$;

c) $P(X_n = c) = p_n \to 1$.

Exercise 2: Suppose that $X_n \overset{P}{\to} X$ ase Y. Show that $X_n \overset{P}{\to} Y$.

Hint: for real numbers a,b,c, $|a - b| \geq \varepsilon$ implies $|a - c| \geq \varepsilon/2$ or $|c - b| \geq \varepsilon/2$.

Definition: *Let A be the set where $\{X_n\}$ converges pointwise to X. If $P(A) = 1$, then $\{X_n\}$ converges almost surely (a.s.) to X. Write $X_n \overset{as}{\rightarrow} X$.*

Exercise 3: Do the "truncate" sequences converge a.s.?

The next theorem was proved in Lesson 3, Part II. The hypothesis of the corollary which follows it defines convergence "a.s. completely" but we do not pursue this further. The succeeding example and exercise illustrate the idea of a "convergence rate" in $X_n \overset{as}{\rightarrow} X$.

Theorem: *The sequence $\{X_n\}$ converges a.s. to X iff*

$$\text{for each } \varepsilon > 0, \quad \lim_{m \to \infty} P(\cup_{n=m}^{\infty} \{ |X_n - X| \geq \varepsilon \}) = 0.$$

Corollary: *If for each $\varepsilon > 0$, $\sum_{n=1}^{\infty} P(|X_n - X| \geq \varepsilon) < \infty$, then $X_n \overset{as}{\rightarrow} X$.*

Proof: Since $P(\cup A_n) \leq \Sigma P(A_n)$,

$$P(\cup_{n \geq m} \{ |X_n - X| \geq \varepsilon \}) \leq \Sigma_{n \geq m} P(|X_n - X| \geq \varepsilon).$$

This sum is the "remainder" term of the convergent series (of real numbers) in the hypothesis so its limit as $m \to \infty$ is 0. The sufficient condition of the theorem is satisfied and the conclusion follows.

Example: Let $P(X_n = n) = n^{-3/2}$, $P(X_n = 0) = 1 - n^{-3/2}$. Then, for each $\varepsilon > 0$, $\Sigma_n P(|X_n| \geq \varepsilon) = \Sigma_{n \geq \varepsilon} n^{-3/2}$ which is a remainder term of a convergent "p" series and is finite. Therefore, $X_n \overset{as}{\rightarrow} 0$.

Exercise 4: Show that $X_n \overset{as}{\to} X$ is equivalent to

$$Y_m = \sup_{n \geq m} |X_n - X| \overset{P}{\to} 0.$$

Hint: $\cup_{n \geq m} \{ |X_n - X| \geq \epsilon \} = \{ \sup_{n \geq m} |X_n - X| \geq \epsilon \}$.

Definition: *Let* $\{X_n\}$, X *be such that for some* $p \in [1,\infty)$, $E[|X_n|^p]$, $E[|X|^p]$ *are all finite;* $\{X_n\}$ *converges in pth-mean to* X *iff* $\lim_{n \to \infty} E[|X_n - X|^p] = 0$. *When* $p = 1$, $\{X_n\}$ *converges in mean to* X; *when* $p = 2$, $\{X_n\}$ *converges in quadratic mean.*

Example: Let $\Omega = R$ with a probability such that $P(\{0\}) > 0$. Let X be the identity function on Ω and let

$$X_n(\omega) = X(\omega) \cdot I_{[-n,n]}(\omega) .$$

 a)   For each ω, $X_n(\omega) \to X(\omega)$ so $X_n \overset{as}{\to} X$; consequently, $X_n \overset{P}{\to} X$.

 b)   Let the probability on Ω be symmetric about 0 with $E[|X|] < \infty$. The $E[X_n] \to 0$ (because they actually $= 0$) yet there are no subsequences X_{n_k} which converge to 0 even in probability. In fact,

$$\{\omega : X_n(\omega) \to 0\} = \{0\} \text{ has positive probability.}$$

Exercise 5: Show that when $\{X_n\}$ converges in mean to X, $E[X_n] \to E[X]$.

Exercise 6: Let $\{X_n\}$ be such that

$$P(X_n = n^\alpha) = 1/n^\beta = 1 - P(X_n = 0).$$

Find values of p, α, and $\beta > 0$ such that the sequence converges:
 a)   in pth mean to 0

   b)   a.s. to 0. Note the differences.

Definition: *For the RVs* $X, X_1, X_2, \cdots$, *let the CDFs be*
$F(x) = P(X \leq x)$, $F_n(x) = P(X_n \leq x)$, $n = 1(1)\infty$. *The sequence*
$\{X_n\}$ *converges in distribution to* X *iff* $\lim_{n \to \infty} F_n(x) = F(x)$ *for each*
x *in the continuity set of* F, *say*

$$C(F) = \{x : F \text{ is continuous at } x\}.$$

Write $X_n \overset{D}{\to} X$.

   We shall have more to say about this restriction to $C(F)$ in
later lessons. For now, we note that when $X_n \overset{D}{\to} X$, $x \in C_F$ and n
is "large", $P(X_n \leq x)$ is approximately equal to $P(X \leq x)$. For
example, if by some means we can show that $X_n \overset{D}{\to} Z$, where Z is
the standard normal RV, then we can approximate probabilties on
X_n by using the normal table. As will be seen in later lessons, the
limiting distribution of many "statistics" is in fact that of Z.

Examples: a) Let

$$F_n(x) = (1/2 + 1/2n) \cdot I\{-n \leq x < n\} + I\{n \leq X < \infty\}.$$

For each $x \in R$, $F_n(x) \to 1/2$. This is not a CDF so $\{X_n\}$ does not
converge in distribution.
   b)   Let X_n be a normal RV with mean zero and variance $1/n$:

$$F_n(x) = \int_{-\infty}^{x} e^{-nz^2/2} dz/\sqrt{(2\pi/n)} = \int_{-\infty}^{x\sqrt{n}} e^{-w^2/2} dw/\sqrt{(2\pi)}.$$

$$G(x) = \lim_{n \to \infty} F_n(x) = \begin{cases} 0 & \text{if } x < 0 \\ 1/2 & \text{if } x = 0 \\ 1 & \text{if } x > 0 \end{cases}.$$

The RV X degenerate at 0 has CDF $F(x) = \begin{cases} 0 & \text{for } x < 0 \\ 1 & \text{for } 0 \leq x \end{cases}$.

Here $G(x) = F(x)$ except at the point "0" which $\notin C(F)$;
hence, $X_n \overset{D}{\to} X$.

c)   Let $F_n(x) = I_{[1/n,\infty)}(x)$ and $F(x) = I_{[0,\infty)}(x)$. Note

$$C(F) = R - \{0\}.$$

For $x \neq 0$, $F_n(x) \to F(x)$ but $F_n(0) = 0$ does not converge to

$F(0) = 1$. Nevertheless, $X_n \overset{D}{\to} X$ as in b).

d)   Let $F_n(x) = (1 - (1 - x/n)^n) \cdot I_{[0,n)}(x) + I_{[n,\infty)}(x)$.

For each x, $F_n(x) \to F(x) = (1 - e^{-x}) \cdot I_{[0,\infty)}(x)$. Of course, this

F is continuous everywhere and $X_n \overset{D}{\to} X$ which happens to
be an exponential RV.

Exercise 7: Let X be a Bernoulli RV with

$$P(X = 1) = 1/2 = P(X = 0).$$

Let $Y = 1 - X$ and $X_n = Y$ for all $n \geq 1$.

a)   Show that $X_n \overset{D}{\to} X$. b) Does $X_n \overset{P}{\to} X$?

Inconveniently, convergence in distribution can not be
obtained thru the PDF; that is , $\lim \int f_n$ may not equal $\int \lim f_n$ as
in the following example. This matter will be cleared up in
Lesson 13.

Example: Let $P(X_n = 1 + 2/n) = 1$; the PDF and the CDF are

$$f_n(x) = \begin{cases} 0 \\ 1 \\ 0 \end{cases} \quad F_n(x) = \begin{cases} 0 & \text{for } x < 1 + 1/n \\ 1 & \text{for } x = 1 + 1/n \\ 1 & \text{for } 1 + 1/n \leq x \end{cases}.$$

Obviously, $\lim_{n\to\infty} f_n(x) = 0$ for all real x so that any integral of lim

f_n is 0. But, $G(x) = \lim_{n\to\infty} F_n(x) = \begin{cases} 0 & \text{for } x \leq 1 \\ 1 & \text{for } x > 1 \end{cases}$ while the RV X degenerate at 1 has CDF $F(x) = \begin{cases} 0 & \text{for } x < 1 \\ 1 & \text{for } x \geq 1 \end{cases}$. Hence, $X_n \overset{D}{\to} X$.

Exercise 8: Let F be a CDF.
  a) Define $F_n(x) = F(x - 1/n)$ for $n \geq 1$.

    i) Show that $F_n \to F$ on C(F).

    ii) What is $\lim_{n \to \infty} F_n(x)$ for $x \notin$ C(F).

  b) Suppose that F is continuous and define
      $G_n(x) = F(x/n)$ for $n \geq 1$. Is $\lim_{n \to \infty} G_n(x)$ a CDF ?

There is an interesting and useful (especially in statistics) inter–relation of convergence in distribution and convergence in probability to a constant. The last part of the following theorem can be extended to rational functions of RVs which converge to constants.

Theorem *(Slutsky): Let $X, X_1, X_2, \cdots, Y_1, Y_2, \cdots$ be RVs such that $X_n \overset{D}{\to} X$, $Y_n \overset{P}{\to} b$ (constant). Then,*

  a) $X_n \pm Y_n \overset{D}{\to} X \pm b$;

  b) $X_n \cdot Y_n \overset{D}{\to} X \cdot b$

  c) $X_n / Y_n \overset{D}{\to} X/b$ when $b \neq 0$.

Partial proof: Consider "+" in a). Let the CDF of X be F_X. Fix x and $\varepsilon > 0$ such that x, x–b–ε, x–b+ε are all in $C(F_X)$. We want the limiting distribution of $X_n + Y_n$. The conditions $x_n + y_n \leq x$ and $|y_n - b| \leq \varepsilon$ partition the (x_n, y_n) plane:

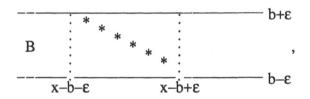

$\leftarrow x_n + y_n = x$

The region of interest is $P(X_n + Y_n \leq x) = P(A \cup B \cup C)$.
Since $A \cup C \subset \{|Y_n - b| > \varepsilon\}$, $P(A \cup C) \to 0$; therefore, we consider

$$P(B) = P(X_n + Y_n \leq x, |Y_n - b| \leq \varepsilon).$$

From the diagram

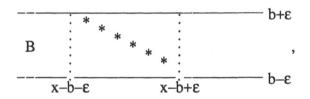

we see that

$$P(B) \leq P(X_n \leq x{-}b{+}\varepsilon, |Y_n - b| \leq \varepsilon) \qquad (*)$$

$$P(B) \geq P(X_n \leq x{-}b{-}\varepsilon, |Y_n - b| \leq \varepsilon). \qquad (**)$$

Now $P(X_n \leq x{-}b{+}\varepsilon) - P(X_n \leq x{-}b{+}\varepsilon, |Y_n - b| \leq \varepsilon)$

$$= P(X_n \leq x{-}b{+}\varepsilon, |Y_n - b| > \varepsilon) \leq P(\{|Y_n - b| > \varepsilon\})$$

which $\to 0$.

It follows that the limit of the probability on the right of $(*)$ is the limit of $P(X_n \leq x{-}b{+}\varepsilon)$ which is $F_X(x{-}b{+}\varepsilon)$. The limit of the probability on the right of $(**)$ is $F_X(x{-}b{-}\varepsilon)$ so that

$$F_X(x{-}b{-}\varepsilon) \leq \liminf_{n \to \infty} P(B) \leq \limsup_{n \to \infty} P(B) \leq F_X(x{-}b{+}\varepsilon).$$

Let $\varepsilon \to 0$, keeping $x \pm b - \varepsilon$ in $C(F_X)$. The final conclusion is

$$\lim_{n \to \infty} P(X_n + Y_n \le x) = F_X(x-b) = P(X \le x - b)$$

$$= P(X + b \le x) \,.$$

Exercise 9: Complete the proof of Slutsky's Theorem. Hint: draw some pictures.

Exercise 10: Let b,m be constants. Let the RVs be such that $Z_n \overset{D}{\to} Z$, $Y_n \overset{D}{\to} b$, $X_n \overset{D}{\to} m$.

a)   Show that $X_n Z_n + Y_n \overset{D}{\to} mZ + b$.

b)   Suppose that Z is the standard normal RV; approximate $P(X_n Z_n + Y_n \le x)$ for the values:

m	1	1	2	2
b	2	2	1	3
x	2	−1	2	−1

LESSON *7. OTHER RELATIONS IN MODES OF CONVERGENCE

As in the previous lesson, we take all RVs on a fixed probability space and finite a.s. Here we will discuss the following relationships:

$$X_n \xrightarrow{P} X \text{ implies } X_n \xrightarrow{D} X;$$

$$X_n \to X \text{ in pth mean implies } X_n \to X \text{ in qth mean, } 1 \le q < p;$$

$$X_n \to X \text{ in pth mean implies } X_n \xrightarrow{P} X.$$

The last theorem in this lesson composes "continuity" with each of the three convergences: a.s., P, D.

When "A implies B", we say that A is "stronger" than B or B is "weaker" than A. For example, since (Lesson 8, Part II) $X_n \xrightarrow{as} X$ implies $X_n \xrightarrow{P} X$, convergence a.s. is stronger than convergence in P. The next exercise contains a more involved version of the example in that lesson showing that convergence in P is strictly weaker than convergence a.s.

Exercise 1: Let $\Omega = [0,1]$ with the Borel σ–field and uniform probability (Lebesgue measure). For each $n = 1(1)\infty$, let $m(n)$ be the integer such that $2^{m(n)} \le n < 2^{m(n)+1}$ so that

$$n = 2^{m(n)} + j \text{ for some } j \in \{0, 1, \cdots, 2^{m(n)-1}\}. \text{ Let}$$

$$X_n(\omega) = I\{j/2^{m(n)} \le \omega \le (j+1)/2^{m(n)+1}\}.$$

Show that $X_n \xrightarrow{P} 0$ but $\{X_n(\omega)\}$ does not converge except for $\omega = 0$.

We now get "the best possible result" going from convergence in probability to convergence a.s.

Theorem: *Let $\{X_n\}$ be a sequence of RVs. If $X_n \xrightarrow{P} X$, then there is a subsequence which converges a.s. to the same limit function X.*

Proof: Suppose that X is finite everywhere. Then $X_n - X \to 0$; we may as well consider $Y_n = X_n - X \to 0$. (This is the formal expression of a common kind of simplification: "Without loss of generality, we may consider $X = 0$." We could retain the symbol X_n.) Then, $\lim_{n \to \infty} P(|Y_n| > 1/2^k) = 0$ for all positive integers k. For each such k, there is an N_k such that

$$P(|Y_{N_k}| > 1/2^k) \le 1/2^k$$

and so

$$\Sigma_k P(|Y_{N_k}| > 1/2^k) \le \Sigma_k 1/2^k = 1.$$

By the Borel-Cantelli Lemma (Lesson 2, Part II),

$$P(\{|Y_{N_k}| > 1/2^k\} \text{ i.o.}) = 0.$$

This means that with probability 1, $|Y_{N_k}| > 1/2^k$ for at most a finite number of k, say $k \le K$; the other way around, with probability 1, $|Y_{N_k}| \le 1/2^k$ for all k > K. Since

$|Y_{N_{k+1}}| \le 1/2^{k+1}$ implies $|Y_{N_{k+1}}| \le 1/2^k$, the probability is 1

that $|Y_{N_k}| \le 1/2^K$ for all k > K or $|Y_{N_k}| \to 0$ a.s.

Exercise 2: Prove the following "converse" of the last theorem. If every actual subsequence Y_n of a sequence of RVs X_n has a further subsequence Z_n which converges a.s. to the RV X, then

$$X_n \overset{P}{\to} X.$$

We turn now to the first implication listed at the beginning of this lesson.

Theorem: *Let $X, X_1, X_2, X_3, \cdots$ be random variables with* *corresponding CDFs* $F, F_1, F_2, \cdots$. *If* $X_n \overset{P}{\to} X$, *then* $X_n \overset{D}{\to} X$.

Proof: For all $x \in R$,

$$\{X_n \leq x\} = [\{X_n \leq x\} \cap \{X > x\}] \cup [\{X_n \leq x\} \cap \{X \leq x\}];$$

more simply,

$$\{X_n \leq x\} = \{X_n \leq x, X > x\} \cup \{X_n \leq x, X \leq x\}.$$

Similarly,

$$\{X \leq x\} = \{X \leq x, X_n > x\} \cup \{X \leq x, X_n \leq x\}.$$

Then, $F_n(x) = P(X_n \leq x, X > x) + P(X_n \leq x, X \leq x)$

and $F(x) = P(X \leq x, X_n > x) + P(X \leq x, X_n \leq x)$.

It follows that

$$|F_n(x) - F(x)| = |P(X_n \leq x, X > x) - P(X \leq x, X_n > x)|$$

$$\leq P(X_n \leq x, X > x) + P(X \leq x, X_n > x). \qquad (*)$$

For each $\varepsilon > 0$, $\{X_n \leq x, X > x\}$ is equal to

$$\{X_n \leq x, X > x, |X_n - X| \geq \varepsilon\} \cup \{X_n \leq x, X > x, |X_n - x| < \varepsilon\}$$

$$\subset \{|X_n - X| \geq \varepsilon\} \cup \{x < X < x + \varepsilon\}.$$

Similarly,

$$\{X_n > x, X \leq x\} \subset \{|X_n - X| \geq \varepsilon\} \cup \{x - \varepsilon < X < x\}.$$

Therefore,

$$P(X_n \leq x, X > x) \leq P(|X_n - X| \geq \varepsilon) + P(x < X < x + \varepsilon),$$

$$P(X \leq x, X_n > x) \leq P(|X_n - X| \geq \varepsilon) + P(x - \varepsilon < X < x)$$

and

$$P(X_n \leq x, X > x) + P(X \leq x, X_n > x)$$

$$\leq 2P(|X_n - X| \geq \varepsilon) + P(x - \varepsilon < X < x + \varepsilon).$$

By hypothesis, $P(|X_n - X| \geq \varepsilon) \to 0$ as $n \to \infty$. Applying all this to the inequality in (*) yields

$$\lim_{n\to\infty} |F_n(x) - F(x)| \le P(x-\varepsilon < X < x+\varepsilon) =$$

$$F(x+\varepsilon + 0) - F(x-\varepsilon).$$

The left hand side of this inequality is free of ε. (Its value will not change under the limit as $\varepsilon \to 0$.) For $x \in C(F)$, the limit of the right hand side as $\varepsilon \to 0$ is 0. In other words,

$$\lim_{n\to\infty} F_n(x) = F(x) \text{ for } x \in C(F).$$

Exercise 3:   Convince yourself that the two "Similarly,$\cdots$" statements in the proof above are in fact correct.

In general, convergence "in D" is weaker than convergence "in P". The next example shows that the converse of the above theorem is false while the succeeding theorem leads to the conclusion that these two convergences are equivalent in a special case.

Example: Let $X_1, X_2, X_3, \cdots$ be IID Bernoulli RVs with

$$P(X_n = 1) = 1/2 = P(X_n = 0).$$

$P(X_n \le x) = P(X_1 \le x)$ so that obviously and trivially, $X_n \overset{D}{\to} X_1$. However, for $0 < \varepsilon < 1$,

$$P(|X_n - X_1| \ge \varepsilon) = P(|X_n - X| = 1)$$

$$= P(X_n = 1, X_1 = 0) + P(X_n = 1, X_1 = 0)$$

$$= (1/2)^2 + (1/2)^2 = 1/2.$$

Therefore $X_n \overset{P}{\to} X$ is false.

Theorem: *For the sequence of RVs $\{X_n\}$ with CDFs $\{F_n\}$, suppose that $X_n \overset{D}{\to} c$(onstant); then $X_n \overset{P}{\to} c$.*

Proof: For any $\varepsilon > 0$, $P(|X_n - c| \ge \varepsilon)$

$$= P(X_n \ge c + \varepsilon) + P(X_n \le c - \varepsilon)$$

$$= 1 - F_n(c + \varepsilon + 0) + F_n(c - \varepsilon).$$

The CDF of "c" is $F(x) = I_{[c,\infty)}(x)$ and $C(F) = R - \{c\}$. Since $c - \varepsilon$ and $c + \varepsilon$ are in $C(F)$,

$$\lim_{n \to \infty} F_n(c - \varepsilon) = F(c - \varepsilon) = 0$$

and

$$\lim_{n \to \infty} F_n(c + \varepsilon) = F(c + \varepsilon) = 1.$$

To get corresponding results for convergence "in mean", we need the following basic lemma. Note that the inequality is trivial when its right–hand side is infinite.

Lemma: *Let X be a real (finite) RV for $[\Omega, \mathcal{B}, P]$. Let g be an even non-negative (measurable) function non-decreasing on R. For each $\varepsilon > 0$,*

$$P(|X| \geq \varepsilon) \leq E[g(X)]/g(\varepsilon).$$

Proof: Since g is non–negative, so is g(X) and its integral exists; since g is even, $g(X) = g(-X) = g(|X|)$. Let $A = \{\omega : |X(\omega)| \geq \varepsilon\}$. On A, $g(X) \geq g(\varepsilon)$ so

$$\int_A g(X) \geq \int_A g(\varepsilon) = g(\varepsilon) \cdot P(A);$$

in any case, $\int_{A^c} g(X) \geq 0$. Putting these inequalities in

$$E[g(X)] = \int_A g(X) + \int_{(A^c)} g(X),$$

yields $E[g(X)] \geq g(\varepsilon) \cdot P(A)$ from which the conclusion follows.

If we take $g(x) = |x|^p$ for $p > 0$, we get

$$P(|X| \geq \varepsilon) \leq E[|X|^p]/\varepsilon^p.$$

The form of the inequality just above is attributed to Markov; the special case $p = 2$ is Chebyshev's:

$$P(|X| \geq \varepsilon) \leq E[X^2]/\varepsilon^2.$$

Exercise 4: Prove the following

Corollary: *Let* $X, X_1, X_2, X_3, \cdots$ *be RVs. If* $\{X_n\}$ *converges to X in pth mean, then* $X_n \overset{P}{\to} X.$

The converse of the corollary is not true as the next example demonstrates. However, there is a partial converse following that.

Example: Define X_n such that

$$P(X_n = 0) = 1 - 1/n^2, \ P(X_n = n) = P(X_n = -n) = 1/2n^2.$$

For each $\varepsilon > 0$, $P(|X_n| \geq \varepsilon) = 1/2n^2$ which does indeed tend to 0 as $n \to \infty$. But, $E[X_n^2] = 1$ for all $n \geq 1$; thus, this $\{X_n\}$ does not converge in quadratic mean (p = 2) to 0.

Theorem: *Let the sequence of RVs* $\{X_n\}$ *be uniformly bounded a.s., that is, for some finite M,* $P(|X_n| \leq M) = 1$ *for all* $n \geq 1$. *Then,*

$$X_n \overset{P}{\to} X \text{ implies } \lim_{n \to \infty} E[|X_n - X|^p] = 0.$$

Proof: a) Let c and d be real; since $|(|c|-|d|)| \leq |c-d|$, $|c-d| < \varepsilon$ implies $|d| < |c|+\varepsilon$. With X_n and X, this gives

$$\{|X_n - X| < \varepsilon\} \subset \{|X| < M + \varepsilon\} \text{ whence}$$

$$P(|X_n - X| < \varepsilon) \leq P(|X| < M + \varepsilon).$$

The left–hand side of the last inequality tends to 1 as $n \to \infty$, so the right–hand side is also 1. This means that $X_n - X \overset{P}{\to} 0$; of course, $X_n - X$ is uniformly bounded (by 2M). Without loss of generality, we may take X = 0 in the rest of the proof.

b)   Since $|X_n| \leq M$ a.s., $E[|X_n|^p] < M^p < \infty$ for all p > 0.

On $A = \{\omega : |X(\omega)| < \varepsilon\}$, $|X_n|^p < \varepsilon^p$ and on A^c, $|X_n|^p \leq M^p$. Since

$$E[\,|X_n|^P] = \int_A |X_n|^P dP + \int_{A^c} |X_n|^P dP,$$

using these inequalities makes

$$E[\,|X_n|^P] \leq \epsilon^P + M^P \cdot P(\,|X_n| \geq \epsilon).$$

Letting $n \to \infty$, obtains $E[\,|X_n|^P] \leq \epsilon^P$; letting $\epsilon \to 0$, obtains the final result.

Exercise 5: Let each $\{X_n\}$ be a sequence of RVs; investigate convergence a.s., in probability, in distribution, and in quadratic mean.

a) $P(X_n = n^2) = 1/n^2 = 1 - P(X_n = 0)$.

b) $P(X_n = n^{1/4}) = 1/n = 1 - P(X_n = 0)$.

Exercise 6: Suppose that $X_n \overset{P}{\to} 0$, $|X_n| \leq Y$ a.s., and $E[\,|Y|^P] < \infty$. Show that $X_n \to 0$ in pth mean. Hint: show that $\int_{A_n} Y\, dP \to 0$ whenever $P(A_n) \to 0$.

Exercise 7: Let $X_n \to X$ in quadratic mean. Show that:

a) $E[\,|X_n - X|] \to 0$ by making use of the CBS inequality;

b) $E[X_n] - E[X] \to 0$

c) $E[X_n^2] - E[X^2] \to 0$.

In ordinary measure theory, "in pth mean" results are referred to as "L_p" theorems; we continue only with the special case of probability. The consequence is the much used result that $E[X^2] < \infty$ implies $E[X]$ is finite.

Theorem: *If $X_n \to X$ in pth mean, then*

$$X_n \to X \text{ in qth mean for } 1 \le q < p.$$

Proof: Lesson 11, Part II contains two results we need:

$|X|^q$ is integrable whenever $|X|^p$ is integrable;
(Hölder's inequality) for $r > 1$ and $1/r + 1/s = 1$,

$$E[|XY|] \le (E[|X|^r])^{1/r} (E[|Y|^s])^{1/s}.$$

With appropriate substitutions, Hölder's inequality yields

$$E[|X_n - X|^q]$$

$$\le \left[E[(|X_n - X|^q)^{p/q}] \right]^{q/p} \left[E[1^{p/(p-q)}] \right]^{(p-q)/p}$$

$$\le \left[E[|X_n - X|^p] \right]^{q/p} \text{ which converges to 0.}$$

Exercise 8: Verify the substitutions for Hölder's inequality.

Finally, we combine "convergence" and "continuity".

Theorem: *Let* $X, X_1, X_2, \cdots$ *be real RVs and let* $g : R \to R$.
Suppose that $P(X^{-1}(A)) = 1$ *where A is the set of R on which g is
continuous. If* $X_n \to X$ *in one of the modes a.s., in P, in D, then
$g(X_n)$ converges to g(X) in the same mode.*

Proof: a) Let B be the set on which $X_n \to X$ with $P(B) = 1$. Then
$P(X^{-1}(A) \cap B) = 1$. For

$$\omega \epsilon X^{-1}(A) \cap B, \; g(X_n(\omega)) \to g(X(\omega))$$

by continuity.

b)   Suppose that $X_n \overset{P}{\to} X$ but $g(X_n) \overset{P}{\to} g(X)$ is false. Then for
some $\epsilon > 0$ and $\delta > 0$, there is a subsequence $\{X_{n(k)}\}$ such
that

$$P(|g(X_{n(k)}) - g(X)| > \epsilon) > \delta \text{ for all } k \ge 1. \qquad (*)$$

But $X_n \overset{P}{\to} X$ implies $X_{n(k)} \overset{P}{\to} X$ which implies (the best possible result) that some subsequence $X_{n(k)'} \overset{as}{\to} X$. By part a), $g(X_{n(k)'}) \overset{as}{\to} g(X)$ and so $g(X_{n(k)'}) \overset{P}{\to} g(X)$. But this condition contradicts (*) so that $g(X_n) \overset{P}{\to} g(X)$ does hold.

c)   Let the RV X_n have CDF F_n; let U be a RV uniformly distributed on [0,1]. Define the inverse functions by
$$F^{-1}(y) = \inf \{x : F(x) \geq y\}, \quad F_n^{-1}(y) = \inf \{x : F_n(x) \geq y\}.$$
Since $Y_n = F_n^{-1}(U)$ has the same distribution as X_n and
$Y = F^{-1}(U)$ has the same distribution as X, "$Y_n \overset{P}{\to} Y$".
Hence part b) applies and $g(Y_n) \overset{P}{\to} g(Y)$. But then
$g(Y_n) \overset{D}{\to} g(Y)$.

Exercise 9: Supply the details to verify that $Y_n \overset{P}{\to} Y$ in part c) above.

LESSON 8. LAWS OF LARGE NUMBERS

In this lesson, we examine some relationships between relative frequency (Lesson 1, Part I) and probability, specifically expectation, (Lesson 11, Part II). In particular, we examine different modes of convergence of the simple arithmetic mean of real random variables defined on a fixed probability space $[\Omega, \mathscr{B}, P]$; each mode has its own law. "Large" refers to the number of RVs, not their values.

We begin with a *weak law of large numbers*. The rather cumbersome proof (omitted here) of the first theorem can be found in books like Chung (1974). On the other hand, the existence of the variance makes the proof of the second theorem almost trivial.

Theorem: *Let* $\{X_n\}$ *be a sequence of pairwise independent and identically distributed real random variables with finite mean* μ; *let* $\bar{X}_n = \sum_{k=1}^{n} X_i/n$. *Then,* $\bar{X}_n \overset{P}{\to} \mu$.

Note: Each $\bar{X}_n(\omega)$ is a simple arithmetic mean. "Pairwise independent" means that X_i, X_j are independent for any $i \neq j$; "identically distributed" means X_i, X_j have the same distribution. "Independent and identically distributed" is abbreviated as IID.

Theorem: *Let* $\{X_n\}$ *be a sequence of IID RVs with mean* μ *and finite variance* σ^2. *Then,* $\bar{X}_n \to \mu$ *in quadratic mean (and hence also in probability).*

Proof: $E[(\bar{X}_n - \mu)^2] = E[(\Sigma(X_i - \mu)/n)^2] = E[\Sigma(X_i - \mu)^2]/n^2$ because, when $i \neq j$, $E[(X_i-\mu)(X_j-\mu)] = 0$ by independence. Hence, $E[(\bar{X}_n - \mu)^2] = \sigma^2/n$ which tends to 0 as $n \to \infty$.

Example: a) (Bernoulli 1713)

$$P(X_n = 1) = \theta = 1 - P(X_n = 0).$$

X_n, the relative frequency of "1" in the sample, converges in probability to θ, the relative frequency (probability) of "1" in the population.

b)   Let the sequence of independent RVs $\{X_n\}$ have the common CDF F. For each $x \in R$, let

$$Y_j = I\{-\infty < X_j \le x\}.$$

Then, $\{Y_n\}$ is a sequence of IID RVs with finite mean and finite variance: $\mu_Y = E[Y_j] = P(X_j \le x) = F(x)$

$$\sigma_Y^2 = E[Y_j^2] - (E[Y_j])^2 = E[Y_j] - (F(x))^2 = F(x)(1-F(x)).$$

The sample mean $\overline{Y}_n = \sum_{j=1}^{n} Y_j/n = \sum_{j=1}^{n} I\{-\infty < X_j \le x\}/n$

$$= F_n(x),$$

the sample CDF (or empirical distribution function) of $X_1, X_2, \cdots, X_n$. It follows that $F_n(x) \xrightarrow{P} F(x)$.

c)   Let each X_n be a Cauchy RV with density $1/\pi(1+x^2)$ for all real x. Then $E[X_n] = \infty$ so that the weak law of large numbers does not apply to such IID RVs. It will be seen later that the distribution of $\overline{X}_n$ in this case is also Cauchy!

Part b) of the previous example allows the sample CDF to be used to estimate the probabilities $\{F(x) : x \text{ is real}\}$ for any distribution. Also, for each x, $\{Y_j\}$ is just a sequence of Bernoulli RVs with $p = F(x)$ so that parts a) and b) are really equivalent.

Exercise 1: Assume the hypotheses of the second theorem.

a)   Apply Chebyshev's inequality to $\overline{X}_n - \mu$.

b)   Prove $X_n \xrightarrow{P} \mu$ from a).

c)   Suppose that $\sigma^2 = 2$. How large a sample size "n" is needed to have

$$P(|\overline{X}_n - \mu| \leq 1) \geq .95 ? \text{ Hint: look at } \frac{\sigma^2/n}{\varepsilon^2} \leq .05 .$$

d)   Suppose that $\sigma = 4$. How large a sample is needed to have $\overline{X}_n$ within .1 units of μ with probability at least .99?

The next two theorems make up Kolmogorov's *strong law of large numbers*; a general proof for the first (omitted) can be found in books like Chung (1974). Again, by assuming a finite second moment, we get a simpler proof, though not as trivial as in the case of the weak law. Results of the following exercise are also needed.

Exercise 2: Let $\alpha(n) = [\sqrt{n}]$, the greatest integer less than or equal to the square root of the positive integer n. Show that for $n \geq 2$, $n - 2\sqrt{n} + 1 \leq \alpha^2(n) \leq n$ and hence that $(\alpha(n))^2/n \to 1$ as $n \to \infty$.

Theorem: *For* $\{X_n\}$ *an IID sequence of RVs with finite mean* $\mu = E[X_n]$, $\overline{X}_n \xrightarrow{as} \mu$.

Partial proof: a) Assume that $E[X_n^2] < \infty$. Then, so is $E[X_n]$; without loss of generality, we assume $E[X_n] = 0$. Let

$$S(n) = \sum_{j=1}^{n} X_j \text{ and } \alpha(n) = [\sqrt{n}].$$

b)   Next we show that $S(n^2)/n^2 \xrightarrow{as} 0$ as $n \to \infty$. By Chebyshev's inequality,

$$P(S(n^2)/n^2 \geq \varepsilon)$$

$$= P\left[|\sum_{j=1}^{n^2} X_j| \geq \varepsilon n^2\right] \leq E\left[\left[\sum_{j=1}^{n^2} X_j\right]^2\right]/(\varepsilon^2 n^4) .$$

Since the X_j's are independent, the right–hand side reduces to $\sigma^2/\epsilon^2 n^2$. Then,

$$\sum_{n=1}^{\infty} P(S(n^2)/n^2 \geq \epsilon) \leq \sum_{n=1}^{\infty} \sigma^2/\epsilon^2 n^2 < \infty$$

and so by the Borel–Cantelli lemma (Lesson 2 Part II),

$$P(\{S(n^2)/n^2 \geq \epsilon\} \text{ i.o.}) = 0.$$

c)   Now, $S(\alpha^2(n))/n = \left[S(\alpha^2(n))/\alpha^2(n)\right] \cdot \left[\alpha^2(n)/n\right]$ and $\alpha^2(n)/n \to 1$ as $n \to \infty$. By part b),

$$S(\alpha^2(n))/\alpha^2(n) \overset{as}{\to} 0 \text{ as } \alpha(n) \to \infty.$$

Hence, $S(\alpha^2(n))/n \overset{as}{\to} 0$ as $n \to \infty$.

d)   Finally, $S(n)/n \overset{as}{\to} 0$ if $S(n)/n - S(\alpha^2(n))/n \overset{as}{\to} 0$. But,

$$P(|S(n)/n - S(\alpha^2(n))/n| \geq \epsilon)$$

$$= P\left[|X_{\alpha^2(n)+1} + \cdots + X_n| \geq n\epsilon\right] \leq (n - \alpha^2(n))\sigma^2/n^2\epsilon^2$$

$$\leq 2\sigma^2\sqrt{n}/n^2\epsilon^2 = 2\sigma^2/\epsilon^2 n^{3/2}.$$

As in part b), $P(\{|S(n)/n - S(\alpha^2(n))/n| \geq \epsilon\} \text{ i.o.}) = 0$. Hence, the strong law of large numbers does hold.

Exercise 3: Fill in the details for part d) of the proof.

We can give a proof of the following "converse ".

Theorem: *Let $\{X_n\}$ be an IID sequence of RVs. If*

$$\sum_{j=1}^{n} X_j/n \overset{as}{\to} \mu \text{ (finite), then } E[|X_n|] < \infty \text{ and } \mu = E[X_n].$$

Proof: Let $S(n) = \sum_{j=1}^{n} X_j$. It suffices to show that

$$E[|X_n|] < \infty$$

for then, by the previous theorem, $S(n)/n \overset{as}{\rightarrow} E[X_n]$ and

$E[X_n] = \mu$ because a.s. limits are unique. It follows from hypothesis that

$$X_n/n = S(n)/n - ((n-1)/n)S(n-1)/(n-1) \overset{as}{\rightarrow} 0.$$

Let $A_n = \{\omega : |X_n(\omega)|/n \geq 1\}$; then by the corollary in Lesson 2, Part III, $P(\{A_n\} \text{ i.o.}) = 0$. Since $\{A_n\}$ are independent, the converse to the second part of the Borel–Cantelli lemma implies that $\sum_{n=1}^{\infty} P(A_n) < \infty$. By the "interesting result" in Lesson 11, Part II, this implies that $E[|X_n|/n]$ is finite so that $E[X_n]$ is also.

The strong law applies to the Bernouilli RVs of the first example because $E[X_j] = \theta = E[X_j^2]$. To see something of the difference in the two laws, we need to emphasize the functional aspect: for each $\omega \ \varepsilon \ \Omega$, $\{X_n(\omega)\}$ has the partial sums

$$S_n(\omega) = \sum_{j=1}^{n} X_j(\omega) \text{ and the sequence}$$

$$S_1(\omega)/1, S_2(\omega)/2, \cdots, S_n(\omega)/n, \cdots$$

is an *orbit* of relative frequencies. The weak law says that, for large n, the probability is close to one that $S_n(\omega)/n$ is close to θ, but says nothing about an individual orbit or its $\lim_{n\to\infty} S_n(\omega)/n$. The strong law says that for almost all orbits, when n is large, $S_n(\omega)/n$ is close to θ, in fact, $\lim_{n\to\infty} S_n(\omega)/n = \theta$. The strong law in particular shows that expectation is an intimate part of probability theory.

Exercise 4:   Let $\{X_n\}$ be a sequence of IID RVs with

$$\mu_k = E[X_n{}^k] \text{ finite for some positive integer k.}$$

$$\text{Let } S(n;k) = \sum_{j=1}^{n} X_j^k/n.$$

a) Show that $S(n;k) \overset{as}{\to} \mu_k$.

b) If $k \geq 2$, show that

$$\sum_{j=1}^{n} (X_j - \mu_1)^2/n \overset{as}{\to} \mu_2 - \mu_1{}^2 = \sigma^2$$

and $\displaystyle\sum_{j=1}^{n} (X_j - S(n;1))^2/n = S(n;2) - (S(n;1))^2 \overset{as}{\to} \sigma^2.$

Exercise 5: (Borel's strong law, 1909) Let $S(n;k)$ be as in the previous exercise; let $\{X_n\}$ be an IID sequence of Bernouilli RVs with $\theta \in (0,1)$.

a) First show that for each $\varepsilon > 0$,

$$P(|S(n^2;1) - \theta| \geq \varepsilon) \leq \theta(1-\theta)/\varepsilon^2 n^2.$$

b) For each positive integer n, let m(n) be such that $(m(n))^2 \leq n < (m(n)+1)^2$. Show that

$$|S(n;1) - \theta| \leq |S(n^2;1) - \theta| + 4/m(n).$$

Conclude that $S(n;1) \overset{as}{\to} \theta$.

The following lemma and exercise from analysis are of interest in themselves (being at the start of summability theory) but here they are applied in probability (Exercise 7).

Toeplitz Lemma: *Let $\{a_n\}$, $\{x_n\}$ be sequences of real numbers such that $a_n > 0$ and, as $n \to \infty$, i) $\displaystyle\sum_{j=1}^{n} a_j \uparrow \infty$, ii) $x_n \to x$. Then,*

$$\left[\sum_{j=1}^{n} a_j\right]^{-1} \sum_{j=1}^{n} a_j x_j \to x \text{ as } n \to \infty.$$

Proof:

$$\left[\sum_{j=1}^{n} a_j \right]^{-1} \sum_{j=1}^{n} a_j x_j = \left[\sum_{j=1}^{n} a_j \right]^{-1} \sum_{j=1}^{n} a_j (x_j - x + x)$$

$$= \left[\sum_{j=1}^{n} a_j \right]^{-1} \sum_{j=1}^{n} a_j (x_j - x) + x .$$

Since $x_j \to x$ as $j \to \infty$, for each $\varepsilon > 0$, there is an N_ε such that for $j > N_\varepsilon$, $|x_j - x| < \varepsilon$. Consequently,

$$\left| \left[\sum_{j=1}^{n} a_j \right]^{-1} \sum_{j=1}^{n} a_j (x_j - x) \right|$$

$$\leq \left[\sum_{j=1}^{n} a_j \right]^{-1} \left[\sum_{j=1}^{N_\varepsilon} a_j |x_j - x| + \sum_{j=N_\varepsilon+1}^{n} a_j \varepsilon \right]$$

$$\leq \left[\sum_{j=1}^{n} a_j \right]^{-1} \sum_{j=1}^{N_\varepsilon} a_j |x_j - x| + \varepsilon .$$

Here, the quantity in brackets tends to ∞ as $n \to \infty$, but the other sum is finite; therefore, the product tends to 0 and the conclusion follows.

Exercise 6: Let $\{\alpha_n\}$ be a sequence of real numbers such that $\sum_{n=1}^{\infty} \alpha_n < \infty$; let the sequence $\{b_n\}$ be postive and tend to ∞ as $n \to \infty$. By using Toeplitz' lemma, show that

$$\sum_{j=1}^{n} b_j a_j / b_n \to 0 \text{ as } n \to \infty.$$

Exericse 7: Let $\{X_n\}$ be a sequence of real RVs such that $X_n \overset{as}{\to} 0$. Show that $\overline{X}_n = \sum_{j=1}^{n} X_j / n \overset{as}{\to} 0$.

INE–Exercises:

1.   Let $\{X_n\}$ be a sequence of RVs such that

$$\sum_{n=1}^{\infty} P(|X_n| < \varepsilon) < \infty$$

for each $\varepsilon > 0$. Show that $X_n \overset{as}{\to} 0$. Hint: Let $A = \cap_{k=1}^{\infty} \cup_{m=1}^{\infty} \cap_{n=m}^{\infty} \{\omega : |X_n(\omega)| \leq 1/k\}$. Use the Borel–Cantelli lemma to show that $P(A) = 1$.

2.   Let the RVs $\{X_n\}$ be pairwise independent with means 0 and a.s. uniformly bounded: $P(|X_n| \leq M) = 1$. Prove a strong law of large numbers: for

$$S(n) = \sum_{j=1}^{n} X_j, \quad S(n)/n \overset{as}{\to} 0.$$

Hint: for each m, there is an n such that

$$n^2 \leq m < (n+1)^2.$$

Write $S(m)/m = S(n^2)/m + [S(m) - S(n^2)]/m$. Use the ideas in the proof of the second part of Kolmogorov's strong law.

LESSON 9. CONVERGENCE OF SEQUENCES OF DISTRIBUTION FUNCTIONS

There were examples in lesson 6 demonstrating that a sequence of probability distribution functions need not converge to a probability distribution function. Here we will introduce notation and techniques to enable us to discuss this phenomenon more thoroughly..

Definition: *Let the function $F : R \to [0,1]$. F is a distribution function (DF) iff F is right-hand continuous and non-decreasing. F is a (probability) cumulative distribution function (CDF) if also*

$$F(-\infty) = \lim_{x \to -\infty} F(x) = 0 \text{ and } F(+\infty) = \lim_{x \to +\infty} F(x) = 1.$$

Example: Let G be a CDF. Let $F_n(x) = G(x + n)$ for $n = 1(1)\infty$. Then each F_n is also a CDF but $\lim_{n \to +\infty} F_n(x) = 1$ for all $x \in R$. The function identically 1 is a DF but not a CDF.

It is worthwhile to note that the continuity set of a DF, say F, is also countable; indeed, the proof in Lesson 3, Part II is still valid since here $0 \le F(x) \le 1$.

Exercise 1: A RV X or its probability P_X is *non-atomic* if $P(X = x) = 0$ for all $x \in R$. Show that the corresponding CDF F is continuous and visa–versa.

Definition: *A sequence of DFs $\{F_n\}$ converges weakly to a DF, say F, (write $F_n \overset{w}{\to} F$) iff $F_n(x) \to F(x)$ for all $x \in C(F)$, the continuity set of F. The sequence converges completely to F $(F_n \overset{c}{\to} F)$ if also $F_n(\pm\infty) \to F(\pm\infty)$.*

Example: The sequence $F_n(x) = I_{[n,+\infty)}(x) \to \begin{cases} 0 & \text{for } x < +\infty \\ 1 & \text{for } x = +\infty \end{cases}$.

This F_n converges weakly to $F \equiv 0$ but not completely.

Now we need some more analysis. The set of real numbers, R, can be constructed by taking the limits of all sequences of rational numbers, Q; we say that Q is dense in R. Although Q is countable and R is uncountable, the following is also true.

Lemma: *The complement of any countable subset B of R is dense in R.*

Proof: The set $A = R - B$ is uncountable. For each real r, the interval $(r,r+1)$ must contain points of A; pick one, say x_1. The interval $(r,(r+x_1)/2)$ must contain points of $A \neq x_1$; pick one, say x_2; the interval $(r,(r+x_2)/2)$ must contain $\cdots$. In this way, we generate a sequence $\{x_n\}$ which converges to r. In other words, every real number is a limit of a (decreasing) sequence of points in B^c; that is, B^c is dense in R.

One substantial import of denseness is the following:

Lemma: *Let F and G be right-hand continuous functions on R; let A be dense in R. If $F(x) = G(x)$ for all x in A, then $F(x) = G(x)$ for all x in R.*

Proof: For any real x in R, let $\{a_n\}$ be a sequence in A converging down to x. Then $F(x_n) = G(x_n)$ yields the limit $F(x) = G(x)$.

These two lemmas can be combined to get the following characteristic property of sequences of RVs.

Exercise 2: Let $F, F_1, F_2, \cdots$ be the CDF corresponding to the RVs $X, X_1, X_2, \cdots$. Prove that $F_n \overset{W}{\to} F$ iff $X_n \overset{D}{\to} X$.

Exercise 3: Suppose that $X_n \overset{D}{\to} X$ which has CDF F. Show that

for each $a < b$ in $C(F)$, $\lim_{n \to \infty} P(a < X_n \leq b) = F(b) - F(a)$.

The following theorem brings out the special character of $\pm \infty$ in these convergences.

Theorem: *Suppose that the sequence of DF $\{F_n\}$ converges weakly to the DF F. Then $F_n \overset{C}{\to} F$ iff*

$$F_n(+\infty) - F_n(-\infty) \to F(+\infty) - F(-\infty).$$

Proof: $F_n(-\infty) \le F_n(x) \le F_n(+\infty)$ for all x, in particular, for $x \in C(F)$. For such x, $\lim F_n(x) = F(x)$ and so

$$\limsup F_n(-\infty) \le F(x) \le \liminf F_n(+\infty). \qquad (*)$$

Letting $x \to +\infty$ in (*) yields $F(+\infty) \le \liminf F_n(+\infty)$;

letting $x \to -\infty$ in (*) yields $\limsup F_n(-\infty) \le F(-\infty)$. Therefore,

$$F(+\infty) - F(-\infty) \le \liminf F_n(+\infty) - \limsup F_n(-\infty)$$

$$= \liminf \{F_n(+\infty) - F_n(-\infty)\}.$$

It is clear that $F_n \overset{C}{\to} F$ iff $\liminf \{F_n(+\infty) - F_n(-\infty)\} = 0$.

In a similar spirit, the next theorem shows the importance of "dense".

Theorem: *A sequence of DF $\{F_n\}$ converges weakly iff*

it converges on a dense set of R.

Proof: a) Necessity: Suppose that $F_n \overset{W}{\to} F$. Then $F_n \to F$ on C(F) which is dense in R.

b)   Sufficiency: Let D be a dense set of R on which $\{F_n\}$ converges. Then the following are immediate:

$\lim_{n\to\infty} F_n(x) = F(x)$ defines a function on D;

for each $x \in R$, there is a sequence $\{x_n\}$ in D which converges to x.

From this x and $\{x_n\}$, define $G(x) = \lim_{n\to\infty} F(x_n)$. Since each

F_n is a DF, so are F and G; moreover, $G(x) = F(x)$ on D.

Now let $x \in C(G)$; take y and $z \in D$ such that $y < x < z$. Then

$$G(y) = F(y) = \lim_n F_n(y) \leq \liminf_n F_n(x)$$

$$\leq \limsup_n F_n(x) \leq \lim_n F_n(z) = F(z) = G(z).$$

Keeping both x and y in D, let y "increase" to x and let z "decrease" to x. Then $G(x) = \lim_n F_n(x)$ for $x \in C(G)$.

In the proof above, G is called an *extension* of F from D to R; that is, i) the domain of G includes that of F & ii) G = F on the common part. Many writers use only one symbol F.

As noted in Lesson 6, convergence in distribution might not be obtained by convergence of the corresponding sequence of densities. Here are some more cases.

Example: a) $f_n(x) = 1$ for $-n-1 < x < -n$, = 0 elsewhere.

$$F_n(x) = \int_{-\infty}^{x} f_n = 0 \qquad \text{for } x < -n-1$$
$$= x+n+1 \quad \text{for } -n-1 \leq x$$
$$= 1 \qquad \text{for } -n \leq x \ .$$

$$\lim f_n(x) = 0 \text{ for all x so } \int \left[\lim f_n\right] \equiv 0$$

but $\lim_n F_n(x) = 1$ for all x.

b) $f_n(x) = 1/n$ for $-n/2 < x < n/2$, = 0 elsewhere.

$$F_n(x) = 0 \qquad \text{for } x < -n/2$$
$$= x/n + 1/2 \quad \text{for } - n/2 \leq x < n/2$$
$$= 1 \qquad \text{for } n/2 \leq x \ .$$

Again $f_n \to 0$ but $F_n(x) \to 1/2$ for all x.

c)   Let $F' = f$ be any fixed density; take f_n as in b). Then

$$h_n(x) = (f_n(x) + f(x))/2 \to f(x)/2 \text{ but } \int_{-\infty}^{x} h_n \to 1/4 + F(x)/2.$$

Exercise 4: Find $\lim f_n$ and $\lim F_n$ when the RV X_n is normally distributed with:

a)   mean $-n$, variance 1 b) mean 0, variance n^2.

A resolution of this predicament involves complete convergence and is given in the following.

Theorem: *(Scheffé) If the sequence of PDFs $\{f_n = F'_n\}$ converges a.e. to a PDF $f = F'$, then $F_n \overset{c}{\to} F$.*

Proof: a) Since we are assuming $F_n(\pm\infty) = F(\pm\infty)$, it suffices to show that $F_n \overset{W}{\to} F$; the following argument shows that $F_n \to F$ *uniformly* for $x \in R$.

b)   Let $g_n(x) = f(x) - f_n(x)$ for $f(x) \geq f_n(x)$; let $g_n(x) = 0$ elsewhere. Since $f_n \to 0$ a.e., $g_n \to 0$ a.e. Since $f_n \geq 0$, $|g_n(x)| \leq f(x)$ for all x and all $n = 1(1)\infty$. By the dominated convergence theorem (Lesson 11, Part II),

$$\lim \int_R g_n(x)\, dx = 0.$$

c)   Let $A(n) = \{x : f(x) - f_n(x) \geq 0\}$. Since f_n and f are densities, $0 = \int_R (f(x) - f_n(x))dx$

$$= \int_{A(n)} (f(x) - f_n(x))dx + \int_{A(n)^c} (f(x) - f_n(x))dx$$

which implies

$$-\int_{A(n)^c} (f(x) - f_n(x))dx = \int_{A(n)} (f(x) - f_n(x))dx = \int_R g_n(x)\, dx.$$

Substituting these in the right–hand side of

$$\int_R |f_n(x) - f(x)|\, dx$$

$$= \int_{A(n)} (f(x) - f_n(x))dx - \int_{A(n)^c} (f(x) - f_n(x))dx$$

makes the left–hand side

$$\int_R |f_n(x) - f(x)|\, dx = 2\int_R g_n(dx).$$

d) $|F_n(x) - F(x)| = |\int_{(-\infty,x]} (f_n(y) - f(y))dy|$

$\leq \int_{(-\infty,x]} |f_n(y) - f(y)| dy \leq \int_R |f(y) - f_n(y)| dy$

$= 2\int_R g_n(y)dy.$

The right–hand side of this inequality converges to 0 independently of x.

Exercise 5: The Student T (family of distributions indexed by the parameter n > 0) has corresponding density

$$f_n(x) = \left[\frac{\Gamma(n/2 + 1/2)}{\Gamma(n/2)\sqrt{(n\pi)}} \right] (1 + x^2/n)^{-(n+1)/2} \quad \text{for x real.}$$

It is obvious that the factor involving x^2 converges to $e^{-x^2/2}$. Use Stirling's approximation (Lesson 4, Part II) for the gamma functions to show that the factor in large parentheses converges to $1/\sqrt{(2\pi)}$.

Application of Scheffé's theorem in the exercise above will show that the limiting distribution of "T" is normal; specifically,

$$T = T_n \overset{D}{\to} Z \quad \text{or} \quad P(T \leq x) \to P(Z \leq x)$$

where Z is the standard normal RV. It should also be pointed out that this example and, indeed the theorem, are valid for n increasing to ∞ thru any set of values not just integers. The next example shows that the hypotheses in Scheffé's theorem are sufficient but not necessary.

Example: Let $f_n(x) = (1 - \cos(2n\pi x)) \cdot I\{0 < x < 1\}$. Then,

$$\begin{aligned}
F_n(x) &= 0 & \text{for } x < 0 \\
&= x - (\sin(2n\pi x))/2n\pi & \text{for } 0 \leq x < 1 \\
&= 1 & \text{for } 1 \leq x .
\end{aligned}$$

The sequence $\{f_n(x)\}$ does not converge for x ε (0,1). But,

$$\begin{aligned}
& 0 & \text{for } x < 0 \\
F_n(x) \to \; & x & \text{for } 0 \leq x < 1 \\
& 1 & \text{for } 1 \leq x.
\end{aligned}$$

This is the CDF for X uniformly distributed on the interval [0,1].

For a proof of the next theorem, we refer to Gnedenko and Kolmogorov, 1954. The Glivenko–Cantelli Theorem is obtained when $\{F_n\}$ is the sequence of empirical CDFs:

$X_1, X_2, X_3, \cdots, X_n$ are IID with CDF F and $F_n(x) = \#(X_i \leq x)/n$. This result is used much in "non–parametric" statistics.

Theorem: *(Pólya) If the sequence of DF $\{F_n\}$ converges completely to a DF F which is continuous, then the convergence is uniform:* $\lim\limits_{n \to +\infty} \sup\limits_{x \varepsilon R} |F_n(x) - F(x)| = 0.$

Example: a) Let $\{q_n\}$ be a sequence of positive rational numbers converging to $\lambda > 0$. The following are all continuous CDFs.

$$F_n(x) = (1 - \exp(-q_n x)) \cdot I\{0 \leq x < +\infty\},$$

$$F(x) = (1 - \exp(-\lambda x)) \cdot I\{0 \leq x < +\infty\}.$$

Since $F_n \to F$ pointwise, $F_n \to F$ uniformly.

b)   For $n = 1(1)\infty$, $F_n(x) = I\{1 - 1/n \leq x < +\infty\}$. Let

$F(x) = I\{1 \leq x < \infty\}$; $F_n \overset{C}{\to} F$ but $\sup\limits_{x \varepsilon R} |F_n(x) - F(x)| = 1$

so that the convergence is not uniform. Of course, F is not continuous.

Exercise 6: Let $Q = \{q_n\}$ be an indexing of all rational numbers. Then $p(q_n) = 6/\pi^2 n^2$ defines a PDF on Q and for $x \varepsilon R$,

$$F(x) = \sum_{q_n \leq x} p(q_n) \text{ is a CDF.}$$

Show that $D(F) = R - C(F)$ is also dense in R.

Exercise 7: For $n = 1(1)\infty$, let $F_n(x) = F(x + n)$ where F is a DF with a jump point x_0: for all $\varepsilon > 0$, $F(x_0 - \varepsilon) < F(x_0 + \varepsilon)$. Show that: $F(-\infty) < F(+\infty)$; F_n converges weakly but not completely to F ; F is not left–hand continuous .

LESSON 10. CONVERGENCE OF SEQUENCES OF INTEGRALS

In this lesson, we will specialize the results of the previous lesson to consider sequences of integrals like $\int g(x)dF_n(x)$ as opposed to the convergence theorems of Lesson 11, Part II, which dealt with integrals like $\int g_n(x)dF(x)$. The results to follow were all inspired by and usually named for, two early discoverers, Helly and Bray.

Theorem: *(Weak compactness). Every sequence of DF contains at least one subsequence which converges weakly to a DF.*

Proof: a) In virtue of the "dense" theorem of the previous lesson, it suffices to show that there exists a subsequence converging on a dense set of R, in particular, the rational numbers $Q = \{q_n\}$.

b)   Let $\{F_n\}$ be the sequence of DFs. The sequence of bounded real numbers $\{F_n(q_1) : n = 1(1)\infty\}$ contains a convergent subsequence. (We have used this Bolzano–Weierstrass property of real numbers in other lessons.) Reduce $\{F_n\}$ to the subsequence for which $F_n(q_1)$ converges. Then the sequence $\{F_n(q_2)\}$ has a convergent subsequence which induces a further sub– sequence of $\{F_n\}$. Continue in this way, obtaining a $\cdots$ subsubsequence which converges for all points in Q.

Example: Let Q^+ be the set of positive rational numbers, say $\{q_n\}$; let the sequence be $F_n(x) = (1 - \exp(-q_n x)) \cdot I_{[0,\infty)}(x)$.

If $\{q_{n'}\}$ is a subsequence converging to $\lambda > 0$, then

$$F_{n'}(x) = (1 - \exp(-q_{n'}x)) \cdot I_{[0,\infty)}(x) \overset{c}{\to} F(x)$$

$$= (1 - \exp(-\lambda x))I_{[0,\infty)}(x)$$

which is a continuous (exponential) distribution. However, if

$\lambda = 0$, $F_{n'}(x) \to 0$ for all $x \in R$ so that $F_{n'} \overset{W}{\to} G$ which is identically 0. This $\{F_{n'}\}$ does not converge completely to G.

Corollary: *If every weakly convergent subsequence of a sequence of DFs $\{F_n\}$ converges to the same F, then $F_n \overset{W}{\to} F$.*

Proof: Otherwise, there is some $a,b \in C(F)$ such that

$$r_n = F_n(b) - F_n(a) \text{ does not converge to } r = F(b) - F(a).$$

Then some subsequence $\{F_{n'}\}$ converges weakly. By hypothesis $r_{n'} = F_{n'}(b) - F_{n'}(a) \to r$ and by assumption $r_{n'} \to$ not r. This contradiction means that the assumption must be false.

Exercise 1: Justify the sentence in quotation marks in the proof above. Hint: Bolzano–Weierstrass can be used again.

Lemma I: *Let the sequence of DF $\{F_n\}$ converge weakly to the DF F. Let g be a continuous function mapping R into R. Then for all $a < b$ such that $F_n(a) \to F(a)$ and $F_n(b) \to F(b)$,*

$$\lim_{n \to \infty} \int_{(a,b]} g(x)dF_n(x) = \int_{(a,b]} g(x)dF(x).$$

Proof: a) Let $a = y_{0,m} < y_{1,m} < \cdots < y_{k_m,m} = b$, $m = 1(1)\infty$, be subdivisions of [a,b] such that $y_{j,m} \in C(F)$ for $j = 0(1)m$ and

$$\sup_{0 \le j \le k_m} (y_{j,m} - y_{j-1,m}) \to 0 \text{ as } m \to \infty.$$

Let $g_m(x) = \sum_{j=1}^{k_m} g(y_{j.m}) \cdot I\{y_{j-j,m} < x \le y_{j,m}\}$. For each $x \in [a,b]$, there is a $j \ge 1$ such that $y_{j-1} < x \le y_{j,m}$; then

$$g_m(x) = g(y_{j,m}) \text{ so that } |g(x) - g_m(x)| \to 0 \text{ as } m \to \infty.$$

Since [a,b] is closed and bounded, the convergence is in fact uniform:

$$S_m = \sup_{a \le x \le b} |g(x) - g_m(x)| \to 0 \text{ as } m \to \infty.$$

b)   It follows that

$$\left| \int_{(a,b]} g_m(x)dF_n(x) - \int_{(a,b]} g(x)dF_n(x) \right| \le S_m(F_n(b) - F_n(a))$$

and

$$\left| \int_{(a,b]} g_m(x)dF(x) - \int_{(a,b]} g(x)dF(x) \right| \le S_m(F(b) - F(a)).$$

Since these DFs are bounded (by 1), the right–hand side of these inequalities tends to 0 as m → ∞; or, for the ordinary Riemann–Stieltjes integrals, m → ∞ implies

$$\int_{(a,b]} g_m dF_n \to \int_{(a,b]} g dF_n \text{ and } \int_{(a,b]} g_m dF \to \int_{(a,b]} g dF .$$

c)   Fix a subdivision, m and j. Then as n → ∞,

$$\int_{(y_{j-1,m}, y_{j,m}]} dF_n(x) = F_n(y_{j,m}) - F_n(y_{j-1,m})$$

$$\to F(y_{j,m}) - F(y_{j-1,m}) = \int_{(y_{j-1,m}, y_{j,m}]} dF(x)$$

and

$$\int_{(a,b]} g_m(x)dF_n(x) = \sum_{j=1}^{k_m} g(y_{j,m})(F_n(y_{j-1,m}) - F(y_{j-1,m}))$$

$$\to \sum_{j=1}^{k_m} g(y_{j,m})(F(y_{j-1,m}) - F(y_{j,m})) = \int_{(a,b]} g_m(x)dF(x).$$

d)   Now $\int_{(a,b]} g(x)dF_n(x) - \int_{(a,b]} g(x)dF(x)$ \hfill (1)

$$= \int_{(a,b]} g(x)dF_n(x) - \int_{(a,b]} g_m(x)dF_n(x) \tag{2}$$

$$+ \int_{(a,b]} g_m(x)dF_n(x) - \int_{(a,b]} g_m(x)dF(x) \tag{3}$$

$$+ \int_{(a,b]} g_m(x)dF(x) - \int_{(a,b]} g(x)dF(x). \tag{4}$$

The difference on line (3) tends to 0 as $n \to \infty$ by result c). The differences on lines (2) and (4) tend to 0 as $m \to \infty$ by result b). It follows that the difference on line (1) tends to 0 as $n \to \infty$ and this is equivalent to the conclusion.

The first extension of the lemma is so simple we need only state it: the conclusion of the lemma will hold if
i)   g is a continuous function on the interval [a,b] and
ii)   a,b ∈ C(F).
Then the integrals may be over (a,b) or (a,b] or [a,b) or [a,b]. The second and third extensions (Lemmas II and III below) put additional conditions on g.

Definition: *g belongs to the class of functions $C_0[R]$ on R if g is real valued and continuous with $g(\pm\infty) = 0$. g belongs to the class of functions $C_B[R]$ on R if g is real valued, continuous and bounded.*

Exercise 2:   Show that each $g \in C_0[R]$ is bounded. Hint: first examine the limit conditions $\lim_{x \to \infty} g(x) = 0 = \lim_{x \to -\infty} g(x)$.

Lemma II : *Let $F, F_1, F_2, \cdots$ be DFs with $F_n \overset{w}{\to} F$. Let $g \in C_0(R)$. Then,*

$$\lim_{n \to \infty} \int_R g(x)dF_n(x) = \int_R g(x)dF(x).$$

Proof: By exercise 2, $|g(x)| \le B < \infty$ for all real x. Adding and subtracting appropriate integrals as in the previous proof, we get for any a,b ∈ C(F),

$$\left| \int_R g(x)dF_n(x) - \int_R g(x)dF(x) \right| \tag{1'}$$

$$\leq \left| \int_R g(x)dF_n(x) - \int_{(a,b]} g(x)dF_n(x) \right| \tag{2'}$$

$$+ \left| \int_{(a,b]} g(x)dF_n(x) - \int_{(a,b]} g(x)dF(x) \right| \tag{3'}$$

$$+ \left| \int_{(a,b]} g(x)dF(x) - \int_R g(x)dF(x) \right|. \tag{4'}$$

The absolute value on line (2') equals $\left| \int_{(a,b]^c} g(x)dF_n(x) \right|$

which is bounded by $\sup\limits_{x \notin (a,b]} |g(x)|$; since $g(\pm\infty) = 0$, this

sup $\to 0$ as $a \to -\infty$ and $b \to +\infty$. Similarly, the absolute value on line (4') tends to 0 as $a \to -\infty$ and $b \to +\infty$.

Finally, $\left| \int_{(a,b]} g(x)dF_n(x) - \int_{(a,b]} g(x)dF(x) \right|$ on line (3') will

tend to 0 as $n \to \infty$ by Lemma I. The conclusion is equivalent to the absolute value on line (1') having limit 0 which it now does.

Exercise 3: Give the details for the limit of line (4').

Lemma III : *Let* $F, F_1, F_2, \cdots$ *be DFs such that* $F_n \overset{c}{\to} F$. *Let* $g \in C_B[R]$. *Then,*

$$\lim_{n \to \infty} \int_R g(x)dF_n(x) = \int_R g(x)dF(x).$$

Proof: By hypothesis, there is a $K < \infty$ such that $|g(x)| \leq K$ for all real x. Since these DFs are also bounded, this condition on g guarantees that all the integrals exist and are finite. We get lines $1', 2', 3', 4'$ as in the previous proof. As $n \to \infty$, the quantity in line (3') tends to zero by Lemma I. The quantity in line (2') is bounded by

$$K(F_n(a) - F_n(-\infty) + F_n(+\infty) - F_n(b));$$

as $n \to \infty$, this tends to $K(F(a) - F(-\infty) + F(+\infty) - F(b))$ because of complete convergence. As $a \to -\infty$ and $b \to +\infty$, this last sum also tends to 0. Similarly, the quantity in line (4') tends to 0 as $n \to \infty$, $a \to -\infty$, $b \to \infty$. Thus the quantity in line (1') tends to 0 as $n \to \infty$.

Exercise 4: Fill in the details for the limit of line (4') in the last proof.

Exercise 5: a) Let F, F_1, F_2, $\cdots$ be CDFs with $F_n \overset{W}{\to} F$. Show that for all $g \in C_B[R]$, $\lim_{n \to \infty} \int_R g(x)dF_n(x) = \int_R g(x)dF(x)$.

b)   State and prove a theorem involving $X_n \overset{D}{\to} X$, $g \in C_B[R]$.

The following theorem is a partial converse to Lemma II.

Theorem: *Let $\{F_n\}$ be a sequence of DFs. Suppose that for all $g \in C_0[R]$, $\int_R g \, dF_n$ converges. Then there is a DF F such that $F_n \overset{W}{\to} F$.*

Proof: By the first theorem in this lesson, there is a subsequence $\{F_{n'}\}$ converging weakly to some DF F. By Lemma II,

$$\int_R g \, dF_{n'} \to \int_R g \, dF.$$

But all subsequences of the convergent sequence $\int_R g \, dF_n$ must converge to the same limit. Hence, $\int_R g \, dF_n \to \int_R g \, dF.$

INE–Exercises: Let $F, F_1, F_2, F_3, \cdots$ be DFs on R; let g be continuous on R Then g is *uniformly integrable wrt* $\{F_n\}$ iff

$$\lim_{a \to \infty} \int_{\{|x| > a\}} |g(x)| \, dF_n(x) = 0 \text{ uniformly in } n.$$

1. Suppose that F_n are CDF of RVs X_n and $g(x) = |x|^r$, $0 < r < \infty$. Show that g is uniformly integrable wrt $\{F_n\}$ iff

$$\sup_{n \geq 1} \int_{\{|x_n| > a\}} g(x_n(\omega)) dP(\omega) \to 0 \text{ as } a \to \infty.$$

2. Suppose that DFs $F_n \overset{W}{\to} F$. Show that g is uniformly integrable wrt $\{F_n\}$ iff $\int |g| dF_n \to \int |g| dF$.

3) Let $X, X_1, X_2, X_3, \cdots$ be the corresponding RVs. Suppose that $X_n \overset{P}{\to} X$ and for some $0 < r < \infty$, $E[|X_n|^r] < \infty$ for all $n \geq 1$. Show that $E[|X_n|^r] \to E[|X|^r]$ iff $g(x) = |x|^r$ is uniformly integrable.

4. Show that if $\lim \int_{-\infty}^{\infty} g \, dF_n = \int_{-\infty}^{\infty} g \, dF$ for all $g \in C_B[R]$, then $F_n \overset{C}{\to} F$.

LESSON 11. ON THE SUM OF RANDOM VARIABLES

In "statistics", an experiment yields outcomes $x_1, x_2, x_3, \cdots$; a common model is that these are observed values of random variables $X_1, X_2, X_3, \cdots$. As in the simple cases of testing and estimation in Part I, what becomes meaningful are functions of the observations; such (measurable) functions are called *statistics*. One of the most commonly used functions is the simple sum

$$X_1 + X_2 + \cdots + X_n$$

(or mean $(X_1 + \cdots + X_n)/n$). In this lesson, we examine general CDFs and PDFs of such sums, mostly when $n = 2$.

Example: Let a joint distribution of (X, Y) be given by the joint density:

Y			
2	6/36	9/36	3/36
1	2/36	1/36	6/36
0	2/36	4/36	3/36
	0	1	2 X

The distribution of $Z = X + Y$ is:

z P(Z = z)

0 P(X = 0, Y = 0) = 2/36

1 P(X = 0, Y = 1)+P(X = 1, Y = 0) = 2/36 + 4/36

2 P(X = 0, Y = 2)+P(X = 1, Y = 1)+

    P(X = 2, Y = 0) = 6/36 + 1/36 + 3/36

3 P(X = 1, Y = 2)+P(X = 2, Y = 1) = 9/36 + 6/36

4 P(X = 2, Y = 2) = 3/36

In each case, $P(Z = z)$ is the sum of the probabilities $P(X = x, Y = y)$ over values (x,y) such that $x + y = z$. Symbolically,

$$P(Z = z) = \sum_{x+y\,=\,z} P(X = x, Y = y)$$

or, with obvious substitutions,

$$P(Z = z) = \sum_{x} P(X = x, Y = z - x)$$

$$= \sum_{y} P(X = z - y, Y = y) .$$ (1)

Exercise 1: Let the distribution be given by the table:

Y			
4	1/9	1/6	1/18
3	1/6	1/4	1/12
0	1/18	1/12	1/36
	0	2	3 X

a)   Show that X and Y are independent; that is, show that

$$(X = x, Y = y) = P(X = x) \cdot P(Y = y) \text{ for all } x, y .$$

b)   Find the density of $Z = X + Y$.

Example: Let (X,Y) be a pair of continuous type Rvs with joint density g; let $W = X + Y$. Then

$$H(z) = P(Z \le w) = P(X + Y \le w) = P(X \le w - Y)$$

$$= \int_{-\infty}^{\infty} \left[\int_{-\infty}^{w-y} g(x,y) \, dx \right] dy$$

is the "volume" over the region:

Because of the continuity, the PDF is

$$H'(w) = \int_{-\infty}^{\infty} g(w - y, y) \, dy = \int_{-\infty}^{\infty} g(x, w - x) \, dx$$

in complete analogy with the sums in (1).

As you may have noted in exercise 1, some simplification occurs when X and Y are independent, say

$$f(x,y) = f_X(x) \cdot f_Y(y).$$

Then (1) becomes

$$P(Z = z) = \sum_x f_X(x) \cdot f_Y(z - x) =$$

$$\sum_y f_X(z - y) \cdot f_Y(y) . \qquad (2)$$

If, correspondingly, $g(x,y) = g_X(x) \cdot g_Y(y)$, then

$$H'(w) = \int_{-\infty}^{\infty} g_X(w - y) \cdot g_Y(y) \, dy =$$

$$\int_{-\infty}^{\infty} g_X(x) \cdot g_Y(w - x) \, dx . \qquad (3)$$

These forms are used in other areas of mathematics and are referred to as the *convolution* of the densities. A common symbol used is "*" ; thus,

$$f_X * f_Y(z) = \sum_x f_X(x) \cdot f_Y(z - x)$$

$$g_X * g_Y(w) = \int_{-\infty}^{\infty} g_X(x) \cdot g_Y(w - x) \, dx.$$

From (2) and (3), we see that

$$f_X * f_Y = f_Y * f_X \quad \text{and} \quad g_X * g_Y = g_Y * g_X .$$

Moreover, when X_1, X_2, X_3 are msi, from

$$P(X_1 + X_2 + X_3 \le z) = P((X_1 + X_2) + X_3 \le z) =$$

$$P(X_1 + (X_2 + X_3) \le z),$$

it follows that

$$(f_{X_1} * f_{X_2}) * f_{X_3} = f_{X_1} * (f_{X_2} * f_{X_3}) ,$$

or, correspondingly,

$$(g_{X_1} * g_{X_2}) * f_{X_3} = g_{X_1} * (g_{X_2} * g_{X_3}), \text{ etc.}$$

Of course, this can be extended to an arbitrary finite number of independent RVs by induction.

Example: To simplify notation, let X have density f, Y have density g; suppose that X and Y are independent.

a)   Let $f(x) = g(x) = \lambda e^{-\lambda x} I\{0 < x < \infty\}$. Because $g(y) = 0$ for $y \leq 0$,

$$f_* g(x) = \int_{-\infty}^{\infty} f(x - y) g(y) \, dy = \int_{0}^{\infty} f(x - y) \lambda e^{-\lambda y} \, dy \, .$$

Moreover, $f(x - y)$ will be zero for $x \leq y$ so that the integral is further reduced to $\int_{0}^{x} \lambda e^{-\lambda(x - y)} \lambda e^{-\lambda y} \, dy$

which equals $\lambda^2 x e^{-\lambda x} I\{0 < x < \infty\}$.

b)   For $f(x) = g(x) = I_{(0,1)}(x)$,

$$f_* g(x) = \int_{-\infty}^{\infty} f(x-y) g(y) \, dy = \int_{0}^{1} f(x-y) dy$$

since $g(y) = 0$ for $y \leq 0$ or $y \geq 1$. Since, $f(x-y) = 0$ for $x \leq y$ or $x - y \geq 1$, the region of non–zero density is the "diamond" determined by four boundaries in the x–y plane:

For $0 < x < 1$, we have $0 < y < x$ and

$$\int_0^1 f(x - y)dy = \int_0^x dy = x.$$

For $1 < x < 2$, $x - 1 < y < 1$ and

$$\int_0^1 f(x - y)dy = \int_{x-1}^1 dy = 2 - x.$$

Hence, $f_* g(x) = x \cdot I\{0 < x < 1\} + (2 - x) \cdot I\{1 < x < 2\}$.

Exercise 2: Let X and Y be IID RVs with the distribution indicated. First find an integral for the CDF of $Z = X + Y$; then find the PDF by differentiation.

a)   X and Y have the standard normal distribution. Hint: integrate $e^{-(z - y)^2/2} \, e^{-y^2/2}$ by completing the square in y.

b)   Another solution for example b) above. X and Y have the uniform distribution on $(0,1)$. Hint: for $0 < z < 1$, use the integral over "A"; for $1 < z < 2$, use the integral over "B":

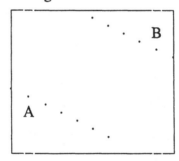

Riemann–Stieltjes theory allows us to represent the CDF of the sum of any two independent RVs in one convolution:

$$P(X + Y \le z) = \int_{-\infty}^{\infty} F_X(z - y) \, dF_Y(y) = F_X * F_Y(z)$$

where F_X and F_Y are the corresponding CDFs. Because the Fubini theorem is still valid, commutativity (and associativity) also hold here:

$$F_{X}*F_{Y} = F_{Y}*F_{X} \text{ , etc.}$$

Exercise 3: Let $F(x) = I\{ 1 < x < \infty\}$,
$G(x) = (1 - e^{-x}) \cdot I\{0 < x < \infty\}$. Show that

$$F_{*}G(x) = (1 - e^{-(x - 1)}) \cdot I\{1 < x < \infty\}.$$

Exercise 4: Show that when F and G are CDFs so is $F_{*}G$; that is, verify montonicity, right–hand continuity, $F_{*}G(+\infty) = 1$, and $F_{*}G(-\infty) = 0$.

An interesting phenomenon occurred in exercise 3: the RV represented by F was discrete and the RV represented by G was continuous, but the sum RV respresented by F*G was continuous. The proof that this is true in general uses slightly different techniques.

Theorem: Let X and Y be independent Rvs; take X to be discrete with CDF F; take Y to be continuous with CDF G such that $G' = g$ a.e. Then

$$Z = X + Y \text{ is continuous with PDF } \int_{-\infty}^{\infty} g(z-y)dF(y).$$

Proof: By earlier remarks, we get

$$H(z) = P(X + Y \le z) = \int_{-\infty}^{\infty} G(z-y)dF(y).$$

Consider points $a \le z_0 \le z_1 \le \cdots \le z_n \le b$. By hypothesis, G is absolutely continuous (Lesson 3) so that for $\varepsilon > 0$, there is a $\delta > 0$, such that

$$\sum_{k=1}^{n} |G(z_k-y) - G(z_{k-1}-y)| < \varepsilon$$

whenever

$$\sum_{k=1}^{n} |(z_k-y) - (z_{k-1}-y)| < \delta.$$

Then

$$\sum_{k=1}^{n} |H(z_k) - H(z_{k-1})|$$

$$\leq \int_{-\infty}^{\infty} \sum |G(z_k-y) - G(z_{k-1}-y)|\, dF(y) < \varepsilon$$

and therefore H' exists. Then

$$H'(z) = \lim_{\delta \to 0} \int_{-\infty}^{\infty} [(G(z+\delta) - G(z))/\delta]\, dF(y)$$

can be evaluated by taking the limit under the integral sign (Lesson 11, Part II):

$$H'(z) = \int_{-\infty}^{\infty} g(z-y)dF(y).$$

Example: Let X and Y be independent RVs and $Z = X + Y$. Take X to be a Bernoulli RV with

$$P(X = 0) = \theta = 1 - P(X = 1);$$

take Y to be uniformly distributed with

$$g(y) = I\{-1 < y < 1\}/2. \text{ Then the PDF}$$

$$h(z) = \int_{-\infty}^{\infty} g(z-y)dF(y) = g(z)\theta + g(z-1)(1-\theta).$$

This will be different from 0 when $-1 < z < 1$ and/or

$$-1 < z-1 < 1;$$

thus,

$$h(z) = \begin{cases} \theta/2 & \text{for } -1 < z < 0 \\ \theta/2 + (1-\theta)/2 & \text{for } 0 < z < 1 \\ (1-\theta)/2 & \text{for } 1 < z < 2 \end{cases} \quad \text{and 0 elsewhere.}$$

Exercise 5: Let X,Y,Z be as in the example above except that Y is uniformly distributed on the interval $(0,1)$. Find the CDF and PDF of Z.

Example: Let X be a normal RV with mean μ, variance σ^2; let Y be uniformly distributed on $(-1,1)$. If X and Y are independent, then the density of $Z = X + Y$ is

$$\int_{-\infty}^{\infty} e^{-(z-y-\mu)^2/2\sigma^2} I_{(-1,1)}(y)\, dy \, / \, \sqrt{(2\pi\sigma)}$$

$$= \int_{-1}^{1} e^{-(z-y-\mu)^2/2\sigma^2} dy \, / \, \sqrt{(2\pi\sigma)}$$

$$= \int_{(z-1-\mu)/\sigma}^{(z+1-\mu)/\sigma} e^{-w^2/2} dw \, / \sqrt{2\pi}$$

which can be evaluated with a table of the standard normal CDF.

Exercise 6: Let X be $N(0,\sigma^2)$; let $P(Y = 0) = 1$; let Z be a RV independent of X with CDF G.

a)   Show that $F_*G \to G$ as $\sigma^2 \to 0$. Then show that

$X + Z \overset{D}{\to} Z$ by using Slutsky's theorem.

b)   Show that as $\sigma^2 \to 0$, $F_X(x) \to \begin{cases} 0 & \text{for } x < 0 \\ 1/2 & \text{for } x = 0 \\ 1 & \text{for } x > 0 \end{cases}$.

Conclude that $X \overset{D}{\to} Y$.

LESSON 12. CHARACTERISTIC FUNCTIONS–I

The characteristic function (CF, defined below) is very useful in many areas of mathematics, not only probability and statistics. Usually, this term is used for what we have called the indicator function and our CF is called a bilateral Fourier transform; needless to say, there is also variation in notation. We begin this lesson with some remarks about the set of complex numbers, $\mathscr{C}$, and complex valued functions.

The imaginary unit is $i = \sqrt{-1}$ (but called j by some). If a,b,c,d are real numbers then $z = a + ib$ and $w = c + id$ are complex numbers with:

sum $z + w = (a + c) + i(b + d)$;

product $zw = (ac - bd) + i(bc + ad)$;

quotient $z/w = (ac + bd)/(c^2 + d^2) + i(bc - ad)/(c^2 + d^2)$

when $c^2 + d^2 > 0$;

conjugate $\bar{z} = \overline{(a + ib)} = a - ib$; z is real iff $z = \bar{z}$;

norm (modulus, absolute value)

$$\|z\| = (a^2 + b^2)^{1/2} = |z|; \quad \|z\|^2 = \bar{z}z;$$

exponential $e^z = e^{a + ib} = e^a \cdot e^{ib} =$

$e^a\{\cos(b) + i\sin(b)\} = \exp(z)$;

exp-product $e^z \cdot e^w = e^{(a + c) + i(b + d)} = e^w \cdot e^z$.

By taking the limit processes to be $\{\lim_t a(t)\} + i\{\lim_t b(t)\}$ for the complex function $z(t) = a(t) + ib(t)$, one can establish most of the rules of analysis and we will assume these as needed. The major difficulties are in some integrations which we do not need herein.

For X,Y real valued measurable functions on Ω of a probability space $[\Omega, \mathscr{B}, P]$, $Z(\omega) = Y(\omega) + iX(\omega)$ is a *complex valued random variable*. If t is real, $e^{itX(\omega)} = \cos(tX(\omega)) + i\sin(tX(\omega))$. Obviously, $\|\exp(itX(\omega))\| = 1$ so that this function is integrable and the following makes sense.

Definition: *The real random variable X has characteristic function (CF)* φ_X *mapping R into $\mathscr{C}$ such that*

$$\varphi_X(t) = E[e^{itX}] = E[cos(tx)] + iE[sin(tX)].$$

A RV X for $[\Omega, \mathscr{B}, P]$ induces the space $[R, \mathscr{B}_1, P_X]$ with $X^{-1}(B_1) \in \mathscr{B}$ and $P_X(B_1) = P(X^{-1}(B_1))$ for all $B_1 \in \mathscr{B}$. The corresponding CDF is

$$F_X(x) = P_X(X \le x) = P(X^{-1}(-\infty, x]).$$

Then φ_X has the equivalent representations

$$\int_\Omega e^{itX(\omega)} dP(\omega) = \int_R e^{itx} dP_X(x) = \int_R e^{itx} dF_X(x).$$

If X is continuous with $F_X' = f_X$, $\varphi_X(t) = \int_{-\infty}^{\infty} e^{itx} f_X(x)\, dx$.

If X is discrete with PDF f_X and support A,

$$\varphi_X(t) = \sum_{x \in A} e^{itx} f_X(x).$$

Example: a) When X is binomial with parameters n and θ, the support A = $\{0,1,2,\cdots,n\}$ so

$$\varphi_X(t) = \sum_{x=0}^{n} e^{itx} \begin{bmatrix} n \\ x \end{bmatrix} \theta^x (1 - \theta)^{n-x} = [\theta e^{it} + 1 - \theta]^n.$$

The last equality is a form of the binomial theorem.

b)   A Poisson RV X has support A = $\{0,1,2,3\cdots\}$. The CF is

$$\varphi_X(t) = \sum_{x=0}^{\infty} e^{itx} e^{-\lambda} \lambda^x / x! =$$

$$e^{-\lambda} \sum_{x=0}^{\infty} (\lambda e^{it})^x / x! = e^{-\lambda} (e^{\lambda e^{it}}).$$

c)   An exponential RV with PDF

$$f_X(x) = \alpha e^{-\alpha x} \cdot I\{0 \le x < \infty\} \text{ has } \alpha > 0$$

and

$$\varphi_X(t) = \int_0^\infty e^{itx} \alpha e^{-\alpha x}\, dx = \int_0^\infty \alpha e^{-(\alpha - it)x}\, dx \qquad (1)$$

$$= \int_0^\infty \alpha e^{-\alpha x} \cos(tx)\, dx + i \int_0^\infty \alpha e^{-\alpha x} \sin(tx)\, dx. \qquad (2)$$

In many cases, one can treat the complex parameters in intergals almost as if they were real, as we now indicate (with the next exercise). The integral on the right of (1) is

$$-\alpha e^{-(\alpha - it)x}/(\alpha - it)\Big]_0^\infty = \alpha/(\alpha - it) \quad \text{because}$$

$$\left| -\alpha e^{-\alpha x} e^{itx}/(\alpha - it) \right| \le \left| \alpha/(\alpha - it) \right| e^{-\alpha x}$$

which $\to 0$ as $x \to \infty$. The integrals in (2) can be evaluated using integration by parts (tables) to get

$$\alpha(\alpha + it)/(\alpha^2 + t^2).$$

Rationalizing the numerator in this fraction will show that the two answers are equal.

Exercise 1: Complete the algebra in parts a) and c) of the example above.

Exercise 2: Find the CF when X is distributed as :
   a)   a gamma RV
   b)   a uniform RV on (–b,b)
   c)   a RV degenerate at b

   d) $N(0, \sigma^2)$. Hint: to integrate $e^{itx} e^{-x^2/2\sigma^2}$, complete the square in x as if "it" were real.

The fact that the CF of a distribution always exists and is bounded by 1 is apparent from the discussion before the definition; also, $\varphi_X(0) = 1$. As we will see, the most useful property of the CF is the one–to–one correspondence between a CDF and its CF. This allows us to work with whichever one is more amenable: given one, we can find (theoretically) the other. This is expressed in the

Theorem: *(Uniqueness) Two CDFs F_1 and F_2 are identical iff their corresponding CFs φ_1 and φ_2 are identical.*

Proof: i) If $F_1 = F_2$, then the definition of the CF makes $\varphi_1 = \varphi_2$.

ii) For the converse, it suffices to show that a CDF F is uniquely determined by its CF $\varphi_F(t) = \int_{-\infty}^{\infty} e^{itx} \, dF(x)$. First note that

$$e^{-ist}\varphi_F(s) = \int_{-\infty}^{\infty} e^{is(x - t)} \, dF(x) \, .$$

Now let the CDF G have CF $\varphi_G(u) = \int_{-\infty}^{\infty} e^{iuy} \, dG(y) \, .$
Then,

$$\int_{-\infty}^{\infty} e^{-ist}\varphi_F(s) \, dG(s) = \int_{-\infty}^{\infty} \varphi_G(x - t) \, dF(x) \, . \qquad (*)$$

If G is the CDF of the normal distribution with mean 0 and variance $1/h$, then, from exercise 2, its CF is $e^{-t^2/2h}$.
Substitution makes (*) equal

$$\int_{-\infty}^{\infty} e^{-ist}\varphi_F(s) \, e^{-hs^2/2} \, ds \, / \, \sqrt{(2\pi/h)} = \int_{-\infty}^{\infty} e^{-(x-t)^2/2h} \, dF(x),$$

equivalently

$$\int_{-\infty}^{\infty} e^{-ist}\varphi_F(t) \, e^{-hs^2/2} \, ds \, / \, 2\pi = \int_{-\infty}^{\infty} \frac{e^{-(x-t)^2/2h}}{\sqrt{(2\pi h)}} \, dF(x) \, .$$

From exercise 6, Lesson 11, we see that the right–hand side is the density of the sum of X and an independent normal RV, say N, with mean 0 and variance h. Since $N \overset{P}{\to} 0$ as $h \to 0$, it follows that $T = X + N \overset{D}{\to} X$ as $h \to 0$. This implies that as $h \to 0$, the CDF

$$\int_{-\infty}^{W}\left[\int_{-\infty}^{\infty}e^{-ist}\varphi_F(s)e^{-hs^2/2}\ ds\ /\ 2\pi\right]dt$$

converges to F(w); in this way, the CF φ_F determines F.

Exercise 3: Complete the argument for (*) above.

By taking other distributions for G, one may obtain other "inversion" formulas (see Feller 1971).

Example: With G the uniform distribution on (–T,T) and $|\varphi_F|$ integrable, it can be shown that

$$F'(x) = \int_{-\infty}^{\infty}e^{-itx}\varphi_F(t)\ dt/2\pi\ .$$

In particular, since

$$\int_{-\infty}^{\infty}e^{-|t|}dt = \int_{-\infty}^{0}e^{t}dt + \int_{0}^{\infty}e^{-t}dt = 2\ ,$$

the corresponding PDF is $\int_{-\infty}^{\infty}e^{-itx}e^{-|t|}dt/2\pi$

$$= \int_{-\infty}^{0}e^{-itx\ +\ t}dt/2\pi\ +\ \int_{0}^{\infty}e^{-itx\ -\ t}dt/2\pi$$

$$= e^{t(1-ix)}/(1 - ix)/2\pi\ \Big]_{t=\ -\infty}^{t=\ 0}\ +$$

$$e^{t(-1-ix)}/(-1 - ix)/2\pi\ \Big]_{t=\ 0}^{t=\ \infty}$$

$$= 1/(1 - ix)2\pi\ +\ 1/(1 + ix)2\pi = 1/\pi(1 + x^2).$$

This is the Cauchy PDF on (–∞,∞). This integration also gives

$$\int_{-\infty}^{\infty}e^{itx}e^{-|x|}dx/2 = 1/(1 + t^2)$$

as the CF of the (Laplace) PDF $\{e^{-|x|}/2\} \cdot I\{-\infty < x < \infty\}$.

The import of the term "identical" in the uniqueness theorem is emphasized by the following which shows that if two CFs are identical only on a finite interval of R, the distributions need not be equivalent.

Exercise 4: The first RV has PDF given by

$$P(X_1 = 0) = 1/2; \; P(X_1 = (2n-1)\pi) = 2/(2n-1)^2 \pi^2$$

for $n = \pm 1, \pm 2, \pm 3, \cdots$. The second RV has PDF given by
$$f_X(x) = (1 - \cos x)/x^2 \cdot I\{x \neq 0\}.$$

a)   Show that

$$\varphi_{X_1}(t) = 1/2 + (4/\pi^2)\sum_{n=1}^{\infty} \{\cos(2n-1)t\pi\}/(2n-1)^2.$$

b)   Show that $\varphi_X(t) = 1 - |t|$ for $|t| \leq 1$.
$$\qquad\qquad\qquad =0 \qquad\qquad \text{for } |t| > 1$$

c)   Show that $\varphi_{X_1}(t) = \varphi_X(t)$ for $|t| \leq 1$.

In the rest of this lesson, we will examine some properties of the CF φ of a RV X with CDF F.

Theorem: *The CF φ is uniformly continuous on R.*

Proof: For s and t real,

$$|\varphi(t) - \varphi(s)| \leq \int_R |e^{itx} - e^{isx}| \, dF(x) \leq \int_R |e^{i(t-s)x} - 1| \, dF(x).$$

Since $|e^{i(t-s)x} - 1| \leq 2$ which is an integrable function, we may take the limit under the intergal sign (Lesson 11, Part II):

$$\lim_{t-s \to 0} \int_R |e^{i(t-s)x} - 1| \, dF(x) = \int_R \lim_{t-s \to 0} |e^{i(t-s)x} - 1| \, dF(x)$$

which is 0 without any other conditions on s and t.

Having discovered the continuity, it is natural to ask about differentiability; the following theorems are not quite converses of each other. We use the notation $\varphi^{(k)}(0)$ for $\varphi^{(k)}(t)$ at $t = 0$; as always,

$$\varphi^{(0)}(t) = \varphi(t) \text{ and } \varphi' = \varphi^{(1)}, \varphi'' = \varphi^{(2)}, \text{ etc.}$$

Theorem: *If the RV X has moments up to order K (a positive integer), then its CF φ has derivatives up to order K.*

Proof:

$$\frac{\varphi(t + h) - \varphi(t)}{h}$$

$$\leq \int_R \left| \frac{e^{i(t+h)x} - e^{itx}}{h} \right| dF(x)$$

$$\leq \int_R \left| \frac{e^{ihx} - 1}{h} \right| dF(x).$$

The (L'Hopital) limit of the integrand in the last integral is $|x|$ so that we may take the limit under the integral sign when $\int_R |x| dF(x)$ is finite. Then $\varphi'(t) = \int_R ix\, e^{itx}\, dF(x)$. Similarly, when $\int_R |x|^2 dF(x)$ is finite,

$$\varphi''(t) = \lim_{h \to 0} (\varphi'(t+h) - \varphi'(t))/h = \int_R (ix)^2 e^{itx}\, dF(x).$$

The proof is completed by induction.

Theorem: *If the CF φ has a finite derivative of even order 2m at $t = 0$, then all moments of order up to 2m of X are also finite.*

Proof: Under the hypothesis, Taylor's formula from analysis yields:

$$\varphi(t) = \varphi(0) + \varphi^{(1)}(0)t + \varphi^{(2)}(0)t^2/2! + \cdots +$$

$$\varphi^{(2m)}(0)t^{2m}/(2m)! + o(|t|^{2m}),$$

where the last "remainder" term means that

$\lim_{t \to 0} o(|t|^{2m}|)/|t|^{2m} = 0$. For m = 1,

$$\varphi(t) - \varphi(0) = \varphi'(0)t + \varphi''(0)t^2/2 + o(|t^2|)$$

so that

$$\varphi(t) - 2\varphi(0) + \varphi(-t) = \varphi''(0)t^2 + o(|t|^2). \qquad (1)$$

By L'Hopital again,

$$\int_R x^2 \, dF(x) = \int_R 2 \lim_{h \to 0} \frac{1 - \cos(hx)}{h^2} \, dF(x). \qquad (2)$$

By one of Fatou's results (Lesson 11, Part II), this integral is at most

$$\liminf_{h \to 0} \int_R 2 \frac{1 - \cos(hx)}{h^2} \, dF(x). \qquad (3)$$

A little algebra turns $-\int_R 2 \dfrac{1 - \cos(hx)}{h^2} \, dF(x)$ into

$$\int_R \frac{e^{ihx} - 2 + e^{-ihx}}{h^2} \, dF(x)$$

$$= \frac{\varphi(h) - 2\varphi(0) + \varphi(-h)}{h^2}. \qquad (4)$$

By (1), this is $\varphi''(0) + o(|h|^2)/h^2$ with limit $\varphi''(0)$ as $h \to 0$. Putting (2),(3),(4) in proper order yields $E[X^2] = -\varphi''(0)$. The proof is completed by (a much more involved) induction.

Exercise 5: Check the algebra and analysis in (1) above; in particular, show that the sum of the "small oh" functions is again "small oh".

The fact that the proof of the second theorem above works only for even moments is demonstrated by a traditional example wherein $\varphi'(0)$ exists but $E[X]$ does not. The RV X has support $\pm2, \pm3, \pm4, \cdots$; for some constant c,

$$P(X = x) = c/x^2 \ln|x| \quad \text{and} \quad \varphi(t) = \sum_{x=2}^{\infty} 2c\{\cos(tx)\}/x^2 \ln(x).$$

Since $\lim_{x \to \infty} x\{1/x\ln(x)\} = 0$, $\sum_{x=2}^{\infty} \{\sin(tx)\}x/x^2\ln(x)$ is uniformly

convergent and so $\sum_{x=2}^{\infty} 2c\{\sin(tx)\}/x\ln(x)$ is the derivative of

$\varphi(t)$. (See Titchmarsh, 1949, page 6). Then

$\varphi'(0) = 0$. But, $\sum_{x=2}^{\infty} x/x^2\ln(x) = \infty$ so that $E[X]$ does not exist.

Exercise 6:   Continue the example above. Verify that
   a)   such a c exists;
   b) $\varphi(t)$ is as given;
   c)   Titchmarsh's lemma applies.

Let the distribution be such that for some positive integer K, $E[X^K]$ is finite; then Taylor's formula for φ can be written as

$$\varphi(t) = \sum_{k=0}^{K} \varphi^{(k)}(0)t^k/k! = \sum_{k=0}^{K} E[X^k](it)^k/k! + o(|t|^K)$$

$$= \sum_{j=0}^{K-1} E[X^k](it)^k/k! + \theta_K \cdot E[|X|^K]t^K/k!$$

where θ_K is a complex quantity with modulus at most 1. If X has

all moments, $\varphi(t) = \sum_{k=0}^{\infty} E[X^k](it)^k/k!$ for $|t| < 1/\limsup_{k \to \infty}$

$\{E[X^k]/k!\}^{1/k}$ which is the radius of the circle of convergence of the series. It follows that when this limsup is finite knowing all the moments will allow one to determine the distribution, at least in terms of the CF. In general, the moments do not determine the distribution.

Exercise 7: (Traditional) Prove the following proposition including the existence of the integrals. Every distribution in the family indexed by $0 < \alpha < 1$ with PDF

$$f(x) = e^{-x^{1/4}}(1 - \alpha \sin(x^{1/4})) \cdot I\{0 < x < \infty\}/24$$

has $E[X^k] = \int_0^\infty w^{4k+3} e^{-w} dw/6$.

Exercise 8: Let the RV X be normal with mean 0 and variance 1. Let m be a positive integer. Show that:

a) $E[X^k] = \begin{cases} (2m)!/2^m m! & \text{for } k = 2m \\ 0 & \text{for } k = 2m + 1 \end{cases}$. Hint: use

separate inductions for k even, k odd.

b) $\sum_{k=0}^\infty E[X^k](it)^k/k! = e^{-t^2/2}$ and conclude that this is

the CF of X .

LESSON 13. CHARACTERISTIC FUNCTIONS–II

In this lesson, we discuss additional basic properties of CFs and some related variations.

Of course, it is necessary to deal with random vectors and their joint characteristic function; the same kind of argument as for real RVs shows that this function exists for all distributions. Notationally, it is convenient to have the *dot product*:

$$\text{for } t = (t_1, t_2, \cdots, t_n) \text{ and } x = (x_1, x_2, \cdots, x_n) \text{ in } R^n,$$

$$t \cdot x = \sum_{k=1}^{n} t_k x_k .$$

Definition: *Let the random vector $X = (X_1, X_2, \cdots, X_n)$ have real components and CDF F_X. For t and x as above, the (joint) characteristic function of X is*

$$\varphi_X(t) = E[e^{it \cdot X}] = \int_{R^n} e^{it \cdot x} \, dF_X(x)$$

$$= \int_{R^n} e^{i(t_1 x_1 + \cdots + t_n x_n)} dF_X(x_1, x_2, \cdots, x_n).$$

As in the case $n = 1$, the calculations are more familiar when the variables are all discrete or all continuous type; e.g., when $n = 3$, the CF is

$$\sum_{x_1} \sum_{x_2} \sum_{x_3} \exp(it_1 x_1 + it_2 x_2 + it_3 x_3) f(x_1, x_2, x_3) \text{ or}$$

$$\int_{-\infty}^{\infty} \int_{-\infty}^{\infty} \int_{-\infty}^{\infty} \exp(it_1 x_1 + it_2 x_2 + it_3 x_3) \, f(x_1, x_2, x_3) \, dx_1 \, dx_2 \, dx_3$$

where "f" is the corresponding PDF.

Marginal CFs are obtained from the joint CF by setting appropriate $t_k = 0$. Again when $n = 3$,

$$\varphi_{X_1, X_2}(t_1, t_2) = \varphi_{X_1, X_2, X_3}(t_1, t_2, 0)$$

because

$$\int_{R^3} e^{i(t_1 x_2 + t_2 x_2)} dF_{X_1, X_2, X_3}(x_1, x_2, x_3)$$

$$= \int_{R^2} e^{i(t_1 x_1 + t_2 x_2)} dF_{X_1, X_2}(x_1, x_2).$$

Similarly, $\varphi_{X_2}(t_2) = \varphi_{X_1, X_2, X_3}(0, t_2, 0)$. Etc.

The conclusions of the next theorem are often presented individually as if they were unrelated results.

Theorem: *Let $X = (X_1, X_2)$ have CDF F and CF φ ; let the marginal CFs be φ_{X_1} and φ_{X_2}. Let c and d be constants. Let the real RV $W = cX_1 + dX_2$. Then,*

a) $\varphi_W(t) = \varphi(tc, td)$;

b) *so that knowing the distribution of all linear combinations W allows us to calculate the bivariate φ and, of course, visa-versa;*

c) *in particular,* $\varphi_{X_1 + X_2}(t) = \varphi(t, t)$;

d) *if X_1 and X_2 are independent,* $\varphi(t_1, t_2) = \varphi_{X_1}(t_1) \cdot \varphi_{X_2}(t_2)$ *and* $\varphi_W(t) = \varphi_{X_1}(tc) \cdot \varphi_{X_2}(td)$;

e) *if $X_2 = 1$ a.s. ,* $\varphi_W(t) = \varphi_{cX_1 + d}(t) = \varphi_{X_1}(tc) e^{itd}$.

Proof: a) and b) The relation

$$\varphi_W(t) = E[e^{it(cX_1 + dX_2)}]$$

$$= E[e^{itcX_1 + itdX_2}] = \varphi(tc, td)$$

is always true so that the bivariate distribution is determined by the distribution of its half–planes.

c) is obvious.

d) $E[e^{it_1 X_1 + it_2 X_2}] = E[e^{it_1 X_1}] \cdot E[e^{it_2 X_2}]$ because (Lesson 13, Part II) the product rule for expectation holds for functions of independent RVs. When $t_1 = tc$ and $t_2 = td$,

$$E[e^{it(cX_1 + dX_2)}] = \varphi_{X_1}(tc) \cdot \varphi_{X_2}(td).$$

e)   When $X_2 = 1$ a.s. , $\varphi_{X_2}(t_2) = E[e^{it_2 X_2}] = e^{it_2 d}$. Then,

$W = cX_1 + d$ a.s. so

$$\varphi_W(t) = E[e^{it(cX_1 + d)}] = \varphi_{X_1}(tc) \cdot \varphi_{X_2}(td) = \varphi_{X_1}(ct)e^{itd}.$$

(Recall that a constant RV is independent of any RV except itself.)

Exercise 1: Let μ be a real number and let σ be a positive real number. Let Z be the standard normal RV. Find the density function of $W = \sigma Z + \mu$. Hint: first find the CF; then use the uniqueness theorem.

Parts a and b) of the theorem imply that there is a corresponding uniqueness theorem for bivariate CFs. From this it follows that part d) has the following partial converse.

Theorem: *Suppose that the CF of $X = (X_1, X_2)$ is factorable as the product of two independent functions:*

$$\varphi(t_1, t_2) = \alpha(t_1) \cdot \beta(t_2).$$

Then X_1 and X_2 are independent random variables.

Proof: The marginal CFs are

$$\varphi_{X_1}(t_1) = \varphi(t_1, 0) = \alpha(t_1) \cdot \beta(0)$$

and

$$\varphi_{X_2}(t_2) = \varphi(0, t_2) = \alpha(0) \cdot \beta(t_2) \; ;$$

also, $1 = \varphi(0,0) = \alpha(0) \cdot \beta(0)$. It follows that

$$\varphi(t_1, t_2) = [\varphi_{X_1}(t_1)/\beta(0)] \cdot [\varphi_{X_2}(t_2)/\alpha(0)]$$

$$= \varphi_{X_1}(t_1) \cdot \varphi_{X_2}(t_2) \,.$$

Applying a uniqueness theorem to both sides yields

$$F_{X_1,X_2} = F_{X_1} \cdot F_{X_2}$$

as desired.

By appropriate inductions, we obtain the following generalities. Let the components of $X = (X_1, X_2, \cdots, X_n)$ be real RVs with joint CDF F_X, CF φ_X,

density f_X (when extant)

marginal CDFs F_{X_k}, CFs φ_{X_k},

densities f_{X_k} (when extant), $k = 1(1)n$.

The components are *mutually stochastically independent* (msi) iff

$$F_X = F_{X_1} \cdot F_{X_2} \cdots F_{X_n} \,,$$

or

$$\varphi_X = \varphi_{X_1} \cdot \varphi_{X_2} \cdots \varphi_{X_n} \,,$$

or

$$f_X = f_{X_1} \cdot f_{X_2} \cdots f_{X_3} \quad \text{a.s.}$$

The components are *identically distributed* iff

$$F_{X_1} = F_{X_2} = \cdots = F_{X_n}$$

or

$$\varphi_{X_1} = \varphi_{X_2} = \cdots = \varphi_{X_n} \,,$$

or

$$f_{X_1} = f_{X_2} = \cdots = f_{X_n} \quad \text{a.s.}$$

If the components $\{X_k\}$ are msi and

$$S_n = X_1 + X_2 + \cdots + X_n \,,$$

then the CF of S_n is $\varphi_{X_1}(t) \cdot \varphi_{X_2}(t) \cdots \varphi_{X_n}(t)$. If the $\{X_k\}$ are also identically distributed, then the CF of S_n is $\left[\varphi_{X_1}(t)\right]^n$.

(Cramér–Wold) Let $c, t_1, t_2, \cdots, t_n$ be real and non–random. For $t = (t_1, t_2, \cdots, t_n)$, the CF φ_X determines and is determined by the relation $\varphi_X(ct) = \varphi_{t \cdot X}(c)$, the CF of an n–dimensional hyper–plane $t \cdot X = Z$.

Example: Let the CF of X be real: $\varphi_X(t) = \overline{\varphi_X(t)}$. Then

$$\varphi_X(t) = E[e^{itX}] = \varphi_X(-t) = E[e^{-itX}] = E[e^{it(-X)}] = \varphi_{-X}(t).$$

Applying the uniqueness theorem to both ends, we see that the CDF of X is equal to the CDF of –X. This is precisely the condition that (the distribution of) X be symmetric. The proof of the converse is a bit more involved.

Theorem: *If the real RV X has a distribution which is symmetric about 0, then its CF is real.*

INE–Proof: Symmetry about 0 means that the CDF F satisfies

$$F(x) = P(X \le x) = P(X \ge -x)$$
$$= 1 - P(X < -x) = 1 - F(-x-0).$$

Hence,

$$\varphi_X(t) = \int_{-\infty}^{\infty} e^{itx} dF(x) = \int_{-\infty}^{\infty} e^{itx} d(1 - F(-x-0)). \qquad (1)$$

By a theorem in Lesson 12, Part II, this is equal to

$$-\int_{-\infty}^{\infty} e^{itx} dF(-x-0). \qquad (2)$$

Under the substitution $x = -w-0$, this becomes

$$\int_{-\infty}^{\infty} e^{it(-w-0)} dF(w) = \int_{-\infty}^{\infty} e^{-itw} dF(w)$$

$$= \varphi_X(-t) = \overline{\varphi(t)}. \qquad (3)$$

Putting (1),(2),(3) together, we get $\varphi_X(t) = \varphi_X(-t)$ as desired.

Exercise 2: Suppose that (the distribution of) X is symmetric about 0, that its CF is φ_X, and that $E[|X|] < \infty$. Prove that $E[X] = 0$ by evaluating $\varphi_X'(0)$.

It is also possible to consider $\varphi_X(z) = E[e^{zX}]$ where $z = a + ib$ and a,b are real. When this function is analytic (that is, has a convergent Taylor series in the complex variable z) in some strip $-\infty < -A \leq a \leq A < \infty$ and $-\infty < b < \infty$, then $M_X(a) = E[e^{aX}]$ is finite for a in some neighborhood of the origin and is called the *moment generating function* (MGF) of (the distribution of) X. Note that when M_X exists,
$M_X(it) = \varphi_X(t)$ or $\varphi_X(-it) = M_X(t)$ for $-A \leq t \leq A$; more importantly, all moments of X exist. Formally,

$$M_X(a) = \int_{-\infty}^{\infty} \left[\sum_{k=0}^{\infty} (ax)^k/k! \right] dF(x)$$

$$= \int_{-\infty}^{\infty} dF(x) + a \int_{-\infty}^{\infty} x\,dF(x) +$$

$$a^2 \int_{-\infty}^{\infty} x^2 dF(x)/2 + a^3 \int_{-\infty}^{\infty} x^3 dF(x)/3! \cdots$$

$$= 1 + aE[X] + a^2 E[X^2]/2 + a^3 E[X^3]/3! + \cdots .$$

Differentiating as with the CF, it follows that $M_X(0) = 1$, $M_X^{(1)}(0) = E[X]$, $M_X^{(2)}(0) = E[X^2]$, $M_X^{(3)}(0) = E[X^3], \cdots$. The uniqueness result translates to: if $M_X(t) = M_Y(t)$ for all t in the same neighborhood of 0, then X and Y have the same probability distribution.

Exercise 3: Find the MGF and the first five moments for:

    a)   the exponential PDF $\alpha e^{-\alpha x} \cdot I\{0 < x < \infty\}$;

b)   the Poisson PDF $e^{-\lambda}\lambda^x/x!$, x = 0(1)∞.

c)   the normal RV with mean μ and variance σ^2. Hint: recall that $X = \sigma Z + \mu$ where Z is the standard normal RV.

Exercise 4: Let X be a real RV with MGF M_X . Show that the variance of X is $M_X^{(2)}(0) - \{M_X^{(1)}(0)\}^2$.

Now we look at another variant which has been found very useful in the study of a discrete RV   X   with support {0,1,2,···}: let

$$Q(t) = E[t^X] = \sum_{k=0}^{\infty} t^k \cdot P(X = k).$$

This series converges absolutely uniformly for $|t| \leq 1$. Of course, if the support is actually finite, so is the sum and Q is defined for all real t. For t > 0,

$$Q(t) = E[t^X] = E[e^{X\log(t)}] = \varphi_X(-i\log(t))$$

so that properties of Q (like inversion, uniqueness) can be inferred from those of the CF φ_X .

Consider the following formal expansions:

$$Q(t) = E[t^X] = P(X = 0) + tP(X = 1)$$
$$+ t^2 P(X = 2) + t^3 P(X = 3) + t^4 P(X = 4)$$
$$+ t^5 P(X = 5) + \cdots$$
$$Q^{(1)}(t) = E[Xt^{X-1}] = P(X=1) + 2tP(X=2) + 3t^2 P(X=3)$$
$$+ 4t^3 P(X = 4) + 5t^4 P(X = 5) + \cdots$$
$$Q^{(2)}(t) = E[X(X - 1)t^{X-2}]$$
$$= 2P(X = 2) + 3 \cdot 2tP(X = 3) + 4 \cdot 3t^2 P(X = 4)$$
$$+ 5 \cdot 4t^3 P(X = 5) + \cdots$$
$$Q^{(3)}(t) = E[X(X - 1)(X - 2)t^{X - 3}]$$

$$= 3 \cdot 2P(X = 3) + 4 \cdot 3.2tP(X = 4) + 5 \cdot 4 \cdot 3t^2 P(X=5) + \cdots$$

Etc.

If we evaluate these expressions in terms of "Q and E", we see Q as a *factorial moment generating function* (FMGF):

$Q(1) = E[1] = 1,$

$Q^{(1)}(1) = E[X]$ when this is finite,

$Q^{(2)}(1) = E[X(X - 1)]$ when this is finite,

$Q^{(3)}(1) = E[X(X - 1)(X - 2)]$ when this is finite, $\cdots$.

If we evaluate these expressions in terms of "Q and P", we see Q as a *probability generating function* (PGF):

$Q(0) = P(X = 0),$

$Q^{(1)}(0) = P(X = 1),$

$Q^{(2)}(0) = 2P(X = 2),$

$Q^{(3)}(0) = 3 \cdot 2P(X = 3), \cdots.$

Example: a) Let $Q(t) = e^{\lambda(t - 1)} = e^{-\lambda} \cdot e^{\lambda t}$. Since the series for e^b converges for all b, we get $Q(t) = e^{-\lambda} \sum_{k=0}^{\infty} (\lambda t)^k / k!$ which means that the distribution is Poisson.

b)   Fix $0 < \theta < 1$, r as a positive integer. For

$$Q(t) = (\theta/(1-(1-\theta)t)^r = \theta^r \sum_{k=0}^{\infty} \binom{-r}{k} (-(1-\theta)t)^k$$

$$= \sum_{k=0}^{\infty} \binom{-r}{k} \theta^r (\theta-1)^k t^k.$$

We see the coefficient of t^k as the PDF of the negative binomial distribution.

Exercise 5: Find a simple expression for the PGF when:
  a)   X is uniformly distributed on $\{0,1,2,\cdots,N\}$;
  b)   X is geometric with $0 < \theta < 1$.

LESSON 14. CONVERGENCE OF SEQUENCES OF CHARACTERISTIC FUNCTIONS

In lesson 12, the uniqueness theorem established the one–to–one correspondence between a CDF and a CF. In this lesson we will see that a similar correspondence exists between sequences of CDFs and sequences of CFs. Roughly speaking,

if $X, X_1, X_2, \cdots$ are random variables (vectors), and

$\varphi, \varphi_1, \varphi_2, \cdots$ their corresponding CFs,

then as $n \to \infty$, $X_n \overset{D}{\to} X$ iff $\varphi_n \to \varphi$.

Some of the finer versions of such results will be included in the exercises. Note that the difficult part of the proof of the first theorem is concerned with the "uniformity" and may be skipped.

Theorem: *Let $F, F_1, F_2, \cdots$ be CDFs with corresponding CFs*

$\varphi, \varphi_1, \varphi_2, \cdots$. *If $F_n \overset{w}{\to} F$, then $\varphi_n \to \varphi$; moreover, the*

convergence of $\{\varphi_n\}$ will be uniform on each finite closed interval.

Proof for real RVs: The hypotheses make the weak convergence be complete.

a)   For each $t \in R$, the function mapping R into $\mathscr{C}$ with values e^{itx} is bounded and continuous. Noting that one may apply Helly–Bray Lemma III (Lesson 10) on real and imaginary parts, we get

$$\int_R e^{itx} dF_n(x) \to \int_R e^{itx} dF(x);$$

that is, $\varphi_n(t) \to \varphi(t)$ for all $t \in R$.

INE–b) For any $\lambda > 0$,

$$|\varphi_n(t+h) - \varphi_n(t)| \le \int_{|x| \ge \lambda} |e^{ihx} - 1| dF_n(x)$$

$$+ \int_{|x| < \lambda} |e^{ihx} - 1| dF_n(x). \qquad (*)$$

Not only is $|e^{ihx} - 1| \leq 2$ but also

$$|e^{ihx} - 1| = |i\int_0^{hx} e^{iu}du| \leq |hx| .$$

Using these in the right side of (*) gives it the bound

$$2\int_{|x| \geq \lambda} dF_n(x) + \int_{|x| < \lambda} |hx|\,dF_n(x)$$

$$\leq 2\int_{|x| \geq \lambda} dF_n(x) + |h|\lambda.$$

For a given $\varepsilon > 0$, there is a λ such that

$$\int_{|x| \geq \lambda} dF(x) < \varepsilon/5;$$

then, $F_n \overset{c}{\to} F$ implies convergence of

$$\int_{|x| \geq \lambda} dF_n(x) \text{ to } \int_{|x| \geq \lambda} dF(x)$$

so that for $n > N = N(\varepsilon,\lambda)$,

$$\int_{|x| \geq \lambda} dF_n(x) < 2\varepsilon/5 .$$

For $n = 1(1)N$, there are $\lambda_1, \cdots, \lambda_N$ such that each

$$\int_{|x| \geq \lambda_n} dF_n(x) < 2\varepsilon/5.$$

Increasing any "λ" decreases the value of the corresponding integral which is bounded by $2\varepsilon/5$ anyhow. Therefore, there is a λ^* ($= \max \{\lambda,\lambda_1,\cdots,\lambda_N\}$) such that for all n and all t,

$$|\varphi_n(t+h) - \varphi_n(t)| \leq 4\varepsilon/5 + |h|\lambda^* .$$

Then $|h| < (\varepsilon-4\varepsilon/5)/\lambda^* = \delta$ implies $|\varphi_n(t+h) - \varphi_n(t)| < \varepsilon$ for all n and t. (The sequence $\{\varphi_n\}$ is said to be *equicontinuous*).

c) There is a finite number of open intervals, say,

$J_1, J_2, \cdots, J_c$, each of length at most δ, whose union covers the given closed interval (the Heine– Borel Theorem in analysis). Each t belongs to some J_k ; let t_k be an endpoint of J_k so that $|h| < \delta$. In

$$|\varphi_n(t) - \varphi(t)| \leq |\varphi_n(t) - \varphi_n(t_k)| +$$
$$|\varphi(t_k) - \varphi(t)| + |\varphi_n(t_k) - \varphi(t_k)|$$

we find: the first term on the right $< \varepsilon$ by part b);
the second term on the right $< \varepsilon$ by uniform continuity of φ; the third term on the right $< \varepsilon$ by convergence of $\{\varphi_n\}$.

In the last, we take n large enough to have all "c+1" differences small making n independent of k whence also t. Finally, $|\varphi_n(t) - \varphi(t)| < 3\varepsilon$ for all t in the given interval.

The next theorem is a partial converse to the first theorem

Theorem: *Let $X_1, X_2, X_3, \cdots$ be real RVs with CFs $\varphi_1, \varphi_2, \varphi_3, \cdots$. Let $\{\varphi_n\}$ converge pointwise on R to a function φ. If φ is continuous at 0, then φ is the CF of some RV X and $X_n \overset{D}{\to} X$.*

Proof: By the weak compactness theorem in Lesson 10, there is a subsequence of the CDFs which converges to a DF; say

$$F_{n'} \overset{W}{\to} F. \qquad (**)$$

a)   Suppose that F is in fact a CDF. By the previous theorem, φ is the CF of F. If some other subsequence converged to another CDF, say $F_{n''} \overset{W}{\to} G$, then φ would be its CF and so F = G by the uniqueness theorem. Thus all subsequences would have the same limit and so (Lesson 9), $F_n \overset{W}{\to} F$. It remains to show that F in (**) is in fact a CDF.

b)   For any DF H,

$$\int_R \frac{e^{itx} - 1}{ix} \, dH(x) = \int_R \left[\int_0^t e^{iux} du \right] dH(x)$$

which is, by Fubini's Theorem, equal to

$$\int_0^t \left[\int_R e^{iux} dH(x) \right] du. \text{ Let } \varphi_H(u) = \int_R e^{iux} dH(x).$$

Since $(e^{itx} - 1)/ix \in C_0[R]$ and $F_{n'} \overset{W}{\to} F$, applying Helly–Bray Lemma I

$$\text{yields} \int_R \frac{e^{itx} - 1}{ix} dF_{n'}(x) \to$$

$$\int_R \frac{e^{itx} - 1}{ix} dF(x).$$

This implies that for $u > 0$,

$$(1/u)\int_0^u \varphi_{F_{n'}}(x)dx \to (1/u)\int_0^u \varphi_F(x)dx. \qquad (\text{***})$$

Since $|\varphi_{F_{n'}}| \leq 1$, applying the Lebesgue dominated convergence theorem on $[0,u]$ gives the left–hand side of (***) the limit $(1/u)\int_0^u \varphi(x)dx$. This means that

$$(1/u)\int_0^u \varphi(x)dx = (1/u)\int_0^u \varphi_F(x)dx.$$

Since φ and φ_F are both continuous at 0, the limit of this equation as $u \to 0$ yields $\varphi(0) = \varphi_F(0)$. Finally, $1 = \varphi_{F_{n'}}(0) \to \varphi(0) = \varphi_F(0) = F(+\infty) - F(-\infty)$ implies that

$$F(+\infty) - F(-\infty) = 1$$

which completes the conditions that F be a CDF.

Even after all this work, we still need the two lemmas below in order to apply this directly to the oldest example, the binomial.

Exercise 1: Use induction to complete the proof in the following lemma.

Lemma: *For n = 1(1)∞ and t > 0,*

$$e^{it} = 1 + it/1! + (it)^2/2! + \cdots + (it)^n/n! + \xi_n t^{n+1}/(n+1)!$$

where ξ_n is a complex function of t with modulus at most 1.

Proof: Let $s_1(t) = e^{it} - 1 = i\int_0^t e^{iu}du$; $|s_1(t)| \le \int_0^t du = t$.

Therefore, $s_1(t) = \xi_1 t$ for $\xi_1 = (e^{it} - 1)/t$. Let

$$s_2(t) = e^{it} - 1 - it = i\int_0^t s_1(u)du; \quad |s_2(t)| \le \int_0^t u\, du = t^2/2.$$

Therefore, $s_2(t) = \xi_2 t^2/2$. $\cdots$

Lemma: *If the sequence of real or complex numbers $\{b_n\}$ has a finite limit b as n → ∞, then $\lim\limits_{n\to\infty} (1 - b_n/n)^n = e^{-b}$.*

Partial proof: Take log to be with base e and Taylor's formula for complex z. Then there is a ξ with $|\xi| < |z| < 1$ such that $\log(1 - z) = -z/(1 - \xi)$. Since b is finite, for large n, $|b_n/n| < 1$ and $n\log(1 - b_n/n)$ is well-defined. Then

$$n\log(1 - b_n/n) = n(-b_n/n(1 - \xi_n)) = -b_n/(1 - \xi_n).$$

Since $|\xi_n| < |b_n/n|$, $(1 - \xi_n) \to 1$ as n → ∞. Finally,

$$\log(1 - b_n/n)^n \to -b \text{ implies } (1 - b_n/n)^n \to e^{-b}.$$

Example: Let $\{X_n\}$ be a sequence of IID Bernoulli RVs with

$P(X_n = 1) = \theta = 1 - P(X_n = 0)$. Then $S_n = \sum\limits_{k=1}^{n} X_k$ is the

ordinary binomial RV with mean $n\theta$ and variance $n\theta(1-\theta)$; the

CF of S_n was computed in the first example of Lesson 11. The standardized value of S_n is

$$V_n = (S_n - n\theta)/\sqrt{n\theta(1-\theta)} = \sum_{k=1}^{n} (X_k-\theta)/\sqrt{n\theta(1-\theta)}$$

$$= S_n/\sqrt{n\theta(1-\theta)} - \sqrt{n\theta/(1-\theta)};$$

this is of the form $cS_n + d$ and its CF is

$$e^{itd}\varphi_{S_n}(ct) = \varphi_{V_n}(t)$$

$$= \left[(1-\theta)e^{-it\sqrt{\theta/n(1-\theta)}} + \theta e^{it\sqrt{(1-\theta)/n\theta}}\right]^n.$$

The results of the first lemma can be applied to the exponentials, with remainder terms (in particular, ξ,η,δ) changing each time:

$$e^{-it\sqrt{\theta/n(1-\theta)}} = 1 - it\sqrt{\theta/n(1-\theta)} - (\theta/n(1-\theta))t^2/2$$

$$+ \xi_2(-it\sqrt{(1-\theta)/n\theta})^3/3!.$$

$$e^{it\sqrt{(1-\theta)/n\theta}} = 1 + it\sqrt{(1-\theta)/n\theta} - ((1-\theta)/n\theta)t^2/2$$

$$+ \eta_2(it\sqrt{(1-\theta)/n\theta})^3/3!.$$

Substituting these in φ_{V_n} and using a little algebra yield

$$\varphi_{V_n}(t) = \left[1 - t^2/2n + \delta_3/n^{3/2}\right]^n$$

$$= \left[1 - \frac{t^2/2 + \delta_3/n^{1/2}}{n}\right]^n$$

which has limit $e^{-t^2/2}$ as $n \to \infty$ (using the second lemma above). This is of course the CF of the standard normal RV, say Z.

Exercise 2: Complete the example by filling in the details for application of the two lemmas.

When n is "large", we can approximate the binomial CDF by the normal distribution function (as also in Lesson 5, Part II):

$$P(S_n \le b) = P(V_n \le (b - n\theta)/\sqrt{n\theta(1-\theta)}) \approx$$

$$P(Z \le (b - n\theta)/\sqrt{n\theta(1-\theta)}).$$

How large is "large"? Other theoretical and numerical studies suggest that the approximation is accurate to within 2 or 3 decimal places when $n\theta$ and $n(1-\theta)$ are both at least 5. Of course, the accuracy is greater for larger sample sizes. Perhaps more importantly, the accuracy is improved when the "correction for continuity" is included:

$$P(S_n \le b) \approx P(Z \le (b + .5 - n\theta)/\sqrt{n\theta(1-\theta)}).$$

This device is explained best in a specific case, say $P(S_n = 16)$. The normal approximation replaces the area of the bar of the histogram by the area " under the curve ":

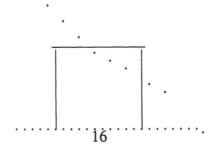

Simple geometry suggests that the approximation must extend from 15.5 to 16.5. Similar pictures can be drawn for other cases.

Example: When n = 50 and θ = .3, $n\theta$ = 15, $n\theta(1-\theta)$ = 10.5.

Then, $P(S_{50} = 16)$ $= P(S_{50} \le 16.5) - P(S_{50} \le 15.5)$

$\approx P(Z \le 1.5/\sqrt{10.5}) - P(Z \le .5/\sqrt{10.5})$

$= P(Z \le .46) - P(Z \le .15)$

$$= .677242 - .559618 = .117624.$$

This can be compared with the value $b(20: 50, .3) = .1147001$ from Lesson 7, Part I.

Although this technique is simple, some care must be exercised. For example,

$$P(16 < S_n \le 25) = P(S_n \le 25.5) - P(S_n \le 16.5),$$

$$P(16 \le S_n \le 25) = P(S_n \le 25.5) - P(S_n \le 15.5),$$

$$P(16 < S_n < 25) = P(S_n \le 24.5) - P(S_n \le 16.5),$$

$$P(16 \le S_n < 25) = P(S_n \le 24.5) - P(S_n \le 15.5); \text{ etc.}$$

For $n = 50$, $\theta = .4$,

$$P(16 < S_n \le 25) \approx P(Z \le 1.59) - P(Z \le -1.01)$$
$$= .944083 - .156248;$$

$$P(16 \le S_n \le 25) \approx P(Z \le 1.59) - P(Z \le -1.30)$$
$$= .944083 - .095098.$$

Exercise 3:   Use the normal approximation for the binomial probabilities:

    a) $n = 50$, $\theta = .4$: $P(S_n = 22)$; $P(S_n \ge 22)$; $P(S_n \le 22)$

    b) $n = 100$, $\theta = .2$: $P(S_n > 22)$; $P(S_n = 20)$;
    $P(18 < S_n \le 23)$.

The following "counter–example" is easier to work out.

Example: For $n = 1(1) \infty$, let X_n be uniformly distributed on the interval $[-n,n]$ so that its PDF is

$$f_n(x) = I\{-n \le x \le n\}/2n;$$

let φ_n be its CF. For $t = 0$, $\varphi_n(t) = 1$; for $t \ne 0$,

$$\varphi_n(t) = \int_{-n}^{n} e^{itx} dx/2n = (e^{itn} - e^{-itn})/2itn. \text{ Then,}$$

$$\varphi_n(t) \to \varphi(t) = \begin{cases} 1 & \text{for } t = 0 \\ 0 & \text{for } t \ne 0 \end{cases}.$$

Note that φ is not continuous at $t = 0$. The corresponding sequence of CDFs is

$$F_n(x) = \begin{cases} 0 & \text{for } x < -n \\ 1/2 + x/2n & \text{for } -n \leq x \leq n \\ 1 & \text{for } n < x \end{cases}.$$

Note that $\lim\limits_{n \to \infty} F_n(x) = 1/2$ for all $x \, \varepsilon \, R$ and, of course,

$F(x) = 1/2$ is not a CDF.

Exercise 4: Let $\{\varphi_n\}$ be a sequence of CFs with

$\lim\limits_{n \to \infty} \varphi_n(t) = \varphi(t)$ for all real t. Show that when φ is continuous, φ is also a CF.

INE–Exercise: Let $\{F_n\}$ be a sequence of CDFs and $\{\varphi_n\}$ be the corresponding sequence of CFs. Show that $\{F_n\}$ converges completely to a CDF F iff $\{\varphi_n\}$ converges uniformly in every finite interval to a function φ which is then the CF of F.

LESSON 15. CENTRAL LIMIT THEOREMS

Actually, we had a central limit theorem (CLT) in lesson 14: the sequence $X_1, X_2, X_3, \cdots$ consists of IID Bernoulli RVs with parameter θ; the partial sum $S_n = \sum_{k=1}^{n} X_k$ is standardized to

$$V_n = (S_n - n\theta)/\sqrt{(n\theta(1-\theta))};$$

the limiting distribution of V_n is normal.

In this lesson, we will see that this scheme is fairly general so we can approximate probabilities for other sample sums S_n or sample means S_n/n by using the standard normal distribution. First an

Exercise 1: "$o(t)$" denotes a function of t such that $\lim_{t \to 0} o(t)/t = 0$. Let $\{a_n\}$ be any sequence of real numbers tending to ∞. Show that for each fixed t, $o(t/a_n) = o(1/a_n)$.

Theorem: *Let $X, X_1, X_2, X_3, \cdots$ be IID RVs with common CDF F, mean $E[X] = \mu, 0 < Var\ X = \sigma^2 < \infty$. Let*

$$S_n = X_1 + \cdots + X_n \text{ and } Z_n = (S_n - n\mu)/\sigma\sqrt{n}.$$

Then $Z_n \xrightarrow{D} Z$ where Z is the standard normal RV; specifically,

for each $z \in R$, $\lim_{n \to \infty} P(Z_n \leq z) = \int_{-\infty}^{z} e^{-w^2/2} dw / \sqrt{(2\pi)}$.

Proof: For $k \geq 1$, $Y_k = (X_k - \mu)/\sigma\sqrt{n}$ has CF

$$\varphi_{Y_k}(t) = E[e^{it(X_k - \mu)/\sigma\sqrt{n}}]$$

$$= \int_{-\infty}^{\infty} e^{it(x-\mu)/\sigma\sqrt{n}} dF(x) = \varphi_{X-\mu}(t/\sigma\sqrt{n}).$$

Now $Z_n = \sum_{k=1}^{n} Y_k$. Since each Y_k is a function of only one of the independent X's, the product rule for expectation (Lesson 13, Part II) yields

$$\varphi_{Z_n}(t) = E[e^{it\Sigma Y_k}]$$

$$= E[e^{itY_1}] \cdots E\{e^{itY_k}] = [\varphi_{X-\mu}(t/\sigma\sqrt{n})]^n.$$

Finiteness of $E[X^2]$ guarantees that all these CF have second derivatives; in particular, from results of Lesson 12,

$$\varphi_{X-\mu}(t) = 1 + iE[X-\mu]t - E[(X-\mu)^2]t^2/2 + o(t^2)$$

$$= 1 - \sigma^2 t^2/2 + o(t^2).$$

Hence, $\varphi_{Z_n}(t) = [\varphi_{X-\mu}(t/\sigma\sqrt{n})]^n = [1 - t^2/2n + o(1/n)]^n$. Using the second lemma in Lesson 14, we get

$$\lim_{n\to\infty} \varphi_{Z_n}(t) = e^{-t^2/2} .$$

Now the second theorem in Lesson 14 applies and the proof is complete.

Exercise 2: Apply L'Hopital's rule to $\lim_{n\to\infty} M_X(t/\sigma\sqrt{n})^n$ when M_X is the MGF (assumed to exist) of a RV X with mean 0 and variance σ^2. What do you conclude ?

Example: a) If we apply the theorem to the Bernoulli case, we get the same limiting normal distribution without all the fuss in lesson 14.

b)   Let $X_1, X_2, X_3, \cdots$ be IID exponential RVs with mean

$$\mu = \alpha > 0. \text{ Then } E[X^2] = \int_0^\infty x^2 e^{-x/\alpha} dx/\alpha = 2\alpha^2 \text{ and}$$

$\sigma^2 = \alpha^2$. For n large,

$$P(X_1 + \cdots + X_n \le b) \approx P(Z \le (b - n\alpha)/\alpha\sqrt{n}).$$

As with the binomial, the accuracy of the approximation depends on all three of n,b,α but note that here there is no "correction for continuity" since the exponential distribution is continuous to begin with.

c) Let $X_1, X_2, X_3, \cdots$ be IID Poisson Rvs with mean = variance = λ. For $S_n = X_1 + \cdots + X_n$,

$$P(S_n \leq b) \approx P(Z \leq (b + .5 - n\lambda)/\sqrt{n\lambda}).$$

The correction for continuity is always needed when a continuous CDF is used to approximate that of a counting RV.

Exercise 3: Recall that the sum S_n of the Poisson RVs in the example has the Poisson distribution with parameter $n\lambda$. For $n = 100$ and $\lambda = .1$, compare the exact and approximate values of $P(S_{100} \leq 10)$.

Exercise 4: Let $\{X_n\}$ be IID standard normal RVs. Show that each of the sequences $\{X_n^2\}$ and $\{X_n^3\}$ satisfy the conditions for the CLT.

When the $\{X_n\}$ are identically distributed, the sample mean $S_n/n = (X_1 + \cdots + X_n)/n$ is usually symbolized as $\overline{X}_n$. Many authors say, incorrectly, that the sample mean is approximately normal. On the contrary, the normal approximation can be applied only to the distribution of the standardized form $V_n = (\overline{X}_n - \mu)/(\sigma/\sqrt{n})$. The following exercise makes this clear.

Exercise 5: Let $X_1, X_2, X_3, \cdots$ be IID standard normal RVs. Show that the distribution of $\overline{X}_n$ is also normal with mean 0 and variance 1/n. Show directly that

$$\lim_{n \to \infty} P(\overline{X}_n \leq w) = \begin{cases} 0 & \text{for } w < 0 \\ 1 & \text{for } w > 0 \end{cases}.$$

Inconveniently, even a slight change in the hypotheses of the first theorem brings about a large change in the proof of a new theorem; we only outline the variations for dropping "identically distributed". Before stating the theorem, we need to explain some other terminology.

Definition: *For each* $n = 1(1)\infty$, $Y_n = (Y_{n,1}, Y_{n,2}, \cdots, Y_{n,k_n})$ *is a vector of real RVs. The family* $\{Y_{n,j}\}$ *is asymptotically negligible (AN) iff for all* $\varepsilon > 0$,

$$\lim_{n \to \infty} \max\left[P(|Y_{n,1}| \geq \varepsilon), P(|Y_{n,2}| \geq \varepsilon), \cdots, P(|Y_{n,k_n}| \geq \varepsilon)\right] = 0.$$

For example, take $Y_{n,j} = (X_j - \mu)/n\sigma^2$, $j = 1(1)n$ with $\{X_j\}$ as in the first theorem. By Chebyshev's inequality,

$$P(|Y_{n,j}| \geq \varepsilon) \leq (\text{Var } Y_{n,j})/\varepsilon^2 = (1/n)/\varepsilon^2$$

which tends to 0 as n tends to ∞. It turns out that something a bit stronger than AN is needed.

Our discussion from here through the Lindeberg–Feller Theorem, will use the following notation. For $n = 1(1)\infty$, X_n are independent real RVs with CDFs F_n, means μ_n, and finite variances $\sigma_n^2 > 0$; $s_n^2 = \sum_{j=1}^{n} \sigma_j^2$; for $j = 1(1)n$,

$$Y_{n,j} = (X_j - \mu_j)/s_n^2.$$

Exercise 6: Use Chebyshev's inequality to find a sufficient condition that the family $\{Y_{n,j}\}$ defined just above be AN.

Definition: *a) Lindeberg condition. For each* $\varepsilon > 0$,

$$L_n(\varepsilon) = \sum_{j=1}^{n} \int_{|x - \mu_j| \geq \varepsilon s_n} (x - \mu_j)^2 dF_j(x)/s_n^2 \to 0 \text{ as } n \to \infty.$$

b)   *Feller condition.* $\lim\limits_{n\to\infty} \max \{\sigma_1/s_n, \ \sigma_j/s_n, \ \cdots, \ \sigma_n/s_n\} = 0.$

c)   *Lyapunov condition. For some* $\delta > 0$,

$$\lim_{n\to\infty} \sum_{j=1}^{n} E[\,|X_j - \mu_j|^{2+\delta}]/s_n^{2+\delta} = 0.$$

Example: a) Let the sequence be identically distributed so that

$F_j = F$, $\mu_j = \mu$, and $\sigma_j = \sigma$. $L_n(\epsilon)$ simplifies to

$$\int_{|x-\mu| \geq \epsilon\sigma\sqrt{n}} (x - \mu)^2 dF(x)/\sigma^2$$

which tends to 0 since the second moment is assumed to be finite.

b)   For $\delta > 0$, the inequality $|x - \mu_j| \geq \epsilon s_n$ implies

$$|x - \mu_j|^{2+\delta} = |x - \mu_j|^2 \cdot |x - \mu_j|^{\delta} \geq |x - \mu_j|^2 \cdot \epsilon^{\delta} s_n^{\delta}.$$

This implies

$$\int_{|x-\mu_j| \geq \epsilon s_n} (x - \mu_j)^2 dF_j(x) \leq \int_{|x-\mu_j| \geq \epsilon s_n} |x - \mu_j|^{2+\delta} dF_j(x)/\epsilon^{\delta} s_n^{\delta},$$

whence,

$$L_n(\epsilon) \leq \sum_{j=1}^{n} \int_{|x-\mu_j| \geq \epsilon s_n} |x - \mu_j|^{2+\delta} dF(x)/\epsilon^{\delta} s_n^{2+\delta}$$

$$\leq \sum_{j=1}^{n} E[\,|X_j - \mu_j|^{2+\delta}]/\epsilon^{\delta} s_n^{2+\delta}.$$

We see that the Lyapunov condition implies the Lindeberg condition.

Exercise 7:   Assume that the sequence $\{X_n\}$ is uniformly bounded; ie., there is a finite number M such that

$$P(|X_j| \leq M) = 1 \text{ for } j = 1(1)\infty.$$

Show that the Lyapunov condition is satisfied. Hint: first show that $|X_j - \mu_j| \leq 2M$ a.s.

Theorem: *(Lindeberg-Feller).* $S_n = X_1 + \cdots + X_n$ *and*
$$V_n = (S_n - E[S_n])/s_n \; ;$$
Z is the standard normal RV. Then the following are equivalent:

 a) $V_n \overset{D}{\to} Z$ *and the Feller condition holds.*

 b) $V_n \overset{D}{\to} Z$ *and the family* $\{Y_{n,j}\}$ *is AN.*

 c)   *The sequence* $\{X_n\}$ *satisfies the Lindeberg condition.*

Proof: See a text like Bauer, 1972 .

In order to extend the basic CLT to sequences of vectors, we need some details about the multivariate normal distribution. Then we will be able to use the Cramér–Wold device to prove the theorem.

Definition: *$Z_1, Z_2, \cdots, Z_k$ are IID standard normal RVs. For $\alpha = 1(1)p$, $\beta = 1(1)k$, $a_{\alpha\beta}$ and μ_α are real constants and*
$$X_\alpha = \sum_{\beta=1}^{k} a_{\alpha\beta} Z_\beta + \mu_\alpha. \text{ Then } X = (X_1, X_2, \cdots, X_p)' \text{ is a}$$
multivariate normal random vector of length p.

Corollary; *For $t' = (t_1, t_2, \cdots, t_p)$ and $\mu' = (\mu_1, \mu_2, \cdots, \mu_p)$, the joint CF of the multivariate normal in this definition is given by*
$$e^{it'\mu} \cdot exp\left[-\sum_{\alpha=1}^{p} \sum_{\gamma=1}^{p} t_\alpha \left[\sum_{\beta=1}^{k} a_{\alpha\beta} a_{\gamma\beta}\right] t_\gamma\right] .$$

Proof: $t'X = \sum_{\alpha=1}^{p} t_\alpha X_\alpha = \sum_{\alpha=1}^{p} t_\alpha \left[\sum_{\beta=1}^{k} a_{\alpha\beta} Z_\beta + \mu_\alpha\right]$

$$= \sum_{\beta=1}^{k} \left[\sum_{\alpha=1}^{p} t_\alpha a_{\alpha\beta}\right] Z_\beta + \sum_{\alpha=1}^{p} t_\alpha \mu_\alpha ,$$

which we rewrite as $\displaystyle\sum_{\beta=1}^{k} b_\beta Z_\beta + t'\mu$.

$$E[e^{it'X}] = E[\exp(i\sum_{\beta=1}^{k} b_\beta Z_\beta) \cdot e^{it'\mu}]$$

$$= e^{it'\mu} E[e^{ib_1 Z_1} \cdots e^{ib_k Z_k}]$$

$$= e^{it'\mu} \cdot e^{-b_1^2/2} \cdot e^{-b_2^2/2} \cdots e^{-b_k^2/2}$$

$$= e^{it'\mu} \cdot \exp(-\sum_{\beta=1}^{k} b_\beta^2/2). \qquad (*)$$

Here

$$\sum_{\beta=1}^{k} b_\beta^2 = \sum_{\beta=1}^{k} \left[\sum_{\alpha=1}^{p} t_\alpha a_{\alpha\beta}\right]^2$$

$$= \sum_{\beta=1}^{k}\sum_{\alpha=1}^{p} t_\alpha a_{\alpha\beta} \cdot \sum_{\gamma=1}^{p} t_\gamma a_{\gamma\beta}$$

$$= \sum_{\alpha=1}^{p}\sum_{\gamma=1}^{p} t_\alpha \left[\sum_{\beta=1}^{k} a_{\alpha\beta} a_{\gamma\beta}\right] t_\gamma .$$

   A generalization of the following matrix example will allow the CF to be written neatly.

Example: For $\quad A = \begin{bmatrix} a_{11} & a_{12} & a_{13} \\ a_{21} & a_{22} & a_{23} \end{bmatrix}$, the transpose

$A' = \begin{bmatrix} a_{11} & a_{21} \\ a_{12} & a_{22} \\ a_{13} & a_{23} \end{bmatrix}$; one product is

$$A' = \begin{bmatrix} a_{11}^2 + a_{12}^2 + a_{13}^2 & a_{11}a_{21} + a_{12}a_{22} + a_{13}a_{23} \\ a_{21}a_{11} + a_{22}a_{12} + a_{23}a_{13} & a_{21}^2 + a_{22}^2 + a_{23}^2 \end{bmatrix}.$$

Now define $\Sigma = \begin{bmatrix} \sigma_{11} & \sigma_{12} \\ \sigma_{21} & \sigma_{22} \end{bmatrix}$ by $\Sigma = AA'$ so that

$$\sigma_{\alpha\gamma} = \sum_{\beta=1}^{k} a_{\alpha\beta} a_{\gamma\beta} \;.\; \text{For } t' = (t_1, t_2), \; t'\Sigma =$$

$$(t_1, t_2) \begin{bmatrix} \sigma_{11} & \sigma_{12} \\ \sigma_{21} & \sigma_{22} \end{bmatrix} \begin{bmatrix} t_1 \\ t_2 \end{bmatrix}$$

$$= t_1 \sigma_{11} t_1 + t_1 \sigma_{12} t_2 + t_2 \sigma_{21} t_1 + t_2 \sigma_{22} t_2$$

$$= \sum_{\alpha=1}^{2} \sum_{\gamma=1}^{2} t_\alpha \cdot \sigma_{\alpha\gamma} \cdot t_\gamma = \sum_\alpha \sum_\gamma t_\alpha \cdot \sum_\beta a_{\alpha\beta} a_{\gamma\beta} \cdot t_\gamma \;.$$

A square matrix Σ is non–singular when there is a matrix Σ^{-1} such that $\Sigma\Sigma^{-1}$ is the identity matrix I; in this example, $I = \begin{bmatrix} 1 & 0 \\ 0 & 1 \end{bmatrix}$.

The CF of the multivariate normal (*) becomes

$$\varphi_X(t) = e^{it'\mu - t'\Sigma t/2}$$

where, in general, t, μ, Σ are p–dimensional. As in the univariate case, this function is infinitely differentiable so that

$$\partial\varphi_X(t)/\partial t_1 = E[e^{it'X}(iX_1)]$$

$$= e^{it'\mu - t'\Sigma t/2}(i\mu_1 - (\sigma_{11} \cdots \sigma_{1p})t);$$

$$\partial^2\varphi_X(t)/\partial t_1 \partial t_3 = E[e^{it'X}(iX_1)(iX_3)]$$

$$= e^{it'\mu - t'\Sigma t/2}(i\mu_1 - (\sigma_{11} \cdots \sigma_{1p})t)(i\mu_3 - (\sigma_{31} \cdots \sigma_{3p})t)$$

$$+ e^{it'\mu - t'\Sigma t/2}(-\sigma_{13}); \text{ etc.}$$

When evaluated at t = 0, we get the moment relations:

$$E[iX_1] = i\mu_1 \ ; \ \ E[(iX_1)(iX_3)] = (i\mu_1)(i\mu_3) - \sigma_{13} \ ; \ \text{etc.}$$

We see that μ is the vector of means and Σ is the covariance matrix.

Short–hand for "X is normally distributed with mean μ and covariance matrix Σ" is $X \cong N(0,\Sigma)$; the "standard normal" is $N(0,I)$. When Σ is non–singular with determinant $|\Sigma|$, the density of X can be shown to be

$$e^{-(x-\mu)'\Sigma^{-1}(x-\mu)/2} \div \{(2\pi)^p |\Sigma|\}^{1/2} .$$

Exercise 7: Rewrite the two–dimensional normal density (Lesson 5) and its CF in terms of vectors and matrices.

Theorem: *For $n = 1(1)\infty$, $X_n = (X_n^1, X_n^2, \cdots, X_n^p)$ is a p-dimensional random vector; $X = (X^1, X^2, \cdots, X^p)$ is also a random vector. The vector $t = (t_1, t_2, \cdots, t_p) \in R^p$. Then $X_n \overset{D}{\to} X$ iff each linear combination $t'X_n$ converges in distribution to $t'X$.*

Proof: If the real RV $t'X_n$ converges in distribution to the real RV $t'X$, then $\varphi_{t'X_n}(r) \to \varphi_{t'X}(r)$ for each $r \in R$. This is the same as

$$\varphi_{X_n}(rt) \to \varphi_X(rt) \text{ whence } \varphi_{X_n}(t) \to \varphi_X(t) \text{ for all } t \in R^p .$$

By the first theorem in Lesson 12, $X_n \overset{D}{\to} X$. As with the earlier form of the Cramér–Wold device, the argument is completely reversible.

Theorem: *(CLT for vectors) Let $X_1, X_2, X_3, \cdots$ be IID random vectors of length p with mean μ, covariance matrix Σ, and CF φ;*

let $\bar{X}_n = \sum_{k=1}^{n} X_k/n$. *Then* $\sqrt{n}(\bar{X}_n - \mu) \overset{D}{\to} Z \cong N(0,\Sigma)$.

Proof: By the previous theorem, the conclusion holds iff

$$t'\sqrt{n}(\bar{X}_n - \mu) \overset{D}{\to} t'Z \text{ for each } t \in R^p.$$

For each $n = 1(1)\infty$, let $Y_n = t'X_n$; an extension of the product rule (Lesson 13, Part II) can be used to show that $Y_1, Y_2, Y_3, \cdots$ are also IID and normally distributed with mean $\nu = t'\mu$, variance $\sigma^2 = t'\Sigma t$. Then $t'\sqrt{n}(\bar{X}_n - \mu) = \sqrt{n}(\bar{Y}_n - \nu)$ is in the same form as in the first CLT so that its limiting distribution is $N(0,\sigma^2)$. This is equivalent to the first sentence of this proof.

Exercise 8: Consider the proof of the last theorem.
  a)   Use the product rule to prove the statement about $\{Y_n\}$.
  b)   Show the truth of the last statement.

REFERENCES

Blyth, C.R. (1986) Approximate binomial confidence limits. *J. Amer. Statist. Assoc.*, 81, 843–855.

Choe, B.R. (1987) An elementary proof of $\sum_1^\infty 1/n^2 = \pi^2/6$. *Amer. Math. Monthly*, 94, 662–663.

Chung, K.L. (1974) *A Course in Probability Theory.* Academic Press.

Diaconis, P. and Freedman, D. (1986) An elementary proof of Stirling's formula. *Amer. Math. Monthly*, 93, 123–125.

Feller, W. (1968, 1971) *An Introduction to Probability Theory and Its Applications.* Vol. I, II. J. Wiley.

Gnedenko, B.V. and Kolmogorov, A.N. (1954) *Limit Distributions for Sums of Independent Random Variables.* Addison–Wesley.

Johnson, N.L. and Kotz, S. (1973) *Urn Models and Their Applications.* J. Wiley.

Johnson, B.R. (1980) An elementary proof of inclusion–exclusion formulas. *Amer. Math. Monthly*, 87, 750–751.

Keller, J.B. (1986) The probability of Heads. *Amer. Math. Monthly*, 93, 191–197.

Leader, S. (1985) The inclusion–exclusion probability formulas by Taylor's thoerem. *Amer. Math. Monthly*, 92, 343–345.

Loève, M. (1963) *Probability Theory.* Van Nostrand.

Neuts, M.F. (1973) *Probability.* Allyn and Bacon.

Tamhane, A.J. (1981) Randomized response techniques for multiple sensitive attributes. *J. Amer. Statist. Assoc.*, 76, 916–923.

Varderman, S.B. (1986) The legitimate role of inspection in modern SQC. *The Amer. Statist.*, 40, 325–328.

Yakowitz, S.J. (1977) *Computational Probability and Simulation*. Addison Wesley.

INDEX

The numbers at the end of a concept refer to the part and the lesson where this concept is first cited.

D

E

F

T

U

V

W